Ecological Studies

Analysis and Synthesis

Edited by

W. D. Billings, Durham (USA) F. Golley, Athens (USA)

O. L. Lange, Würzburg (FRG) J. S. Olson, Oak Ridge (USA)

H. Remmert, Marburg (FRG)

Volume 47

Pollutants in Porous Media

The Unsaturated Zone
Between Soil Surface and Groundwater

Edited by
B. Yaron   G. Dagan   J. Goldshmid

With 119 Figures

Springer-Verlag
Berlin Heidelberg New York Tokyo 1984

B. Yaron
Agricultural Research Organization, The Volcani Center
Institute of Soils and Water
Bet-Dagan, Israel

G. Dagan
Faculty of Engineering, University of Tel-Aviv
Tel-Aviv, Israel

J. Goldshmid
Environmental Engineering and Design Corporation
Tel-Aviv, Israel

ISBN 3-540-13179-5 Springer-Verlag Berlin Heidelberg New York Tokyo
ISBN 0-387-13179-5 Springer-Verlag New York Heidelberg Berlin Tokyo

Library of Congress Cataloging in Publication Data. Main entry under title: Pollutants in porous media. (Ecological studies; v. 47). Papers presented at an international workshop organized by the Institute of Soils and Water of the Israeli Agricultural Organization, and held in Mar. 1983 at Bet-Dagan, Israel. 1. Water, Underground – Pollution – Congresses. 2. Soil absorption and adsorption – Congresses. 3. Water – Analysis – Congresses. I. Yaron, B. (Bruno), 1929–. II. Dagan, G. (Gedeon), 1932–. III. Goldshmid, J. (Jhuda), 1931–. IV. Makhon le-kara' u-mayim (Israel). V. Title: Unsaturated zone between soil surface and groundwater. VI. Series. TD426.P65. 1984. 628.5″5. 84-1255.

Typesetting, printing, and bookbinding: Brühlsche Universitätsdruckerei, Giessen
2131/3130-543210

Preface

The unsaturated zone is the medium through which pollutants move from the soil surface to groundwater. Polluting substances are subjected to complex physical, chemical and biological transformations while moving through the unsaturated zone and their displacement depends on the transport properties of the water-air-porous medium system. Pollution caused by human activities, agriculture, and industry, has brought about a growing interest in the role of the unsaturated zone in groundwater pollution. Due to the complexity and the multidisciplinary nature of the subject, it is being investigated by specialists from various scientific disciplines, such as soil physicists, chemists, biologists, and environmental engineers. This state of affairs has motivated the initiative taken by the Water Quality Commissions of IUPAC (the International Union of Pure and Applied Chemistry) and IAHS (the International Association of Hydrological Sciences) to convene an international workshop, which was organized and hosted by the Institute of Soils and Water of the Agricultural Research Organization in Bet-Dagan, Israel in March 1983. The lecturers at the workshop were an invited group of specialists who are engaged in studying the many facets of the unsaturated zone, and the present book is a selection of their presentations.

Each chapter of the book relates to a different aspect of the unsaturated zone. Due to the limitations of time and extent of the workshop, these chapters are by no means comprehensive discussions of the various topics, as these could only be treated by more numerous symposia and books. They merely bring into focus a few recent developments which necessarily reflect the interest and preferences of the participants at the workshop. The main purpose of the workshop, however, was to bring together a group of selected specialists from different and separate disciplines and to further the exchange of views which is indispensable to a thorough understanding of the pollution process. This is the main reason for our belief that the book is of interest to a wide group of readers: it gives an up-to date picture of recent developments in the many areas of research involved in the study of the behaviour of pollutants in the unsaturated zone. Furthermore, we have attempted to strike a balance between basic research and engineering applications and management. It is our hope that the book will contribute to a better understanding of the subject by the scientific community and by the agencies dealing with pollution control.

We acknowledge the support given to the workshop by the Israeli Ministry of Agriculture, the Water Commissioner, the National Council for Research and Development and the Seagram Center for Soil and Water Research of the Hebrew University. We are grateful to the workshop participants for their willingness to prepare the written contributions and we thank Springer-Verlag for undertaking the publication of the book.

Bruno Yaron
Gedeon Dagan
Jhuda Goldshmid

Contents

Contributors

AVNIMELECH, Y.   Faculty of Agricultural Engineering, Technion, Israel Institute of Technology, Haifa, Israel

BOUWER, H.   USDA-ARS, U.S. Water Conservation Laboratory, Phoenix, AZ, USA

BRESLER, E.   Institute of Soils and Water, ARO, The Volcani Center, Bet-Dagan, Israel

CALVET, R.   Institut National Agronomique de Paris-Grignon, Thiverval-Grignon, France

DAGAN, G.   Faculty of Engineering of Tel-Aviv, Tel-Aviv, and Institute of Soils and Water, ARO, The Volcani Center, Bet-Dagan, Israel

FINE, P.   Institute of Soils and Water, ARO, The Volcani Center, Bet-Dagan, Israel

FYLKNER, A.   Department of Land Improvement and Drainage, Royal Institute of Technology, Stockholm, Sweden

GOLDSHMID, J.   Environmental Engineering and Design Corporation, Tel-Aviv, Israel

HAYES, M. H. B.   Department of Chemistry, University of Birmingham, Birmingham, England

IDELOVITCH, E.   TAHAL Consulting Engineers Ltd., Tel-Aviv, Israel

JACKS, G.   Department of Land Improvement and Drainage, Royal Institute of Technology, Stockholm, Sweden

JURY, W. A.   Department of Soil and Environmental Sciences, University of California, Riverside, CA, USA

KANFI, Y.   Israel Water Commission, Tel-Aviv, Israel

KNUTSSON, G.   Department of Land Improvement and Drainage, Royal Institute of Technology, Stockholm, Sweden

LETEY, J.   Department of Soil and Environmental Sciences, University of California, Riverside, CA, USA

LYNCH, J. M.	Glasshouse Crops Research Institute, Littlehampton, England
MAGARITZ, M.	Department of Isotopes, The Weitzmann Institute of Science, Rehovot, Israel
MATTHESS, G.	Geologisch-Palaeontologisches Institut der Universität Kiel, Kiel, Fed. Rep. of Germany
MAXE, L.	Department of Land Improvement and Drainage, Royal Institute of Technology, Stockholm, Sweden
MERCADO, A.	TAHAL Consulting Engineers Ltd., Tel-Aviv, Israel
METZGER, L.	Institute of Soils and Water, ARO, The Volcani Center, Bet-Dagan, Israel
MINGELGRIN, U.	Institute of Soils and Water, ARO, The Volcani Center, Bet-Dagan, Israel
PEKDEGER, A.	Geologisch-Palaeontologisches Institut der Universität Kiel, Kiel, Fed. Rep. of Germany
PRATT, P. F.	Department of Soil and Environmental Sciences, University of California, Riverside, CA, USA
RAATS, P. A. C.	Institute for Soil Fertility, Haren, The Netherlands
RONEN, D.	Israel Water Commission, Tel-Aviv, Israel
SALTZMAN, SARINA	Institute of Soils and Water, ARO, The Volcani Center, Bet-Dagan, Israel
SCHWILLE, F.	Bundesanstalt für Gewässerkunde, Koblenz, Fed. Rep. of Germany
SHEVAH, Y.	TAHAL Consulting Engineers, Ltd., Tel-Aviv, Israel
STEINNES, E.	Department of Chemistry, University of Trondheim, Dragvoll, Norway
VINTEN, A. J.	Institute of Soils and Water, ARO, The Volcani Center, Bet-Dagan, Israel
WALDMAN, MIRIAM	National Council for Research and Development, Ministry of Science and Development, Jerusalem, Israel
YARON, B.	Institute of Soils and Water, ARO, The Volcani Center, Bet-Dagan, Israel

A. Modeling of Pollutants Transport

Introductory Comments

G. Dagan

The modeling of pollutants transport in the unsaturated zone deals with the formulation of a conceptual framework and subsequent quantitative relationships, which permit one to determine the distribution of pollutants in space and time. Models are therefore needed in order to understand the influence of various mechanisms involved in the transport process on the one hand and in order to predict the pollutants transfer on the other hand. They are also useful for interpretation of tests and determination of physical coefficients of interest.

The modeling of two categories of pollutants has to be discussed separately: those miscible or those immiscible with water.

Most of the material of this book is concerned with miscible transport, which has been studied intensively in the past. The interest in it is also related to other important applications like salt and fertilizer movement in soil. The traditional and well-established procedure to model transport is to solve the flow equation (Richards' equation) first, in order to determine the water velocity as function of space and time and subsequently, the convection-diffusion equation for the solute concentration. Terms related to adsorption, precipitation, or evaporation can be incorporated in the transport equation as well. An up-to-date review of this approach can be found in the recent book by Bresler et al. (Bresler, Mc Neal, and Carter, *Saline and Sodic Soils: Principles-Dynamics-Modeling*. Springer-Verlag, 236 p., 1982) and various computer codes have been developed in the last few years to solve complex cases. The chapters by Raats (Chap. 1) and by Dagan and Bresler (Chap. 2) discuss some frontier problems which depart from the traditional procedure.

In the first part of his chapter, which is still in the realm of the traditional approach, Raats describes an approximate and simplified method of solving the flow and transport equations by using the tracing of parcels. Readers with background in applied mathematics or fluid mechanics will recognize in it the method of characteristics or the Lagrangean description of flow. The second part of chapter 1 deals with the modeling of a few phenomena whose role has been recognized only recently: the representation of water contained in aggregates as immobile water and the solute exchange with the mobile water, the possible instability of wetting fronts, and the secondary flows induced by density variations. The impact of these phenomena upon transport in field conditions has to be yet established, but modelers should be aware of their existence.

The chapter by Bresler and Dagan (Chap. 2) reviews their work in a relatively new area, namely, the modeling of solute transport at field scale. The basic assumption is that large scale fields are heterogeneous, i.e., their hydraulic properties vary in space. This variability has a profound influence upon transport at field scale, which is modeled as a stochastic process to reflect the uncertainty regarding the variation of soil properties in space. The paper describes briefly the methodology developed by the researchers in order to determine the various statistical moments of the concentration. The chapter stresses the need to sample field data as a prerequisite to a realistic modeling of solute transport and emphasizes the limitations of the traditional approach, which regards soil as a homogeneous unit, whenever heterogeneity is present.

The second class of transport problems, namely, the immiscible flow of an organic pollutant in the unsaturated zone has received less attention in the past. The main reason is that the subject is of a limited scope in applications being related mainly to oil spills. The motion of immiscible fluids in a saturated porous medium has been investigated intensively in a different context by reservoir engineers. The chapter of Schwille (Chap. 3), summarizing many years of activity of his institution in this area, deals first with an exposition of the basic physical, chemical and mechanical aspects of motion of liquids immiscible with water through the unsaturated zone and in the underlying groundwater body. This part, in which the complexity of the problem is emphasized, is followed by the description of a few laboratory simulations of the downward seepage of bodies of oil products (lighter than water) and aliphatic chlorohydrocarbons (heavier than water). This comprehensive article is a useful introduction to a subject which is not covered thoroughly in the literature.

The main conclusion of the three studies is that whereas considerable progress has been made in modeling pollutants transport in the unsaturated zone under laboratory conditions, there is a need to further develop the tools necessary for predicting the motion of pollutants under field conditions.

1. Tracing Parcels of Water and Solutes in Unsaturated Zones

P.A.C. RAATS

1.1 Introduction

Convection with water is the chief mechanism for transport of solutes in the unsaturated zone. The fate of parcels of water in rather homogeneous media, subject to rather homogeneous boundary conditions can be determined on the basis of the traditional theory, and such an approach is considered in the first part of this chapter. It yields good estimates of patterns of penetration and of distributions of travel times of solutes. Unfortunately, in practice, numerous factors may complicate the flow patterns, generally leading to more erratic patterns of penetration and wider distributions of travel times. Among these factors are (1) spatial variability of the physical properties and the boundary conditions; (2) coarse structure due to aggregates, cracks, and channels; and (3) secondary flows due to density gradients in the liquid phase and to instability of wetting fronts. Implications of spatial variability are discussed by Dagan and Bresler (Chap. 2). The influences of structure and secondary flows are considered in the second part of this chapter.

1.2 Convective Transport by Simple Flows

1.2.1 Movement of Water in Unsaturated Soils

It is now widely accepted that movement of water in unsaturated soils can be described by the simplest possible balance of mass and balance of momentum, the latter expressed in Darcy's law. Restricting attention to one-dimensional vertical flows, the balance of mass for the water may be written as:

$$\frac{\partial \theta}{\partial t} = - \frac{\partial}{\partial z}(\theta v), \tag{1}$$

where t is the time, z is a vertical coordinate with its origin at the soil surface and taken positive downward, θ is the volumetric water content, and v is the velocity of the water. The volumetric flux θv is given by Darcy's law:

$$\theta v = -k[\theta]\frac{\partial h}{\partial z} + k[\theta], \tag{2}$$

$$= -D[\theta]\frac{\partial \theta}{\partial z} + k[\theta], \tag{3}$$

where h is the tensiometer pressure head, k is the hydraulic conductivity, and D is the diffusivity defined by

$$D = k \frac{dh}{d\theta}.$$ (4)

Symbols enclosed in square brackets denote functional dependence. Unlike the dependence of k upon θ, the dependence of h upon θ tends to be subject to hysteresis. As a consequence, Eq. (3) is, strictly, only valid for monotonic changes in water content from some initial condition with uniform θ and h.

From the point of view of soil physics, the relationships between h and θ and between k and θ define a soil. In theoretical developments, entire classes of soils are sometimes considered at once. For example, in an analysis of drainage of profiles with a deep water table, it is worthwhile to consider the class of soils characterized by linear retention curves and exponential dependencies of the hydraulic conductivity upon the water content (Raats 1976, 1983):

$$h = h_r + \gamma(\theta - \theta_r),$$ (5)

$$k = k_r \exp \beta(\theta - \theta_r).$$ (6)

In Eqs. (5) and (6), the subscript r denotes a reference state and γ and β are empirical constants. The primary relationships (5) and (6) imply the following derived relationships:

$$D[\theta] = D_r \exp \beta(\theta - \theta_r), \quad \text{where } D_r = \gamma k_r,$$ (7)

$$k[h] = k_r \exp \alpha(h - h_r), \quad \text{where } \alpha = \beta/\gamma,$$ (8)

$$D[h] = D_r \exp \alpha(h - h_r).$$ (9)

The reciprocal of the empirical constant α may be regarded as a characteristic length of the soil.

Equation (1) implies that a function q[z, t] exists, such that (Raats 1982):

$$\theta \equiv \frac{\partial q}{\partial z},$$ (10)

$$\theta v \equiv -\frac{\partial q}{\partial t},$$ (11)

i.e., a function q[z, t] exists, such that (1) the space rate of change of q at a certain instant specifies the water content θ and (2) the negative of the time rate of change of q specifies the flux θv. Substitution of Eqs. (10) and (11) into Eq. (1) shows that this definition of q is valid. The name parcel function is appropriate, since q effectively labels all members of a collection of parcels of water. Integration of Eqs. (10) and (11) and suitably combining the results gives an explicit expression for q:

$$q[z, t] = q[0, 0] - \int_0^t \theta_0 v_0 dt + \int_0^z \theta dz,$$ (12)

where $\theta_0 v_0$ is the volumetric flux at the soil surface. On the right hand side of Eq. (12), the first term represents the parcel of water initially at the soil surface, the second term the cumulative flux at the soil surface, and the third term the water stored above depth z. If for a particular flow, the time courses of $\theta_0 v_0$ and of the θ-profile are known, then the parcel function q[z, t] can be calculated from Eq. (12). If at time t, the parcel q[0,0] is still present in the profile, then the depth of penetration z* is defined by

$$q[z^*, t] = q[0, 0], \tag{13}$$

or equivalently,

$$\int_0^t \theta_0 v_0 dt = \int_0^{z^*} \theta dz. \tag{14}$$

1.2.2 Movement of Solutes in Unsaturated Soils

For a solute not subject to reaction or interaction with the solid phase, the balance of mass may be written as

$$\frac{\partial(\theta c)}{\partial t} = -\frac{\partial F}{\partial z}, \tag{15}$$

where c is the concentration and F is the mass flux of the solute. The flux F is assumed to be given by

$$F = \theta v c - \mathscr{D} \frac{\partial c}{\partial z}, \tag{16}$$

where $\mathscr{D}$ is the dispersion coefficient of the solute. Introducing Eq. (16) into (15), expanding the derivatives of θc and $\theta v c$, and using Eq. (1) to eliminate two terms in the resulting equation gives

$$\theta \frac{\partial c}{\partial t}\bigg|_z = \theta v \frac{\partial c}{\partial z} + \frac{\partial}{\partial z} \mathscr{D} \frac{\partial c}{\partial z}. \tag{17}$$

Using Eqs. (10) and (11) to transform (17) from the spatial coordinates (t, z) to the material coordinates (t, q) gives

$$\frac{\partial c}{\partial t}\bigg|_q = \frac{\partial}{\partial q} \theta \mathscr{D} \frac{\partial c}{\partial q}. \tag{18}$$

If $\mathscr{D} = \theta D_m$ and the dispersion occurs in a zone in which the variation of the water content is negligible, i.e., $\theta[z, t] = \theta^*[t]$, then Eq. (18) can be further reduced to

$$\frac{\partial c}{\partial \tau} = \frac{\partial^2 c}{\partial q^2}, \tag{19}$$

where

$$\tau = D_m \int_0^t (\theta^*)^2 dt. \tag{20}$$

1.2.3 Detailed Analysis of Constant Flux Absorption

Consider a semi-infinite, horizontal column with uniform initial conditions $\theta[0,z]=\theta_i$ and $c[0,z]=c_i$ and flux boundary condition $\theta v=\text{constant}=\theta_0 v_0$ (Smiles et al. 1981). It turns out that if one introduces the reduced variables $T=(\theta_0 v_0)^2 t$, $Z=(\theta_0 v_0)z$, $F=(\theta v)/(\theta_0 v_0)$, $\Gamma=(\theta_0 v_0)^2 \tau$, and $Q=(\theta_0 v_0)q$, then the constant flux $\theta_0 v_0$ no longer appears explicitly. The flux-concentration method of Philip and Knight can be used to calculate successive water content profiles. Ignoring complications near $X=0$ and $T=0$, the solution of Eq. (19) for the step change in solution concentration at $Q=0$ can be written as:

$$\frac{c-c_0}{c_i-c_0}=\frac{1}{2}\left\{\operatorname{erf}\frac{Q\Gamma^{-1/2}}{2}+1\right\}. \tag{21}$$

For a Bungendore fine sand with initial water content of approximately 0.1, Fig. 1 shows for three selected values of T calculated profiles of water content and solute concentration. The calculated profiles agreed closely with 12 observed profiles with different values $\theta_0 v_0$ and sampled in sets of 4 at the 3 selected values of T (Smiles et al. 1981).

The sharpness of the solute fronts in Fig. 1 suggests that at least for water in the soil used in the experiments reported by Smiles et al., the principle of impenetrability of continuum mechanics is closely met. Such results support the use of a sharp front approximation in the analysis of implications of random rate of infiltration and variable soil properties by Dagan and Bresler (Chap. 2). In the following two subsections convective displacement during infiltration and during drainage is considered.

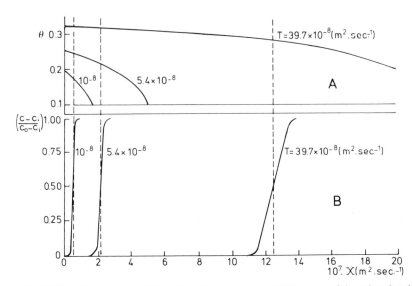

Fig. 1 A, B. Profiles of water content (**A**) and solute concentration (**B**) in terms of the reduced variables T and X. *Vertical broken lines* denote the interface between the aqueous solutions present at $T=0$ and absorbed since that time. The *curves* in **B** were calculated using $\mathscr{D}=1.9\times10^{-9}\times\theta m^2\,\text{sec}^{-1}$. (Adapted from Smiles et al. 1981)

1.2.4 Relative Speeds of Wetting and Solute Fronts

Consider infiltration into a uniform soil with initially a uniform water content θ_i and a corresponding uniform gravitational flux k_i. If at $t=0$, the rate of infiltration is changed to a value $\theta_0 v_0$, such that $k_i < \theta_0 v_0 < k_s$, where k_s is the hydraulic conductivity of the soil at saturation, then for large times, the wetting front approaches a time-invariant, moving profile. The velocity v_p of the wetting front and the flux f_r of the water relative to the wetting front are, respectively, given by:

$$v_p = \frac{k_0 - k_i}{\theta_0 - \theta_i},\tag{22}$$

$$f_r = \frac{\theta_0 k_i - \theta_i k_0}{\theta_0 - \theta_i}.\tag{23}$$

Given the dependence of the pressure head and the hydraulic conductivity upon the water content, the profiles of θ and h and the function q can be easily calculated. If the wetting front is assumed to be sharp, then it follows from Eq. (12) that the parcel function q is given by:

$$q[z,t] = q[z,0] - k_i t, \quad \text{for} \quad z > v_p t,\tag{24}$$

$$q[z,t] = q[z,0] - k_i(z/v_p) - k_0(t-z/v_p), \quad \text{for} \quad z < v_p t.$$

Behind the wetting front the velocity of a parcel of water is then k_0/θ_0, so that the ratio R_{wf} of the velocity of such a parcel to the velocity of the wetting front is given by

$$R_{wf} = \frac{k_0/\theta_0}{(k_0-k_i)/(\theta_0-\theta_i)} = \frac{k_0}{k_0-k_i}\frac{\theta_0-\theta_i}{\theta_0}.\tag{25}$$

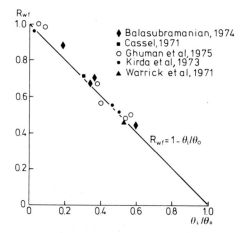

Fig. 2. Comparison of calculated *(solid line)* and experimentally measured *(data points)* ratios R_{wf} of the velocities of parcels of water behind the wetting fronts and of the velocities of the wetting fronts. (Adapted from Rao et al. 1976, where full references to the five sources of data are given)

If $k_i \ll k_0$, then Eq. (25) reduces to

$$R_{wf} = 1 - \theta_1/\theta_0 . \tag{26}$$

Figure 2 shows a comparison of Eq. (26) with experimental data from five different sources.

1.2.5 Drainage to a Deep Water Table

Drainage of uniform profiles with a deep water table requires solving Eqs. (1) and (2) subject to the initial condition $\theta = \theta_i$ for all $z > 0$ and the surface boundary condition $\theta v = \theta_0 v_0$ for all $t \geq 0$. If by dropping the first term on the right hand side of Eq. (2), the effect of capillarity is neglected, then combining Eqs. (2) and (1) gives a so-called kinematic wave equation. According to that equation, water contents in the range $0 < \theta < \theta_i$ will propagate downward from the soil surface at speeds $dk/d\theta$. Neglecting capillarity, Eq. (2) implies that the speed v of parcels of water is k/θ. Hence, if capillarity is neglected, the ratio R_{gf} of the parcel speed and the wave speed is

$$R_{gf} = \frac{k/\theta}{dk/d\theta} . \tag{27}$$

Note the similarity between Eqs. (27) and (25).

For the class of mildly nonlinear soils, with $h[\theta]$ and $k[\theta]$ relationships given by Eqs. (5) and (6), the drainage problem can be solved approximately by either a method of separation of variables (Raats 1976) or a simple iterative method (Parlange 1982). Both solutions yield water content profiles of the form (cf. Raats 1983):

$$\theta = \theta_i + \beta^{-1} \ln \frac{1 + \alpha z}{F[t]} . \tag{28}$$

According to Eq. (28), the successive water content profiles are parallel to each other (Fig. 3). For $\alpha\beta k_i t > 10$, the function $F[t]$ rapidly approaches $\alpha\beta k_i t$. Then the decrease of θ at any depth is logarithmic in time. To the casual observer, it will eventually appear as if the drainage has stopped. This is the basis for the vague notion field capacity (cf. Hadas et al. 1973).

Introducing $\theta_0 v_0 = 0$ and Eq. (28) in Eq. (12) gives

$$q[z, t] = q[0, 0] + \{\theta_i - \beta^{-1}(1 + \ln F[t]\}z$$
$$+ (\alpha\beta)^{-1}(1 + \alpha z)\ln(1 + \alpha z) . \tag{29}$$

Equation (29) shows that as soon as $F[t]$ approaches $\alpha\beta k_i t$, the parcel function q at a given depth z decreases logarithmically in the course of time.

Observations on a draining profile are commonly used to infer the dependence of the hydraulic conductivity upon the water content. Integration of (1) from the soil surface, where $\theta v = 0$, to the depth z, where $\theta v = -k\partial h/\partial z + k$, and solving for k gives

$$k = \frac{\dfrac{d}{dt}\displaystyle\int_0^z \theta dz}{\partial h/\partial z - 1} . \qquad (30)$$

If the time courses of the distributions of θ and h are measured, then $k[\theta]$ can be calculated from Eq. (30). In many cases, the data sets show one or more of the following features (cf. Raats 1983, where further references are given):

1. the gradient of the pressure head is $\ll 1$, i.e., the flow is nearly purely gravitational;
2. the average water content above depth z is a linear function of the water content at depth z;
3. the water content distributions at successive times are parallel.

These features can be understood on the basis of approximate solutions of the form Eq. (28). They are of great interest, since they relax the data requirements and/or simplify the data analysis.

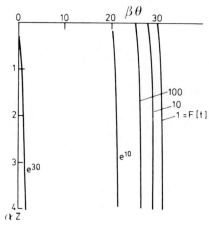

Fig. 3. Successive profiles of water content during drainage to a deep water table. (Adapted from Raats 1983)

1.3 Convective Transport by Complex Flows

1.3.1 Structured Media

In order to account for the structure of a soil, a substance may be considered to be divided between a mobile phase and a stagnant phase. The mobile phase may correspond to a network of large pores, such as the pores between aggregates; the stagnant phase may correspond to a network of small pores, such as the pores in aggregates. Important features in the model are the ratio of the capacities of the two pore systems to store the substance and the finite response time associated with the exchange of the substance between the two pore systems. Another important factor is the mechanism of transport of the substance in the mobile phase.

For any constituent, the combined balances of mass for the mobile and stagnant phases can be expressed as

$$\frac{\partial \varrho_m}{\partial t} = -\nabla \cdot \mathbf{F} - \frac{\partial \varrho_s}{\partial t}, \tag{31}$$

where ϱ_m is the bulk mass density in the mobile phase, $\mathbf{F}$ is the flux in the mobile phase, and ϱ_s is the bulk mass density in the stagnant phase. The bulk mass density ϱ_m in the mobile phase is assumed to be given by

$$\varrho_m = (\theta_m + K)c, \tag{32}$$

where c is the concentration in the mobile phase, θ_m is the volume fraction of the mobile phase, and the constant K describes instantaneous storage capacity in addition to the volume fraction θ_m. The flux $\mathbf{F}$ is assumed to be the sum of a convective component $\theta_m v_m c$ and a diffuse component $-\mathscr{D}\nabla c$:

$$\mathbf{F} = \theta_m v_m c - \mathscr{D}\nabla c, \tag{33}$$

where v_m is the velocity of the mobile phase and $\mathscr{D}$ is the dispersion coefficient.

The mechanism of exchange between the mobile phase and the stagnant phase is assumed to be Fickian diffusion. If the mobile phase is assumed to be a well-stirred fluid and the stagnant phase is assumed to occur in spheres of radius R, then the time rate of change of the bulk mass density ϱ_s in the stagnant phase is assumed to be related to the time rate of change of the concentration c by

$$\frac{\partial \varrho_s}{\partial t} = (\theta_m + K)r \int_0^{t'} \frac{\partial c}{\partial t} \sum_{n=1}^{\infty} 6\exp\{-n^2\pi^2(t'-\tau')\}d\tau', \tag{34}$$

where r is the ratio of the bulk densities in the mobile and stagnant phases at equilibrium, $t'=(D/R^2)t$ is the dimensionless current time, and $\tau'=(D/R^2)\tau$ denotes instants in the past. The combination $n^2\pi^2R^2/D$ denotes a discrete spectrum of relaxation times. The sum $\sum_{n=1}^{\infty} 6\exp\{-n^2\pi^2(t-\tau')\}$ describes the memory of the stagnant phase for changes of concentration at the boundary between the mobile and stagnant phases.

If Eqs. (32)–(34) are introduced in Eq. (31), a linear partial integro-differential equation results. For slow motion, this equation can be converted to either one of two alternative partial differential equations, in which all derivatives with respect to time occur. The derivation involves the use of Laplace transforms and is similar to a procedure long familiar in linear viscoelasticity. The resulting equations are

$$(\theta_m + k)\frac{\partial c}{\partial t} + \theta_m v_m \nabla c - \mathscr{D}\nabla^2 c$$

$$-(\theta_m + K)r\left\{1 - \frac{1}{15}\frac{R^2}{D}\frac{\partial}{\partial t} + \frac{6}{945}\left(\frac{R^2}{D}\right)^2\frac{\partial^2}{\partial t^2} - \ldots\right\}\frac{\partial c}{\partial t} \tag{35}$$

and

$$\left\{1+\frac{1}{15}\frac{R^2}{D}\frac{\partial}{\partial t}-\left(\frac{R^2}{D}\right)^2\frac{\partial^2}{\partial t^2}+\cdots\right\}$$

$$\left\{(\theta_m+K)\frac{\partial c}{\partial t}+\theta_m v_m \nabla c-\mathscr{D}\nabla^2 c\right\}=-(\theta_m+K)r\frac{\partial c}{\partial t}. \tag{36}$$

Neglecting all but the first terms of the infinite series in braces corresponds to assuming instantaneous equilibrium between the mobile and stagnant phases. Retaining the second term of the infinite series in Eq. (35) introduces a second-order derivative of c with a sign opposite from the first-order time derivative. It represents the simplest possible model of the retardation of uptake and release of solute by the stagnant phase. Retaining the second term of the infinite series in Eq. (36) introduces at once three new terms. It corresponds to assuming an equivalent film resistance between the mobile and stagnant phases and thus replacing Eq. (34) by

$$\frac{\partial \varrho_s}{\partial t}=15\frac{D}{R^2}(\varrho_s-r\varrho_m). \tag{37}$$

Simple expressions of this kind have been used in numerous studies.

Retaining only the first-order derivatives in Eqs. (35) or (36) gives:

$$\frac{\partial c}{\partial t}=-\frac{\theta_m v_m}{(1+r)(\theta_m+K)}\nabla c. \tag{38}$$

Substituting the proportional relationship between time and space derivatives implied in Eq. (38) into (35) or (36) gives

$$(1+r)(\theta_m+K)\frac{\partial c}{\partial t}+\theta_m v_m \nabla c-(\mathscr{D}+\mathscr{D}_e)\nabla^2 c=0, \tag{39}$$

where

$$\mathscr{D}_e=\frac{1}{15}\frac{R^2}{D}\frac{\theta_m^2}{(\theta_m+K)}\frac{r}{(1+r)^2}v_m^2. \tag{40}$$

This expression for the equivalent dispersion coefficient $\mathscr{D}_e$ was also obtained by Passioura (1971). Elsewhere I have given a similar analysis for structured media in which the mobile, stagnant, and inert phases occur in layers (Raats 1981b).

1.3.2 Secondary Flows Induced by Density Variations

In pollution problems, one is often dealing with movement of fluid masses embedded in other fluids in porous media. Duyvenbooden and Kooper (1981) described density currents underneath a waste disposal site near the North Sea coast. The density of the percolate was about $1.01\ kg\ dm^{-3}$. Just above the fresh–saline water interface at a depth of approximately 40 m, increased concentrations of HCO_3, NH_3, and Fe were found, resulting in a density of $1.005\ kg\ dm^{-3}$.

Density currents have also been of concern in connection with placement of fertilizers (Raats 1969).

Locally the influence of the change of the density and/or the viscosity can be read at once from Darcy's law. The difference between the actual flux $\mathbf{F}$ and the flux $\mathbf{F}_0$ of a reference liquid with the same distribution of the pressure is given by

$$\mathbf{F} - \mathbf{F}_0 \kappa \frac{\mu - \mu_0}{\mu} \mathbf{F}_0 + \frac{\theta \kappa}{\mu} (\gamma - \gamma_0)\mathbf{g}, \tag{41}$$

where μ is the viscosity, κ the intrinsic permeability, γ the density, and $\mathbf{g}$ the gravitational force per unit mass, while the subscript o denotes the reference liquid. Equation (41) is a key element in an analysis of density currents induced by a line source of salt (Raats 1969), and also in an analysis of the movement of fluid masses embedded in other fluids in porous media (Yih 1963, Raats 1981a).

In a uniform ambient velocity field, the movement of embedded masses of fluid, both in the form of cylinders with elliptical cross section and in the form of ellipsoids, is analogous to the movement of solid bodies in an ideal fluid (Yih 1963). It can be shown that the velocity of such embedded masses can be written in the concise form (Raats 1981a):

$$\mathbf{V} = \mathbf{A}\mathbf{V}_0 + \mathbf{B}\,\kappa(\gamma - \gamma_0)\mathbf{g}, \tag{42}$$

where $\mathbf{V}_0$ is the velocity of the ambient fluid and $\mathbf{A}$ and $\mathbf{B}$ are matrices, which depend on the viscosities and the shape, but not on the size, of the embedded fluid mass. As an example, for a cylindrical mass with a circular cross section and $\mu = \mu_0$, the matrices $\mathbf{A}$ and $\mathbf{B}$ are given by

$$\mathbf{A} = \mathbf{I}, \tag{43}$$

$$\mathbf{B} = \mu \begin{vmatrix} 1 & 0 & 0 \\ 0 & \frac{1}{2} & 0 \\ 0 & 0 & \frac{1}{2} \end{vmatrix}. \tag{44}$$

where $\mathbf{I}$ is the identity matrix. In $\mathbf{B}$, the component μ applies to the direction of the cylinder, while the components $\frac{1}{2}\mu$ apply to directions perpendicular to the cylinder. The matrix $\mathbf{B}$ reflects that the embedded mass drags along some of the ambient fluid. Figure 4 shows the flow pattern around the cylindrical mass and the deformation of initially horizontal lines of parcels in the ambient fluid.

1.3.3 Unstable Wetting Fronts

In almost all studies of movement of water in unsaturated soils, it is tacitly assumed that small perturbations in flow patterns will not have a tendency to grow. In other words, it is usually assumed that the flows are stable. Yet instability of flows has been observed under a wide variety of circumstances (Raats 1973 and references given there; White et al. 1976). Among the causes of instability are abrupt and gradual increases of hydraulic conductivity with depth, compression

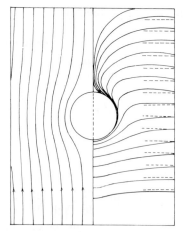

Fig. 4. Flow around a cylindrical mass of heavier liquid based on the solution of an analogous problem by Maxwell (1870). *Left:* flow pattern, *right:* deformation of initially horizontal material lines

Fig. 5. Beach of Ameland, The Netherlands. *Foreground:* barchan dune with evidence of unstable infiltration. (Photo by Corrie de Kuijer-Raats)

of air ahead of the wetting front, redistribution of water when supply from infiltration decreases or stops entirely, and water repellency of the solid phase.

An early, but not quite satisfactory, attempt to derive criteria for instability to occur was based on the Green and Ampt model and the assertion that flows are unstable, if the velocity of the wetting front increases with depth (Raats 1973). Subsequently, Philip (1975a, b) presented a more rigorous analysis based on the Green and Ampt model and the technique from fluid mechanics for analysis of the stability of a fluid motion to small disturbances. This technique was used already in the late 1950's in the context of displacement of oil by water (Wooding and Morel-Seytoux 1976). The fundamental criterion for instability emerging from Philip's analysis is that the gradient of the pressure head behind the front should oppose the flow.

Most observations of instabilities reported in the literature to date have been in laboratory experiments. In practice, erratic flow patterns due to instability may remain unnoticed or may be misinterpreted as being caused by erratic variation of physical properties. When water penetrates an already wet material, then capillarity will tend to stabilize the front. This stabilizing influence is least effective when the front is sharp. Monodisperse, coarse texture, and low initial water content favor a sharp front.

Figure 5 shows unstable penetration of water in a small barchan dune on the beach of the island Ameland at the North Sea coast of The Netherlands. Such dunes are formed when the wind blows long enough at high speed from a fixed direction. During high intensity rainfall, air is trapped between the wetting front and the wet beach sand on which the dunes rest. The increase of the air pressure above atmospheric pressure appears to be the cause of the instability of the wetting front. The instability significantly increases the depth of penetration of the water.

1.4 Conclusions

A full description of transport of solutes in unsaturated soils with time-dependent distribution of water content and water flux is rather complicated. However, in many cases, a transformation from spatial to material coordinates leads to simple results. These results can in turn be used in analyses of implications of random boundary conditions and variable soil properties.

Often the structure of the solid phase causes a lack of local equilibrium within representative volume elements, which are small in comparison with the size of the entire system. Then a distribution of liquid and/or solutes among a mobile and a stationary phase, roughly corresponding to networks of large and small pores can be used. The resulting partial integro-differential equations can be converted to partial differential equations, in which all derivatives with respect to time occur. Commonly used models based on (1) an equivalent film resistance between the mobile and stationary phases; or (2) an effective dispersion coefficient, or (3) equilibrium among the phases, correspond, respectively, to rougher approximations of these partial differential equations. As an illustration, some new results for spherical aggregates have been presented.

In a uniform ambient velocity field, the movement of embedded masses of fluid, both in the form of cylinders with elliptical cross section and in the form of ellipsoids, is analogous to the movement of solid bodies in an ideal fluid. The velocity of such embedded masses can be written as the sum of a vector proportional to the velocity of the ambient fluid and a vector proportional to the difference of the densities of the two fluids.

Instability of wetting fronts has been observed under a wide variety of circumstances. The fundamental criterion for instability is that the gradient of the pressure head behind the front should oppose the flow.

References

Duyvenboden W van, Kooper WF (1981) Effects on groundwater flow and groundwater quality of a waste disposal site in Noordwijk, The Netherlands. Sci Total Environ 21:85–92

Hadas A, Swartzendruber D, Rijtema PE, Fuchs M, Yaron B (eds) (1973) Remarks and discussion. In: Physical aspects of soil water and salts in ecosystems. Ecological Studies, Vol. 4. Springer, Berlin Heidelberg New York, pp 203–204

Maxwell JC (1870) On the displacement in a case of fluid motion. Proc London Math Soc 3:82–87

Parlange JY (1982) Simple method for predicting drainage from field plots. Soil Sci Soc Am J 46:887–888

Passioura JB (1971) Hydrodynamic dispersion in aggregated media. Soil Sci 111:339–344

Philip JR (1975a) Stability analysis of infiltration. Soil Sci Soc Am Proc 39:1042–1049

Philip JR (1975b) The growth of disturbances in unstable infiltration flows. Soil Sci Soc Am Proc 39:1049–1053

Raats PAC (1969) Steady gravitational convection induced by a line source of salt in a soil. Soil Sci Soc Am Proc 33:483–487

Raats PAC (1973) Unstable wetting fronts in uniform and nonuniform soils. Soil Sci Soc Am Proc 37:681–685

Raats PAC (1976) Analytical solutions of a simplified flow equation. Trans Am Soc Agric Eng 19:683–689

Raats PAC (1981a) Movement of fluid masses embedded in other fluids in porous media. EOS Trans Am Geophysical Union 62:807

Raats PAC (1981b) Transport in structured porous media. In: Verruyt A, Barends FBJ (eds) Flow and transport in porous media. Proc Euromech 143, Delft 2–4 September 1981 Balkema, Rotterdam, pp 221–226

Raats PAC (1982) Convective transport of ideal tracers in unsaturated soils. In: NUREG/CP-0030, Proc Symp unsaturated flow and transport modeling. US Nucl Regulat Comm Washington, pp 249–265

Raats PAC (1983) Implications of some analytical solutions for drainage of soil water. Agric Water Manage 6:161–175

Rao PSC, Davidson JM, Hammond LC (1976) Estimation of non-reactive and reactive solute front locations in soils. In: Residual Management by Land Disposal. Proc Hazardous Waste Res Symp Tucson, Arizona EPA-600/9–76–015 pp 235–242

Smiles DE, Perroux KM, Zegelin SJ, Raats PAC (1981) Hydrodynamic dispersion during constant rate absorption of water by soil. Soil Sci Soc Am J 45:453–458

White I, Colombera PM, Philip JR (1976) Experimental study of wetting front instability induced by sudden change of pressure gradient. Soil Sci Soc Am J 40:824–829

Wooding RA, Morel-Seytoux HJ (1976) Multiphase fluid flow through porous media. Annu Rev Fluid Mech 8:233–274

Yih CS (1963) Velocity of a fluid mass embedded in another fluid flowing in a porous medium. Phys Fluids 6:1403–1407

2. Solute Transport in Soil at Field Scale

G. Dagan and E. Bresler

2.1 Introduction

We consider here the transport of solutes and particularly of pollutants by water flowing downward through the unsaturated zone. Let z be a downward vertical coordinate, x and y coordinates in the horizontal plane, t the time, and C the concentration defined, for instance, as mass of solute per volume of solution. The aim of mathematical modeling of the transport phenomenon is to provide a conceptual framework and formulas, which predict C(x,y,z,t) for a given field and given flow conditions. The knowledge of the concentration distribution in space and time permits one to assess the pollution hazards in the unsaturated zone as well as the danger of contamination of the underlying aquifer.

The common approach to determining C(x,y,z,t) is to model the transport by a partial differential equation for C and to solve it for appropriate boundary and initial conditions. The validation of the model, as well as the calibration of various coefficients, is usually carried out by laboratory experiments. Thus, in a common experiment, vertical flow is created in a soil column and solute at a given concentration is introduced at the upper entrance. Subsequently, C(z,t) is measured at the bottom ($z = z_b$), the curve $C(z_b,t)$ being known as the breakthrough curve.

A large body of literature has been devoted in the past to the derivation of the transport equation and to its solution by analytical or numerical means, as well as to its validation under laboratory conditions. Some recent developments have been presented by Raats in chapter 1 of this volume.

The simplest case, which is the one discussed here for the purpose of illustration, is that of an inert solute, which does not affect the matrix properties. The transport is governed then by two mechanisms: convection of solute by water flowing with the macroscopic downward velocity on the one hand and combined molecular diffusion and pore-scale dispersion on the other. In the absence of the latter, the hypothetical C distribution is a step function, which moves downward with the water pore-velocity, this being known also as the piston flow model, whereas diffusion and pore-scale dispersion "smears" the concentration profile. For the sake of convenience and without loss of generality, we normalize the concentration such that it varies between C = 0 (initially) and C = 1 (maximal concentration at the surface). Then, rather than the step function solution $C(z,t) = 1-H(z-z_f)$, where H is the Heaviside function and $z_f(t)$ is the front location, the actual profile has a sigmoid shape with C = 0.5 at $z = z_f$. It is customary to regard the point for which C(z,t) = 0.5 as representing the front in any case. The transition zone around the front caused by dispersion is completely described by the equation of $C(z,t)$, but a rough characterization of its extent is given for instance by the depths $z_{0.1}$ and $z_{0.9}$ for which $C=0.1$ and $C=0.9$, respectively. The transition zone $L = z_{0.1} - z_{0.9}$ grows like $(Dt)^{1/2}$, where D is the diffusion-dispersion coefficient, if the latter is constant.

In agricultural and environmental applications, we are generally interested in pollution at field scale. The horizontal extent of a field is characterized by a length scale, which may vary between tens of meters, in, for example, an infiltrating pond, to kilometers or more in the case of pollution by pesticides. In any case, in most applications of interest, this scale is much larger than the vertical one, i.e., than the thickness of the unsaturated zone which is of the order of meters. The concentration is a function of x,y,z,t, but the transport is essentially a vertical one. Hence, if the soil is homogeneous and the initial and boundary conditions for water flow and solute are uniform in the horizontal plane, the solute distribution in space and time is going to be similar to that in a laboratory column under same conditions; the x, y coordinates are immaterial.

In reality, this is not the case since as a rule soils are heterogeneous, i.e., their hydraulic properties vary in space. This variation occurs with depth, many times in a systematic fashion due to layering. We are concerned here, however, mainly with variations in the plane, which may be quite irregular and of large magnitude. These variations are measured by sampling soil properties at a few locations in the field (we shall describe the results of such sampling for two examples of actual fields in Sect. 2.3). Variations in the plane may also occur in the rate of application of water on the soil surface as it happens for irrigation systems and even for rainfall.

If we regard the field as a collection of homogeneous vertical columns of varying properties in the horizontal x,y plane, the concentration profile for each column is similar to that of a laboratory one, but C depends now on x,y as well, because of heterogeneity. To illustrate schematically this point, we have sketched in Fig. 1 the three surfaces for which $C=0.9$, $C=0.5$, and $C=0.1$, respectively, at a fixed t, for a field for which solute at $C=1$ has been applied on the surface $z=0$ at $t=0$. In Fig. 1a, these are horizontal planes corresponding to an ideal homogeneous field and to uniform conditions, whereas in Fig. 1b, they are the irregular surfaces which one should expect in an actual heterogeneous field.

A complete description of the concentration distribution in a heterogeneous field is achieved if we can determine $C(x,y,z,t)$ and particularly $z_{0.1}(x,y,t)$, $z_f(x,y,t)$ and $z_{0.9}(x,y,t)$. This is a formidable task (Philip 1980, regards it as

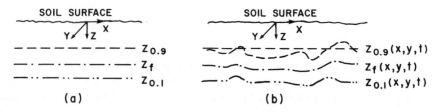

Fig. 1 a, b. Surfaces of equal concentration at time t after beginning of solute transport: **a** ideal homogeneous field, **b** heterogeneous field

pertaining to the realm of "trans-science"), which faces two obstacles at least: first, we do not possess the necessary detailed data on the distribution of soil properties and on boundary conditions in space and secondly, even if the data are available, the computational problem is exceedingly complex. Fortunately, in applications we may seek only some gross properties of the concentration distribution of Fig. 1 b over the entire field. Thus, rather than asking what is the concentration at each x,y,z, and t, it is sufficient to know the percentage of the area of the field at a given depth z and time t for which the concentration is larger than a given value. Figure 1 b depicts such a control horizontal plane and it is seen that the concentration is larger than 0.9 over the area of that plane above the surface $z_{0.9}(x,y,t)$. In other words, this type of information gives us the extent of the contaminated area, but not its detailed shape. The simplest and basic type of information is, of course, the average concentration $\bar{C}$ over a horizontal plane as a function of z and t, which is easily determined from the previous function. This would be enough in practice to determine, for instance, the rate of pollutant leaching to groundwater from the *entire* field (a more systematic characterization of C is discussed in Sect. 2.2).

The traditional approach of transport modeling is to regard the field as an equivalent homogeneous unit and to attempt to predict the average distribution $\bar{C}(z,t)$. This approach raises a few basic questions of relevance in applications like: (1) is it possible to define a homogeneous field of effective properties such that the concentration distribution in it is identical to the average concentration $\bar{C}(z,t)$ in the actual heterogeneous field? (2) if the answer is positive, how can we determine the effective properties? (3) what is the level of confidence in results as a function of the degree of variability and amount of data? (4) what can we say about the variation of C around its mean $\bar{C}$? and (5) do we need accurate and intricate models of transport in a column in order to describe the average concentration in the field?

We have shown (see, e.g., Bresler and Dagan 1981, 1983 a, b) that the answer to the first question is generally negative and therefore there is no answer to (2). As for questions (3) and (4), answers can be obtained only if we take into account the field heterogeneity. Paradoxically at first glance, the answer to question (5) is that a fairly accurate description of the average behavior of the solute transport in the field can be achieved by using simple flow models, precisely as is the case for the macroscopic laws of flow in a porous medium, which can be derived with the aid of relatively simple models of the pores.

The aims of this paper are to present and summarize the results we have obtained so far on modeling solute transport at field scale and to point out some problems yet to be solved.

2.2 Basic Concepts of Stochastic Modeling

We consider first independent variables like soil properties (for instance, hydraulic conductivity at saturation K_s or pore-scale dispersivity λ) or like water rate of application on the surface R and denote these by X,Y,Z, ... These are assumed to vary in the field in the x,y plane, i.e., $X = X(x)$, $Y = Y(x)$ ... where x is a coordinate vector of Cartesian components x,y. These variables are measured at a set of points x_j $(j = 1, ..., M)$ and as a rule, they are found to vary in an irregular manner in the plane. The value of X at any point other than x_j is subjected to uncertainty and its prediction on the basis of X at x_j is error-prone. These errors are of two types: measurement errors associated with instrumentation, interpretation, etc. and interpolation errors, which are related to the spatial variability of X and the scarcity of measurement points.

We shall not distinguish here between these two types of errors and we account for them by regarding X,Y,Z, ... as random functions of x. A general and complete mathematical characterization of such functions is quite complex (see, e.g., Beran 1968) and we shall limit our knowledge to:
– the probability density functions $f(X), f(Y)$ ... and particularly the expectation value

$$\mu_X = \langle X \rangle = \int X f(X) \, dX \qquad (1)$$

and the variance

$$\sigma_X^2 = \langle X'^2 \rangle = \int X'^2 f(X) \, dX; \qquad X' = X - \mu_x \qquad (2)$$

– the two-points covariance

$$C_X(x_1, x_2) = \langle X'(x_1) X'(x_2) \rangle = \sigma_X^2 \varrho_X(x_1, x_2), \qquad (3)$$

where X' is the residual and ϱ_X is the correlation function. It is emphasized that the averaging operation represented by $\langle \ \rangle$ is an ensemble averaging, i.e., it is carried out on all possible fields of heterogeneous structure similar to the actual one.

Generally, $f(X)$, μ_X, and σ_X^2 depend on x, whereas C_X depends on both x_1 and x_2. Our basic assumption is of weak stationarity, i.e., μ_X is constant or has a simple trend, whereas $f(X)$ does not depend on x. Furthermore, $C_X(x_1, x_2)$ or equivalently the variogram of the residuals γ_X defined by

$$\gamma_X(x_1, x_2) = \tfrac{1}{2} \langle [X'(x_1) - X'(x_2)]^2 \rangle \qquad (4)$$

depend on the lag $x_1 - x_2$ rather than x_1 and x_2 separately. If stationarity prevails, we are entitled to make use of the ergodic hypothesis and to exchange ensemble and space averaging. In other words, given enough data, we can identify $\mu_X, f(X')$ and C_X from the actual field by well-known techniques of statistical inference and we shall assume that this is indeed the case.

The next step is to adopt a deterministic model of flow and transport in a vertical homogeneous column, e.g., Richard's equation for water flow and the convection-diffusion equation for solute transport. For given soil properties X, Y and boundary condition represented by Z, the mathematical model will provide solutions for the flow variables, e.g., the moisture content θ and the solute concentration C. These dependent variables can be expressed in principle as follows

$$C(x,y,z,t) = C(z,t;X,Y,Z), \qquad (5)$$

where in the rhs of Eq. (5) we have emphasized that C is a function of z and t in a given profile, but it depends on the horizontal coordinates through the soil and water application variables X,Y,Z. Since the latter are random functions, the same is true for C. Thus, we can characterize C only in probablistic terms, e.g., by its expectation value and variance rather than by deterministic values. The average concentration is therefore given by

$$\langle C(z,t) \rangle = \int \int \int C(z,t; X, Y, Z) f(X) f(Y) f(Z) \, dX \, dY \, dZ, \qquad (6)$$

where it has been assumed, for simplicity, that X, Y, and Z are independent random functions. In a similar way, we can calculate $\sigma_C^2(z,t)$ and any other second-order moment of the concentration field.

By the ergodic argument, we can write

$$\langle C(z,t) \rangle \cong \bar{C}(z,t) = \frac{1}{A} \int_A C(x, y, z, t) dx \, dy, \qquad (7)$$

where A is the field area and $\bar{C}$ is a space average. The application of this fundamental equality is subjected to two conditions: first that the probability density functions $f(X), \ldots$ are known accurately and secondly, that A is of much larger extent that the correlation scales of X,Y, and Z. If these conditions are not fulfilled, $\bar{C}$, regarded as a random variable, has a variance which is different from zero. We have analyzed this topic in a previous work (Russo and Bresler 1981) and we shall not discuss it here. Assuming that Eq. (7) is valid, with a similar equation for the variance

$$\sigma_C^2(z,t) \cong \frac{1}{A} \int_A (C - \bar{C})^2 dx \, dy, \qquad (8)$$

we have answered the basic problem raised in Sect. 2.1, namely, the computation of the gross features of the concentration distribution at field scale.

Summarizing this section, the stochastic model is based on two basic components: models of flow and transport in a homogeneous vertical column on the one hand and the second-order statistical structure of soil properties and boundary conditions on the other.

Among the various questions raised in Sect. 2.1, we wish to make a comment at this point about the last one. Assume that we use an approximate model of vertical flow and transport so that the solution for the concentration is affected by a model error ε_{Cm}, which depends also on x,y through the random functions X,Y,Z. Thus, ε_{Cm} is also a random function and its impact upon $\langle C \rangle$ [Eq. (7)] and upon σ_C^2 [Eq. (8)] stems from its average $\langle \varepsilon_{Cm} \rangle$ and variance σ_ε^2. It may happen, and we have shown that this is indeed the case for the data pertaining to actual heterogeneous fields, that $\langle \varepsilon_{Cm} \rangle$ is close to zero because of mutual cancellations and σ_C^2 is influenced at a larger extent by $\sigma_X^2, \sigma_Y^2 \ldots$ than by σ_ε^2. In other words, at first glance our stochastic model seems to be prohibitively complex, since it requires to solve first the water flow and transport problem in a homogeneous profile, which is a difficult problem in itself, and subsequently, to carry out statistical averages as in Eq. (6). Our comment above suggests, however, that fairly accurate results for $\langle C \rangle$, for instance, can be obtained by using crude models of flow and the computational gains may offset the additional complexity related to heterogeneity.

2.3 Succinct Review of Previous Work

Stochastic modeling of unsaturated flow and of transport in the unsaturated zone has been developed only recently. The review here is confined to our own work in the area, but we shall mention first the article by Biggar and Nielsen (1976), who carried out a leaching experiment in a field of Panoche soil and analyzed the spatial variability of the pore-scale dispersion coefficient D. We interpreted their data (Bresler and Dagan 1981) in order to determine the dispersivity λ, which is in the order of milimeters in a typical laboratory experiment. It has been found that λ is log-normal with an average $\bar{\lambda} \cong 3$ cm and a variance of $\sigma^2_{\ln \lambda} \cong 1.4$. The large value of $\bar{\lambda}$ as compared to laboratory data can be attributed to local heterogeneities resulting from aggregation, cracks, etc., which are present in a soil profile in the field. The finite and relatively large variance reflects the spatial variability of such events both in the vertical profile and among profiles in the plane. Thus, even local heterogeneity leads already in field conditions to values of dispersivity, which are much larger than those determined under standard laboratory conditions.

In our first works on modeling of transport in heterogeneous fields (Bresler and Dagan 1979), we considered a Panoche field soil, whose properties have been measured and analyzed by Warrick et al. (1977). They found that the hydraulic conductivity at saturation K_s can be related to a similarity parameter which is log-normal. With $X = \ln K_s$, it was found that

$$\mu_X = -2.76; \quad \sigma_X = 2.32, \quad (9)$$

where K_s is expressed in cm^2/h. We have adopted some simple power relationships for the dependence of $K(\theta)$ and $\psi(\theta)$ upon θ such that the soil hydraulic properties are characterized by a few parameters besides K_s, e.g., θ_s (moisture content at saturation) and θ_{ir} (irreducible θ) and η (characterizing pore size distribution). Furthermore, we have assumed that only K_s vary throughout the field, while the other parameters have deterministic constant values. Next, to investigate transport, we have adopted the simplest conditions of water flow, namely, we have assumed that flow is caused by application of a recharge Y on the soil surface, which is steady in time, but varies irregularly in the plane. Furthermore, we have assumed that Y is applied long enough to create a gravitational steady flow in each profile. Thus, along the vertical, both the moisture content θ and the pore velocity $v = K/\theta$ are constant, but they vary in the plane because of the variability of X and Y. Transport of a solute has been assumed to start at a given time $t = 0$, when the concentration in the soil is uniform and equal to zero, whereas on the surface $C = 1$, for $t > 0$. The concentration profile has been assumed to have a step function shape, i.e., we have neglected pore-scale dispersion. Thus, the piston flow solution is given by

$$C = 1 - H(z - z_f); \quad z_f = vt, \quad (10)$$

i.e., the front between the invading solute and the native water is moving downward at constant velocity v in time, but varying in the plane, since v is a function of X and Y.

With given probability density functions of X (lognormal) and Y (rectangular) we could calculate in a simple way the various concentration statistical

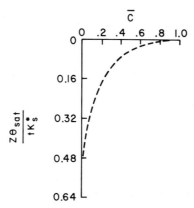

Fig. 2. The average concentration as function of dimensionless depth calculated for Panoche soil field (Bresler and Dagan 1979)

moments. To illustrate the point, we reproduced in Fig. 2 an example of a profile of the concentration expectation value $\langle C(z, t) \rangle$ which is presumably equal to its average in the plane $C(z, t)$ [Eq. (7)]. The selected profile is for the ratio between the average recharge and the average K_s equal to $\bar{Y}/\bar{K}_s \cong 0.05$.

Examination of results (Bresler and Dagan 1979) and of Fig. 2 reveals some interesting points: (1) the shape of the $\bar{C}(z, t)$ profile for a fixed t differs considerably from the step function one would have obtained for the same field by assuming that it behaves as a homogeneous unit with constant recharge over the entire surface; (2) the shape also departs from the sigmoid, which is characteristic of laboratory experiments; (3) the profile expands linearly with time, i.e., the transition zone defined in Sect. 2.1 $L = z_{0.1} - z_{0.9}$ grows linearly with t, whereas in a diffusion process the growth is like $t^{1/2}$; and (4) if we define an equivalent dispersion and dispersivity, λ_{eq} the latter is found to grow with time and to be much larger than pore-scale dispersivity. Thus, for the case of Fig. 2 and for $z_f = 1$ m, $\lambda_{eq} \cong 74$ cm.

Since $\bar{C}$ represents the average concentration at depth z and at time t, these conclusions are relevant to problems of pollution in which we wish to determine the extent of pollutant spread into the soil profile.

In the next stage of our investigations (Bresler and Dagan 1981), we considered in addition to the Panoche soil, the Bet-Dagan field, the statistical properties of which have been reported elsewhere (Russo and Bresler 1981). Thus, the pertinent data were now

$$\mu_X = 2.27 \; ; \quad \sigma_X = 0.8 . \tag{11}$$

Comparison with Eq. (10) shows that the Bet-Dagan soil is much more permeable and less heterogeneous than the Panoche soil, which can be attributed in part to its smaller size. At this stage, we introduced in addition to the two random functions $X = \ln K_s$ and Y also the pore-scale dispersivity $Z = \ln \lambda$. As a result, the concentration in each profile is no more a step function, but displays a gradual transition. The dispersivity was assumed to be lognormal with field values taken from the aforementioned work by Bigger and Nielsen (1976).

$$\mu_Z = 1.41 \; ; \quad \sigma_Z = 1.17 . \tag{12}$$

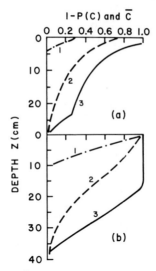

Fig. 3a, b. The average concentration $\bar{C}$ and the probability 1-P(C) that concentration is larger than C=0.1 or C=0.9 as a function of z for fixed t. **a** Panoche soil, t=24 h: *line 1* (C=0.9, λ=3 cm), *line 2* ($\bar{C}$, λ=3), *line 3* (C=0.1, λ=3). **b** Bet-Dagan soil, t=5 h: *line 1* (C=0.9, λ=3 cm), *line 2* ($\bar{C}$, λ=3), *line 3* (C=0.1, λ=3) (Bresler and Dagan 1981)

The main conclusions (Bresler and Dagan 1981) were: (1) the variability of pore-scale dispersivity and recharge have a much lesser impact than that of K_s upon concentration and they can be taken as constant and equal to their average field values, (2) in the case of the Panoche soil, which is of considerable heterogeneity, the influence of pore-scale dispersion is small and the transport mechanism is dominated by the large spatial variations of the convective velocity. In contrast, for the less heterogeneous Bet-Dagan soil, the augmented field pore-scale dispersion (λ=3 cm) has a definite influence on the concentration distribution and (3) in any case, the soil cannot be modeled as a homogeneous unit.

To illustrate the results of this stage, we reproduced in Fig. 3, the probability that the concentration is larger than 0.1 or 0.9, as well as average $\bar{C}$, as a function of z for t=24 h.

The curve of probability that C is larger than 0.1 can be interpreted as representing the percentage of the field area at various z and fixed t for which the concentration is larger than 0.1. Thus, under the conditions specified in Fig. 3 and for the Panoche soil, 50% of the field is at a concentration larger than 0.1 at a depth of 9 cm at t=24 h, whereas in Bet-Dagan, the corresponding figure is 24 cm. Again, such information cannot be obtained from the traditional solution of the transport problem for a homogeneous field.

Finally, in a series of recent articles (Dagan and Bresler 1983 and Bresler and Dagan 1983a, b), we solved the problem of transient water flow (an infiltration and redistribution cycle) and that of solute transport, both being affected by the variability of K_s only. We would like to mention here one of the major conclusions of these studies, namely, that an effective soil conductivity K_{ef}, which would lead to

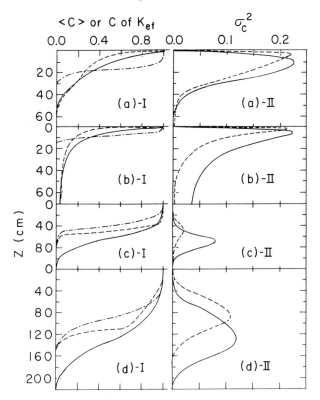

Fig. 4a–d. *I Average concentration* $\bar{C}$ as a function of depth after infiltration (deterministic for uniform soil of effective conductivity, *dashed-dotted line;* for steady gravitational flow in heterogeneous soil, *dashed lines;* for unsteady infiltration *solid line*). *II* Concentration variance σ_c^2 as function of time (symbols as in *I*). **a** Panoche soil, water rate of application $Y = 0.5\,\mathrm{cm/h}$, infiltration time $t^i = 24\,\mathrm{h}$; **b** Panoche soil, $Y = 6.5\,\mathrm{cm/h}$, $t^i = 6\,\mathrm{h}$; **c** Bet-Dagan soil, $Y = 0.5\,\mathrm{cm/h}$, $t_i = 24\,\mathrm{h}$; **d** Bet-Dagan soil, $Y = 6.5\,\mathrm{cm/h}$, $t^i = 6\,\mathrm{h}$

a concentration profile identical to the average concentration $\bar{C}(z, t)$ cannot be defined. This point is illustrated in Fig. 4, in which we represented the average concentration profiles at two different times in the two types of soil and simulated by three different methods (curves I): (1) by assuming that flow takes place in a deterministic uniform field of an assumed effective saturated conductivity (dashed-dotted line); (2) by assuming steady gravitational flow in the heterogeneous field along the lines of the forementioned work (Bresler and Dagan 1981); and (3) by simulation infiltration in the heterogeneous field.

Similarly, the curves II represent the concentration variance [Eq. (8)] computed by methods (2) and (3) above.

Figure 4 reveals again that there is considerable difference between the average concentration profile computed for an equivalent uniform field and for the actual heterogeneous field in the case of Panoche soil (Fig. 4 a, b) and a significant, though lesser one, for the Bet-Dagan field (Fig. 4, c, d). Furthermore, the steady flow model performs similarly to the unsteady one for the more heterogeneous Panoche soil, but not so much for the less heterogeneous Bet-Dagan soil.

Finally, the concentration variance σ_C^2 (Fig. 4II) is zero at the soil surface, because of the assumption that $C=1$ is given deterministically and is also zero at large depth where $C=0$. Its maximum is reached close to the depth for which $\bar{C}=0.5$ and can reach the maximal theoretical value $\sigma_C^2=0.25$.

2.4 Conclusions and Outline of Future Investigations

We would like to emphasize first a few general conclusions of our studies reviewed in Sect. 2.3. First, the soil spatial variability may have a significant impact upon solute distribution at field scale, depending on the degree of heterogeneity and areal extent of the field. Secondly, in a heterogeneous field, the gross properties of the solute distribution (average, variance) can be computed with the aid of simplified models of transport in a homogeneous profile. The third conclusion, stemming from the previous ones, is that in the future emphasis should be put on better knowledge of field data rather than on further refinements of models of transport in homogeneous columns.

Modeling of flow and transport in spatially variable fields has been developed only recently, as witnessed by references here. Many problems are still to be addressed and we would like to mention a few we are concerned with: (1) impact of adsorption and its spatial variability upon solute transport; (2) the feed-back mechanism of change of soil properties due to presence of solutes; (3) the effect of the roots zone upon water and solute distribution; and (4) the influence of spatial variability at depths larger than the few meters of the upper layer and particularly in the zone adjacent to a water table. These and other related subjects pose challenging problems to those engaged in modeling of transport of pollutants in the unsaturated zone.

References

Beran MJ (1968) *Statistical Continuum Theories.* Interscience, New York

Biggar JW, Nielsen DR (1976) Spatial variability of leaching characteristics of a field soil. Water Resour Res 12:78–84

Bresler E, Dagan G (1979) Solute dispersion in unsaturated heterogeneous soil at field scale. II. Applications. Soil Sci Soc Am J 43:467–472

Bresler E, Dagan G (1981) Convective and pore-scale dispersive solute transport in unsaturated heterogeneous fields. Water Resour Res 17:1683–1693

Bresler E, Dagan G (1983a) Unsaturated flow in spatially variable fields. II. Application of water flow models to various fields. Water Resour Res 19:421–428

Bresler E, Dagan G (1983b) Unsaturated flow in spatially variable fields. III. Solute transport models and their application to two fields. Water Resour Res 19:429–435

Dagan G, Bresler E (1983) Unsaturated flow in spatially variable fields. I. Derivation of models of infiltration and redistribution. Water Resour Res 19:413–420

Philip JR (1980) Field heterogeneity: some basic issues. Walter Resour Res 16:443–448

Russo D, Bresler E (1981) Soil hydraulic properties as stochastic processes. I. An analysis of the field spatial variability. Soil Sci Soc Am J 45:682–687

Warrick AW, Mullen GJ, Nielsen DR (1977) Scaling field-measured soil hydraulic properties using a similar media concept. Water Resour Res 13:355–362

3. Migration of Organic Fluids Immiscible with Water in the Unsaturated Zone

F. Schwille

3.1 Introduction

In systemizing fluids potentially harmful to groundwater with regard to their migration behaviour, it has been proven useful in groundwater hydrology to divide these into two main groups

miscible with water – immiscible with water

based on fluid dynamics. This division is indispensable since the simultaneous flow of two (or more) immiscible fluids gives a completely different migration pattern than that of the simultaneous flow of fluids miscible with one another. Accordingly, one must fundamentally differentiate between the two types of flow

one phase flow – two (multi-) phase flow.

For fluids immiscible with water, one is concerned nearly exclusively with organic substances, which one can summarize in two main groups based on their potential for contaminating groundwater:

I: natural organic "products" (mineral oils)

II: organic chemical products (chemicals).

Under mineral oils are understood crude oils as well as refined products which under normal conditions are fluid, such as fuels and heating oils and further refined products, such as motor oils and cutting oils. Also included are products obtained from the coking and carbonization of coal and from the extraction of oil-bearing rock and their fluid refined products, e.g. benzene and toluene. The majority of spills with organic fluids which contaminate groundwater still result from mineral oils.

As chemicals, all "unnatural" anthropogenic substances are included. As representatives of this group, the low molecular solvents may be mentioned, which recently appear to be providing severe competition for the mineral oil products as groundwater contaminants in some areas in industrialized countries.

Although the fluid dynamic principles of multiphase flow were laid down by petroleum scientists about 50 years ago, groundwater research took over their findings on the explanation of the flow behaviour of mineral oil products only 2 decades ago. Only 3 years ago, the principle of multiphase flow was transferred to the subgroup of low molecular solvents as well. As is shown by certain theoretical considerations in connection with laboratory studies, under certain conditions the principle of multiphase flow may be applied fundamentally to all fluids immiscible with water, whereby of course the different physical-chemical properties lead to a multiplicity of modifications and variations of the hypothetical postulate (Wyckoff and Botset 1936, Leverett and Lewis 1941, van Dam 1967, Schwille 1982).

3.2 Multiphase Flow

The migration of fluids underground depends on the type and composition of the rock formation as well as on the type and composition of the fluid. Because of the different kinds of flow paths, one must consider:

porous rocks (media) – fissured (fractured) rocks (media).

Of course, there are differing transitional forms, as for example, the porous fissured rocks, which have a certain permeability and storage capacity in the pore space, but whose groundwater flow occurs mainly in the fissures. The karst rocks are properly assigned to the fissured rocks and are not to be considered as an independent rock group.

The principle of the simultaneous flow of two fluids immiscible with one another in a porous medium is shown in the well-known diagram (going back to Wyckoff and Botset 1936) of relative permeability for a wetting and a nonwetting fluid (Fig. 1). Water is usually the wetting fluid and mineral oils or chlorohydrocarbons, for example, are the nonwetting fluids. Since for each fluid flowing, only a part of the pore space and thus only a part of the cross section A under consideration is available, the discharge Q of each fluid must be lower, corresponding to their proportion of the cross-sectional area A. Two phase flow may then be described by modification of the general Darcy Law, so that

$$Q = -\frac{k\varrho}{\mu} \cdot A \cdot \frac{d\theta}{dx}, \tag{1}$$

(where k represents the permeability, ϱ the density, μ the dynamic viscosity and θ the fluid potential), in which the permeability k, having the dimension of an area, is supplemented for each phase by the factors:

k_{rw} relative permeability for water in the presence of a nonwetting fluid,
k_{rnw} relative permeability for the nonwetting fluid in the presence of water.

Then one obtains

$$Q_w = -\frac{k_{rw}k\varrho_w}{\mu_w} \cdot A \cdot \frac{d\theta_w}{dx}, \tag{2}$$

$$Q_{nw} = -\frac{k_{rnw}k\varrho_{nw}}{\mu_{nw}} \cdot A \cdot \frac{d\theta_{nw}}{dx}. \tag{3}$$

The relative permeabilities according to Eqs. (2) and (3) depend on the degree of saturation (S) expressed as the proportion of pore space of each of the two phases (Fig. 1). It is remarkable that the relationships between the relative permeabilities

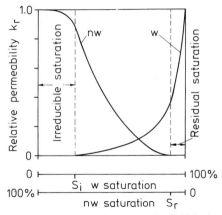

Fig. 1. Relative permeabilities for a wetting (*w*) and a nonwetting fluid (*nw*) as a function of the degree of saturation

and degree of saturation are not in the least linear and that the curves don't begin on the abscissa at $S = 0$ or $S = 100$. Thus, the fluids hinder each other in the pore space to a varying extent; the permeabilities decrease differently related to the one-phase permeabilities (i.e., permeability at $S = 100\%$). Each of the fluids must reach a minimum saturation degree before it can flow at all in the medium in question. Thus, water cannot flow until the degree of saturation S_i is reached, since it is in a state of irreducible saturation. The nonwetting fluids cannot flow until the degree of saturation S_r is reached, since these substances are in a state of residual saturation and under normal pressure conditions not mobilizable. The effect of residual saturation is the real reason why a particular quantitiy of a fluid immiscible with water in a porous medium can only spread out a specific distance and only occupy a certain volume of the medium.

The model for multiphase flow may be understood considerably easier when considering fissured media or a single fissure than when considering porous media. Based on model studies, Romm (1966) came to the conclusion that for two-phase flow in very narrow fissures (fissure with $c \sim 0.1$ mm), the dependence of the relative permeability on the degree of saturation must be linear in character (i.e., $k_{rw} \, k = S_w$ and $k_{rnw} \, k = 1 - S_w$) (Fig. 2). This conclusion should be examined in our opinion. According to our own studies, a type of residual saturation occurs also in fissures with $c < 0.2$ mm with hydraulically rough fissure walls for unsaturated flow, so that after the flow process is finished, isolated droplets of the nonwetting phase remain in the fissure.

If air occurs in the pore space as a third phase – air is the nonwetting fluid relative to water – then one talks about three-phase flow. Although this process is considerably more complicated than two-phase flow, it follows essentially the same principles, the flow equation for air corresponds accordingly to Eq. (3). The presentation of the relationships in the well-known triangle diagram (Fig. 3) shows that now in even larger areas at least one of the fluids is immobile and that all three fluids can flow simultaneously only in an extremely limited saturation region. There is also residual saturation for air, which contributes to a lowering of the permeability of the other phases.

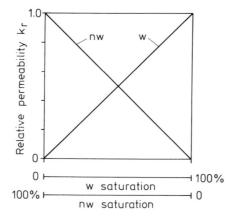

Fig. 2. Relative permeabilities for a wetting (*w*) and a nonwetting (*nw*) fluid as a function of the degree of saturation for flow between parallel planes (Romm 1966)

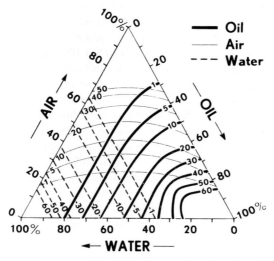

Fig. 3. Relative permeabilities for three-phase flow (Van Dam 1967)

Clear as the model presentation based on two-phase flow may be for general considerations, considerable difficulties arise in quantitative application. The dependence of the relative permeability on the degree of saturation can only be derived in the laboratory with considerable time necessary; the determination of relative permeability is therefore not a routine laboratory procedure. Diagrams for two- or three-phase flow have been derived nearly exclusively for mineral oils.

Apparently the residual saturation S_r is the limiting factor for the estimation or calculation of the spreading out of immiscible fluids for generally limited discharge quantities. In practice, the knowledge of the maximum migration is of incomparably greater interest than a detailed knowledge of the flow processes themselves. The determination of the residual saturation with relatively simple laboratory methods must therefore be considered as a high priority task. In this context, however, differentiation should be made between residual saturation in the unsaturated zone, where the nonwetting fluid is in the pendular stage and residual saturation in the saturated zone, where it is in the insular stage (Dracos 1966, van Dam 1967, Bear 1972, Schwille 1975, Schiegg 1977, Fried et al. 1979).

3.3 Physical Characteristics

Assuming that the substances in the temperature range (0 °–30 °C) in which the migration processes take place beneath the ground are fluid (i.e., neither boil nor solidify nor stop flowing), then only a few physical-chemical characteristic numbers are required in order to evaluate the migration process with sufficient reliability. It is taken for granted that the properties of the rock formation are essentially known.

The miscibility or immiscibility with water, or what is the same, the solubility of the fluid in water, is the decisive determining parameter for the migration process. Fluids miscible with water are fully displaced by water, while immiscible fluids are held back in residual saturation in the pore space and can be further transported by water only according to their solubility.

The density ϱ is the next most important parameter to the extent that the difference in density between fluid and groundwater determines the level in which the migration chiefly takes place relative to the aquifer: in the region of the groundwater table ($\varrho < 1$) or in the region of the impervious bedrock ($\varrho > 1$). Even small density differences of a few tens of grams per kilogram groundwater can be descisive.

The viscosity on the other hand determines the velocity of the flow process, thus, the conductivity of the porous medium for the fluid (fl) concerned.

For $\qquad\qquad S_{fl} = 100\%, \qquad K_{fl} = \dfrac{\varrho k}{\mu} = \dfrac{k}{v},$ $\qquad\qquad$ (4)

where $v = \mu/\varrho$ is the coefficient of kinematic viscosity.

The surface tension σ is responsible for capillary effects as well as for the spreading of fluids on free water surfaces. It is without significance in the first phase of spreading, in which the effect of gravity prevails, but gains increasing importance towards the end of the spreading phase.

Many organic fluids, e.g., solvents or gasoline, possess a high volatility, whose underground effect has been largely underestimated up to now. The evaporation rate of a fluid is mainly dependent on the vapor pressure and the latent heat of evaporation of the fluid. Since the rate of evaporation cannot be calculated from the boiling state and must be determined in each case under specific conditions, it has proven useful to provide relative values, so-called evaporation numbers, which may be determined very easily. Ether is chosen as reference fluid with an evaporation rate set equal to 1. The values of other fluids are given as a multiple of this evaporation number. The determination occurs at 20 °C and 65% relative humidity so that natural conditions are approximated.

The relative density of vapor (relative to air $=1$ at 20 °C) decides whether it sinks in the unsaturated zone and spreads over an impermeable layer or over the capillary fringe (values >1) or whether the vapor rises (values <1).

3.4 The Migration of the Immiscible Phase

The behavior of immiscible fluids underground will now be schematically shown using as examples the above mentioned mineral oil products (oil) and aliphatic chlorohydrocarbons (CHC's). We feel that these two groups are sufficiently representative for by far the largest proportion of organic substances immiscible with water, which are potentially harmful to groundwater, as far as migration behavior is concerned. To begin with, the migration procedure is presented for a porous medium, whereby for a complete understanding, the unsaturated as well as the saturated zone must be treated (Schwille 1981, 1982).

Whether infiltrating oil or CHC's reach the groundwater table or not, depends decisively upon the spilled volume, the infiltration process (e.g., quasi-point or areal source, infiltration rate) and upon the retention capacity of the porous medium. The retention capacity for oil in the unsaturated zone lies in general by 3–$5 \, l \, m^{-3}$ in highly permeable media and at the most by 30–50 $l \, m^{-3}$ in media of low permeability; for CHC's similar values were found. The heterogeneity of the unsaturated zone (e.g., alteration of different permeable layers) and the degree of water saturation of the individual layers (e.g., so-called perched capillary fringes at the boundary of less permeable to more permeable layers) results in a lateral component of the percolation process and consequently a broadening of the oil/CHC-body with increasing depth (Fig. 4). The heterogeneity of the unsaturated

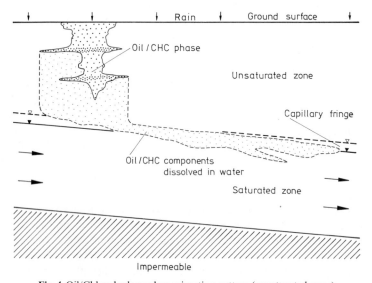

Fig. 4. Oil/Chlorohydrocarbon migration pattern (unsaturated zone)

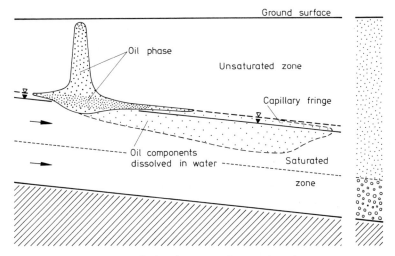

Fig. 5. Oil migration pattern (saturated zone)

zone thus results in general in a considerable reduction of the penetration depth in comparison with that of a homogeneous medium. The retention capacity of the unsaturated zone can therefore only be estimated satisfactorily with a good knowledge of its structure and moisture content. Oil and CHC's behave similarly as a phase in the unsaturated zone, disregarding different percolation velocities as a result of different viscosities.

Oil and CHC's behave fundamentally differently, however, if the infiltrated fluid volume exceeds the retention capacity of the unsaturated zone. Let us first consider the oil (Fig. 5). As soon as the oil reaches the upper boundary of the saturated zone, which lies within the capillary fringe, it spreads out at first downwards as well as laterally, the extent depending on the pressure of the fluid. But the lateral spreading quickly dominates. With decreasing fluid pressure, the oil which was forced beneath the groundwater table tries to reach the surface of the saturated zone again. The tendency of the oil to "swim" up to the surface is so strong that even oil injected beneath the groundwater table does not remain in the saturation zone, except for that part corresponding to the residual saturation.

In principle, the given oil migration patterns (Figs. 4 and 5) are applicable to crude oils, too. The flow characteristics of crude oils are very complicated, however, and depending on their origin, dissimilar. At temperatures at which all crude oil components are dissolved within one another, crude oil is a true Newtonian fluid. When the oil is cooled, substances (paraffins and others) crystallize out and form agglomerates. In this way, the flow behavior of the oil is considerably altered. This microcrystalline precipitate can lead to stoppages in fine-pored media, which can bring the flow of oil to a stop even though the viscosity of the oil itself would allow flow. For crude oils, therefore, the viscosity may be considered only as a first criterion for the velocity of the flow process.

A similar effect, but one caused completely differently, may be found for fresh crude oils with volatile components. Where these components have an opportunity

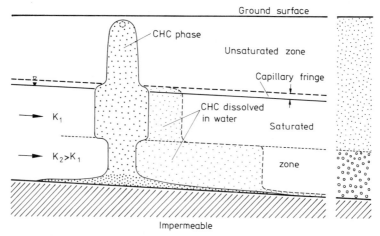

Fig. 6. Chlorohydrocarbon migration pattern (saturated zone)

to evaporate, the crudes become increasingly viscous and flow correspondingly slower.

The flow behavior of certain crude oils cannot therefore be derived from the usual relevant fluid dynamic parameters without reservation. It is recommended that the flow behavior of crude oil be determined in the laboratory in specific media at fixed temperatures which correspond to those occurring in the unsaturated zone and with consideration for the evaporation of gases (Bertsch et al 1979).

The spreading of CHC's is presented schematically in Fig. 6. When the fluid has reached the surface of the saturated zone, the flow is slowed down as would be expected, since groundwater must be displaced, but the laterally effective component is clearly not of such descisive importance. The fluid sinks into the saturated zone. The heterogeneity that is usually present in aquifers results in a lateral spreading on layers, as indicated in the unsaturated zone in Fig. 4. If the retention capacity of this zone is exceeded (this is governed by the residual saturation in the same manner as percolation in the unsaturated zone), then the fluid sinks right to the impervious bedrock and spreads out under formation of a more or less flat mound. The spreading out takes place under the pressure head in the mound and follows the relief of the bedrock as slavishly as the oil over the groundwater table.

In nature, however, the bedrocks are in no way ideal planes as in Fig. 6, but have as a rule a more or less undulating relief. The CHC's keep strictly to the deepest parts of the relief, following the local depth contours, whose course in comparison with the groundwater table is usually unknown or not sufficiently known. When the bedrock is formed of layers of comparatively low hydraulic conductivity, then an estimate of the form and extent of the CHC "puddles" is practically impossible.

The area occupied by the immiscible phase may be calculated for oil for a non-fluctuating groundwater table according to Schiegg (1977). The oil lens would, according to Schiegg, theoretically be fairly thin at its end stage and would have to be correspondingly large. Practice has shown, however, that the areas required by theory are not attained. The reason for this is that because of the seasonally

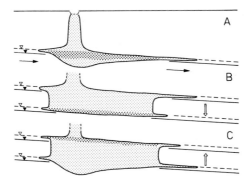

Fig. 7. Oil redistribution in the fluctuation zone of the groundwater table

dependent fluctuations of the groundwater table, the mobile oil still in the spreading stage is distributed in the vertical until eventually, with a certain size of the oil lens, a stage of residual saturation and thus immobility of the oil is attained (Fig. 7).

With the CHC's, which sink to the impervious bed of the aquifer, such a secondary distribution in the vertical is not possible. After the infiltration from above has stopped, the mound so-formed flattens out more and more, until eventually here too residual saturation is reached or until the fluid collects in a depression without an outflow.

3.5 Migration of Dissolved Components

The oil and CHC bodies, which are found only in a condition of residual saturation after reaching the final spreading stage of the fluid phase, are in no way impervious for seepage water and groundwater. The flow velocity of the water is, according to Fig. 1, somewhat reduced, but the bodies remain previous for water. The globule distribution in the pore space provides a very large interface with the water. With the mostly very slow flow velocities, a saturated solution is already reached after a short flow path. It may be therefore assumed that in the lee of the body, a plume with dissolved components is formed ("solution zone"), whose initial concentration corresponds to the solubility or the saturation concentration of the substance in question. The density of these hydrous solutions is only slightly higher than that of the groundwater, so that there is no significant tendency for sinking within the aquifer.

The spreading of the dissolved substances follows the principle of hydrodynamic dispersion, which may here assumed to be known. The extension of the plume and the concentration distribution of the solution within the plume can, assuming a homogeneous medium, be sufficiently accurately calculated by the appropriate determination of the maximum cross section of the oil or CHC body and the specific discharge (Darcy flux) of the groundwater. As an example for ascertaining the maximum cross-sectional area of an oil body (saturated zone) after practically complete immobility has been reached, Fig. 8 will serve. Important in this connec-

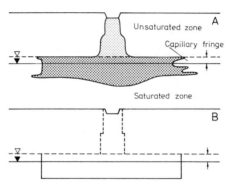

Fig. 8. Cross section through the oil body perpendicular to flow direction

tion is the suggestion that the heterogeneity of the aquifers, especially multilayered aquifers, shapes the plumes to a much greater extent than would hydrodynamic dispersion alone.

The calculation of the dispersion plume assuming a homogeneous porous medium, as shown in Fig. 9, can in most cases simply be considered as a useful model for better understanding of the transport processes. In fissured rocks, such model calculations are misleading rather than helpful. Here the widest open fissures, whose number and course for the most part are unknown, determine the flow path and the discharges (Schwille and Ubell 1982).

For the water-soluble components, the knowledge of the maximum value of the solubility, or, when multicomponent liquids are treated, the saturation concentration, is of the greatest importance. General statements on mixing with water are insufficient. Attention should further be paid to the fact that commercial grade products can vary greatly from the pure product through additions of additives, stabilizers or technically conditioned contaminations. These additions can present a much greater danger to the groundwater than the pure fluid itself.

The mobility of the dissolved substances can be more or less strongly retarded by adsorption and desorption processes on the solid matter as well as the organic matter of the soil. First indications of this are provided by the n-octanol/water distribution coefficients and the polarity (polarization) of the substances in question. The distribution coefficient of a substance between water and the lipophile solvent n-octanol permits one to recognize to what degree the dissolved substance can accumulate in the organic matter of the soil. In this way, the weakly polar CHC-solvents are considerably absorbed by the organic matter. With more polar molecules on the other hand clay minerals of the soil play an increasingly important role in adsorption effects.

3.6 The Migration of Gas Components

The high volatility of gasoline and certain CHC's causes an evaporation of the substances and thus their migration in the unsaturated zone. Around the gasoline CHC-bodies, an envelope of gas forms with decreasing concentration outwards

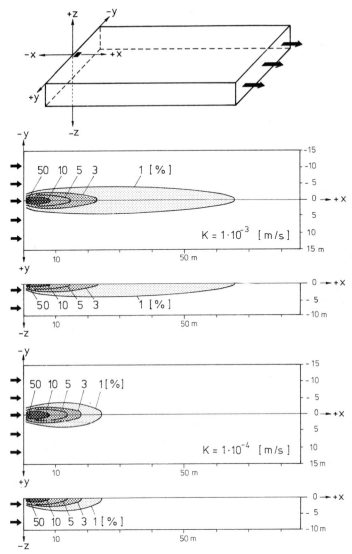

Fig. 9. Concentration distribution of dissolved CHC's as a result of hydrodynamic dispersion (point source) after 3 months

(Fig. 10). Because of the relatively high density, the vapors tend to sink and spread out chiefly over the capillary fringe.

The evaporation underground is advantageous on the one hand as it allows the vapors to be released to some extent to the open air through diffusion. On the other hand, through the formation of vapor in the pore space, an additional contact with the percolating water is made possible through which a correspondingly large area of the groundwater surface can be contaminated.

The effect of diffusion should not be underestimated for highly volatile liquids. Gas components can diffuse even out of the solution plume in the aquifer

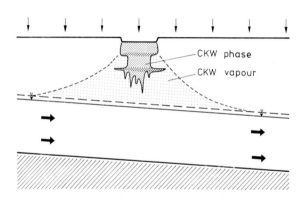

Fig. 10. CHC spreading; formation of a CHC gaseous zone

into the unsaturated pore space, but correspondingly slowly. Even so, by taking advantage of this effect, the course of the solution plume may sometimes be localized by investigating the pore air with suitable gas-tracing instruments in small bore holes. For mineral oil products with higher boiling points, e.g., light heating oil, the evaporation losses are quantitatively unimportant however; the gas-tracer method promises little success in these cases. In fissured rocks, the gas-tracer method may be usefully applied only where the fissured rock is covered by a layer of porous loose rock.

3.7 Microbial Degradation

The degradation of organic substances underground can be a purely chemical as well as a microbial process. Although fluid dynamic processes can have a significant bearing on these processes (e.g., mixing with oxygen, dilution, evaporation), the complex degradation processes will not be treated within the context of this paper oriented towards fluid dynamics.

It appears important to us, however, to point out to the microbiologists that they should consider and evaluate the biodegradation processes, differentiating according to the individual phase zones. For example, one must differentiate when considering oil degradation between oil as a fluid phase in the unsaturated and in the saturated zone as well as between dissolved components in the unsaturated and the saturated zone. In conclusion, the biodegradation of gaseous components in the unsaturated zone must not be ignored.

3.8 Migration in Fissured Media

Basically the same physical laws are valid here as for porous media, of course, with the difference that the finely branched network of the micropore spaces cannot be compared with a fissure space with approximately parallel planes.

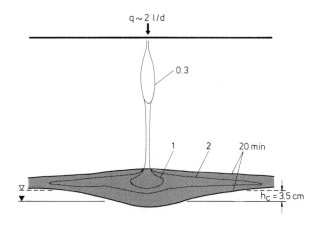

Fig. 11. Fissure model, $c = 0.2$ mm. Kerosine. Stage after infiltration finished

The migration process is explained on the basis of two examples. A fissure model of the type Hele-Shaw is applied, where hydraulically smooth and rough glass plates form the fissure walls.

In Fig. 11, kerosine ($\varrho = 0.75$) was infiltrated into a fissure with width $c = 0.2$ mm and smooth walls. The unsaturated zone is rapidly percolated on a very narrow "hose". As soon as the kerosine reaches the upper boundary of the capillary fringe, the lateral spreading begins, exactly as observed in porous media.

In a further test (Fig. 12), the fissure was narrowed to $c = 0.1$ mm and the infiltrating rate stayed about the same. At first, a relatively broad kerosine body forms and extending roots grow quickly at the lower boundary. Further flow takes place only on these roots. As soon as their tips have reached the upper capillary fringe, the lateral migration begins with a slight displacement of the water in the capillary fringe. When the infiltration was finished, the kerosine body of the unsaturated zone "bled out" in both cases, so that only insignificant kerosine "films" remained on the walls.

In Figs. 13 and 14 tests with tetrachloroethene C_2Cl_4 ($\varrho = 1.62$) in a fissure with hydraulically smooth walls are presented. In the fissure with $c = 0.2$ mm, the infiltrating fluid forms a very narrow hose (Fig. 13). The upper boundary of the capillary fringe causes a lateral spreading at first. However, at the lower boundary of the body, roots form quickly. These extend down to the bed of the aquifer and form a shallow mound there, which expands laterally. In the fissure with $c = 0.1$ mm (Fig. 14), the infiltration proceeds at first similarly to that of kerosine. After the infiltration there remain in both the unsaturated and saturated zones only slight residues. Nearly all the infiltrated fluid collects on the bed of the aquifer.

In order to compare hydraulically smooth and rough fissure walls (the glass plates were roughened by sandblasting), the last two tests were repeated with rough walls. Now even with the 0.2 mm fissure width a considerably more marked branched flow network formed (Fig. 15). After the infiltration, a large

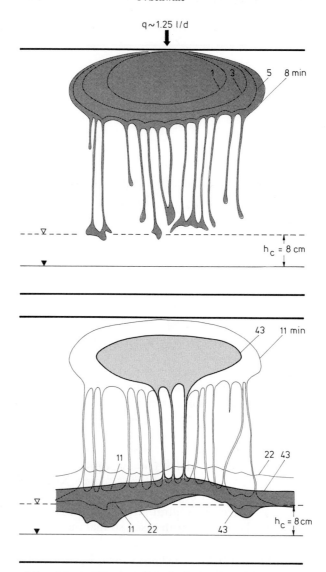

Fig. 12. Fissure model, c = 0.1 mm. Kerosine. *Above:* beginning stage; *below:* after infiltration finished

number of globules remained in the unsaturated zone and in the saturated zone fewer, but larger "spots". From this test, the influence of the nature of the fissure walls is recognized, which in nature certainly are not smooth and vary mineralogically as well. With a fair degree of confidence, one may say that the character of the fissure walls can exert an influence only under 0.5 to 0.2 mm width.

Important here is the geological assertion that in regions appropriated by humans for living areas and transportation routes, the fissured rock is usually covered by a more or less thick layer of soils, which as a porous medium has a considerable retention capacity. From the combination of porous over fissured

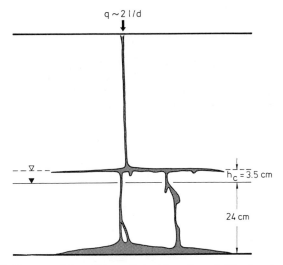

q ~ 2 l/d

$h_c = 3.5$ cm

24 cm

Fig. 13. Fissure model, c − 0.2 mm. Tetrachloroethene. Infiltration stage

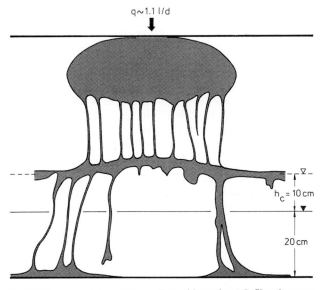

q ~ 1.1 l/d

$h_c = 10$ cm

20 cm

Fig. 14. Fissure model, c = 0.1 mm. Tetrachloroethene. Infiltration stage

media in the unsaturated zone very complicated percolation processes can result, which are not readily comprehensible.

3.9 Conditions in the Unsaturated (Porous) Zone

This general exposition, including the saturated zone, was essential for understanding the behavior of immiscible fluids in the unsaturated zone.

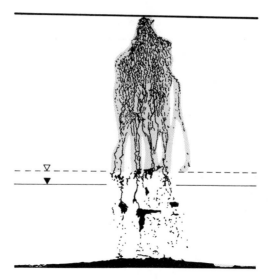

Fig. 15. Fissure model, c = 0.2 mm. Tetrachloroethene. Fissure walls hydraulically rough. Infiltration stage

When a fluid which can contaminate groundwater infiltrates the ground, one is at first interested in how quickly this happens and what quantity of fluid will be retained in the unsaturated zone. In dry soil (e.g., beneath factories), one can use the conductivity for the immiscible fluid K_{nw} for the infiltration or percolation process, if a saturated flow is assumed. The air is displaced to the degree of its residual saturation, as in the well-known case of percolation of water. The velocities of percolation may therefore be considered to be approximately inversely proportional to the kinematic viscosities. A light heating oil with $v = 4$ mm^2 s^{-1} will move 4 times slower and trichloroethene with $v = 0.4$ mm^2 s^{-1} 2.5 times faster than water. Model studies essentially confirm this concept.

As long as the soil moisture content does not exceed the irreducible saturation, the conductivity for the immiscible fluid compared to that by full saturation is only moderately reduced, since the water occupies the least pervious areas in the pore space and thus hinders only slightly the flow of the immiscible phase (cf. Fig. 1). The conductivity is strongly reduced only when higher water content is found, as may be recognized in the steeply declining curve for the nonwetting phase. But as the water contents as a rule are unknown, the relative decline in conductivity can only be evaluated to an order of magnitude.

In this connection, it must be emphasized that the diagrams for 2- and 3-phase flow (Figs. 1 and 2) can only provide initial information on the general course of k_r, but are not suitable for calculation of k_r or K_r.

The decisive parameter in practical groundwater protection remains the retention capacity of a porous medium for an immiscible fluid – the residual saturation. Here there are differences to be expected for the case of a dry medium, where only air and the immiscible fluid are present and for the case of a moist medium with the additional presence of water.

There exist a large number of measurements on the residual saturation of oil in porous media, partly expressed as a percent of the pore volume and partly in liter per cubic meter of the medium. The latter value is more useful in practice. For the low molecular CHC's (solvents), some laboratory determinations have provided values of the same order of magnitude as for oil, which was to be expected on the basis of the physical properties. For all other immiscible groups of substances, we know of no published values.

The high volatility of many organic substances forces one to try to obtain quantitative values of evaporation underground. There appears to be virtually no experience in this direction (Albertsen 1977). From observations of spills with gasoline, it is known that the phase bodies in the unsaturated zone diminish considerably more quickly in comparison with middle distillates (e.g., light heating oil) and finally disappear. In spills with dichloromethane, for example, which has an evaporation number of 2, notably smaller solution plumes were found than with tetrachloroethene with values at least twice as high with comparable infiltration quantities. Apparently the dichloromethane evaporates out of the unsaturated zone much quicker than tetrachloroethene, so that only a correspondingly small proportion of the resulting amount can be dissolved by seepage water.

There is just as little field experience on the rate of evaporation in porous media as on the migration of vapors. Let us return to the example just mentioned. Dichloromethane has a vapor-density ratio of 2.9, tetrachloroethene of 5.8. Presumably dichloromethane sinks slower into the deeper areas of the unsaturated zone for this reason, whereby diffusion to the earth's surface may be favored.

3.10 Measures After Spills of Organic Liquids

The measures to be taken to protect groundwater depend on a reliable evaluation of the migration of a groundwater-contaminating fluid as a phase within the unsaturated zone. The following measures may be considered.

a) The fluid hardly endangers the groundwater. The fluid may remain underground.

b) The groundwater is endangered. It will pay to clean out the contaminated soil and safely dispose of it and to replace it with clean material.

c) Measure (b) is for particular reasons not realizable. One must try to wash out and pump out the fluid in soluble form in the groundwater by seepage of water. To this belongs the method of "fertilizing" with trace elements (N, P, K) and introduction of oxygen to degrade microbial attackable substances. Under certain in conditions, the fluid occurring in residual saturation can be mobilized by introducing surface active substances and pumped out. The details of these measures, for example, the decontamination of the soil or the pumped out groundwater (aeration, application of activated carbon, etc.), storage of soil in disposal sites, cannot be treated here.

All other methods are extremely expensive and should be considered only in special cases. These include enclosing the liquid phase body by diaphragm walls to the bedrock of the aquifer or fixing the fluid by injection of chemicals.

3.11 Practical Investigations

It will be impossible to develop special laboratory tests for all potentially harmful substances in order to research their behavior underground. One must try to study chosen representatives of each group of substances carefully and transfer the knowledge gained with caution to the entire group. In detail, the following investigations are urgent.

a) The simulation of the seepage process with glass troughs or glass cylinders of sufficient size with suitable soil types and in fissure models. Here one is not concerned with model tests in the sense of model technology, but with natural tests on a small scale. Only the physical models, which sometimes have been considered old-fashioned, are in a position to achieve a basic representation of the migration process for the conception of an analog model.

In these studies, the capillary fringe must be included, although it is hydraulically a part of the aquifer and the boundary between the unsaturated zone and the saturated zone runs in nature mostly within the capillary fringe. In connection with the investigations on the behavior of volatile fluids, one should not avoid the term "aerated zone". The explanations of the terms "saturated zone" and "unsaturated zone", as they occur, for example, in a recent glossary (Johnson 1981), should be correspondingly altered.

As an example of the value of physical models, a test in a glass trough may serve (Fig. 16). A comparison of tetrachloroethene (above) and dichloromethane (below) shows under the same conditions a considerably broader percolation area of the dichloromethane compared to the tetrachloroethene, and as a result a considerably slower infiltration in depth. As a result of the lower kinematic viscosity of dichloromethane ($v = 0.32$ mm^2 s^{-1}), one would have expected a quicker and "thinner" seepage than with tetrachloroethene ($v = 0.54$ mm^2 s^{-1}). An essential role is presumably played here by the higher vapor pressure and the higher rate of evaporation of dichloromethane. After the saturation region (capillary fringe) is reached, this effect seems to be of minor importance.

b) The retention capacity in the unsaturated zone is best determined with the aid of glass columns of not too small height and diameter. For different reasons, we use cylinders ot 100 to 150 cm length and 20 cm diameter for our own investigations. We strive to standardize the determination of the retention capacity as far as possible with regard to instrumentation, soil type, water content and method. The significance of the determination will be demonstrated by an example.

Each of two glass cylinders filled with standard sands with a hydraulic conductivity of $K_w = 2.10^{-4}$ m s^{-1} and 2.10^{-3} m s^{-1} were saturated to excess with tetrachloroethene after a drainage time of 8 days. The difference between the infiltrated quantity of tetrachloroethene (V_t) and the amount running out from the bottom of the cylinder (V_d) provides the retention capacity (V_r) (Fig. 17).

The "bleeding out" occurs very quickly; after just a few days the run-off curve flattens out and appears to approach a boundary value asymptotically. A limitation of the test to about 20 days seems justified. The retention capacity shows values of 53 and 49 l m^{-3} after 20 days of drainage.

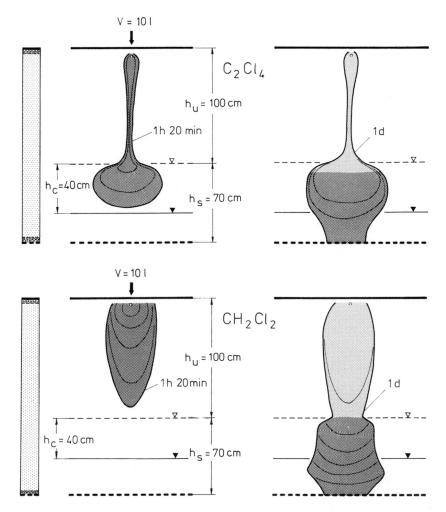

Fig. 16. Model study, porous medium. *Above:* Tetrachloroethene; *below:* dichloromethane; *right:* stage
after infiltration finished

c) For chemicals too, studies should be carried out to determine the relative permeabilities of the water, air and immiscible fluid phases, in order to obtain at least
a few typical diagrams for some representative fluids and soil types. Because of
the considerable efforts involved, one is severely restricted.

d) The evaporation processes in the unsaturated zone require a thorough investigation in suitable model troughs.

e) Model calculations to evaluate gas diffusion from the underground into the atmosphere exemplifying volatile substances should be carried out.

f) Suitable cases of contamination, in which chemicals infiltrate underground,
should be studied as examples. The case history should be documented carefully.

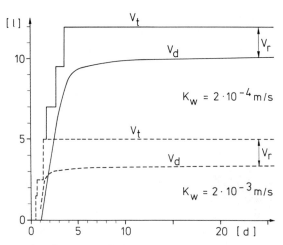

Fig. 17. Retention capacity of porous media for tetrachloroethene with dependence on the drainage time

g) For investigating spills and especially for the sampling of water, soil and soil air, appropriate field methods must be developed.

h) The development of analytical methods and instrumentation that are easy to use and lead quickly to results is necessary.

3.12 Cooperative Research

The above comments show how complex the problem of migration of immiscible fluids in porous and fissured media is. It is clear that hydrogeology, which is to the largest extent concerned with groundwater contaminations, cannot solve the current problems with its own methods, not even with application of groundwater dynamics, which up to now has treated one-phase flow nearly exclusively. Multiphase flow has always been the domain of the petroleum engineer and petroleum physicist, who treated fluid dynamics in its widest sense. They too have met their limitations, as they have recently been confronted with problems of tertiary oil recovery and now seek the aid of chemodynamics.

Future research on the migration behavior of chemicals underground should in our opinion be based on the three disciplines:

groundwater dynamics – fluid dynamics – chemodynamics.

Without the aid of these disciplines, hydrogeology will not be able to overcome the difficult tasks that arise. One is not chiefly concerned here with discovering and solving basic problems, but rather sensibly applying knowledge already obtained for solving groundwater problems. The petroleum sciences and process engineering in chemistry are in a position to provide hydrogeology with the necessary equipment. In conclusion, one should not forget to insure at an early stage the cooperation of microbiology and geochemistry. A thorough and systematic investigation of migration and transport processes of chemicals in the environment can only be successful as an interdisciplinary undertaking.

3.13 Conclusions

Organic chemicals are usually lumped together as substances harmful to groundwater. Until now little has been investigated on their migration behavior underground and especially in groundwater. The frequently observed differing behavior of organic fluids in comparison to hydrous solutions appears to justify the sceptical evaluation. The unexplained behavior may be quickly made clear, however, when one differentiates the organic fluids with regard to their miscibility with water. As a rule, it is the immiscible fluids that deviate the most from the usual groundwater-hydrological concepts.

The extensive research work carried out over the past two decades on the behavior of mineral oil products underground has provided a great deal of knowledge on these fluids immiscible with water, which on the basis of theoretical considerations one may transfer without hesitation to the organic chemicals. From a fluid dynamic point of view, no fundamental differences are to be expected between mineral oil products and chemicals immiscible with water.

Using as an example organic solvents (CHC's), upon which we have persistently applied our experiences with oil, it is shown that the conclusions were right. In our opinion, the basic features of migration behavior of all organic chemicals immiscible with water may be considered as explained. This does not mean at all that no further research is necessary, but simply that one can now attack specific problems in detail. Interdisciplinary cooperation is of course essential if one wishes to achieve a satisfactory solution to the whole problem complex within a reasonable period of time.

The concepts treated here reveal the significance of the unsaturated zone as a protective covering over the groundwater. Tests with the fissure model, for example, show forcefully how susceptible unprotected fissure groundwater can be. The type, structure and extent of the unsaturated zone are decisive for measures of preventive and restorative groundwater protection.

Acknowledgements. The author wishes to thank his colleagues Dr. W. Bertsch, G. Diesler, W. Reif, and S. Zauter for carrying out the model studies, A. Kaczmar for translating Russian literature and R. Andrews for translating this paper.

References

Albertsen M (1977) Labor- und Felduntersuchungen zum Gasaustausch zwischen Grundwasser und Atmosphäre über natürlichen und verunreinigten Grundwässern. Diss Kiel

Bear J (1972) Dynamics of fluids in porous media. American Elsevier, New York

Bertsch W, Schwille F, Ubell K (1979) Versickerungsversuche mit Import-Rohöl und niedrig viskosem schwerem Heizöl. Dtsch Ges Mineralölwiss Kohlechem, Forschungsber 144, Hamburg

Dam J van (1967) The migration of hydrocarbons in a water bearing stratum. – In: Hepple P (ed) The joint problems of the oil and water industries. Inst Petrol London, pp 55–96

Dracos Th (1966) Physikalische Grundlagen und Modellversuche über das Verhalten und die Bewegung von nichtmischbaren Flüssigkeiten in homogenen Böden (Erste Resultate). Mitt Versuchsanst Wasserbau Erdbau ETH Zürich, 72

Fried JJ, Muntzer P, Zilliox L (1979) Groundwater pollution by transfer of oil hydrocarbons. Groundwater Vol 17, 6:586–594

Johnson AI (1981) Glossary Spec Technic Publ 746, Am Soc Testing Materials, Philadelphia

Leverett MC, Lewis WB (1941) Steady flow of gas-oil-water mixtures through unconsolidated sands. Trans AIME 142:107

Romm ES (1966) Die Filtrationseigenschaften der klüftigen Gesteine. „Nedra" Moskau, pp 84–88 (in russisch)

Schiegg HO (1977) Methode zur Abschätzung der Ausbreitung von Erdölderivaten in mit Wasser und Luft erfüllten Böden. Mitt Versuchsanst Wasserbau Hydrol Glaziol ETH Zürich, 22 p

Schwille F (1975) Groundwater pollution by mineral oil products. IAHS-AISH Publ 103:226–240

Schwille F (1981) Groundwater pollution in porous media by fluids immiscible with water. Sci Total Environ 21:173–185

Schwille F (1982) Die Ausbreitung von Chlorkohlenwasserstoffen im Untergrund, erläutert anhand von Modellversuchen. DVGW-Schriftenr Wasser Eschborn 31:203–234

Schwille F, Ubell K (1982) Strömungsvorgänge im vermaschten Kluftmodell. Gwf-wasser/abwasser 123, H 12:585–593

Wyckoff RD, Botset HG (1936) The flow of gas-liquid mixtures through unconsolidated sands. Physics 7:325–345

B. Behavior of Inorganic Chemicals

Introductory Comments

B. Yaron

The main sources of inorganic chemicals in the unsaturated zone are the direct incorporation into the land surface of commercial fertilizers, land disposal of sludge, industrial and domestic effluents, and dispersion of aerosols. In addition, there exists a "background" contribution of biological residues and a release of elements from the solid phase. The properties and the binding ability of the inorganic chemicals determine the extent to which they are absorbed on the solid phase, precipitate or move downward into the unsaturated zone until they may reach the groundwater. Adsorption-desorption-precipitation processes accompany the transport phenomenon and govern the distribution of the inorganic pollutant in the unsaturated zone.

In the second part of this book, the behavior of some of the most important potential inorganic pollutants, such as nitrates, phosphates, and heavy metals is described. The role of acid rain and long distance atmospheric transport is also discussed.

Pratt and Jury (Chap. 4) show that the nitrogen added to land surfaces is distributed in the soil profile, removed in harvested crops, leached as NO_3^- from the root zone, and escaped to the atmosphere by volatilization of NH_3 or as N_2O_2 and N_2. Avnimelech (Chap. 5) treats the fate of phosphorus in soil and discusses its solubility and speciation. Modeling of nitrate and phosphate transport in the unsaturated zone is also discussed in both chapters of Pratt and Jury (Chap. 4) and Avnimelech (Chap. 5). The chemical decomposition (nitrate case) or adsorption (phosphate case) are considered in the prediction of leaching of these chemicals through the unsaturated zone to the groundwater.

The natural and man-induced input of heavy metals in the unsaturated zone is discussed by Matthess (Chap. 6) as a function of abundance, geochemical mobility, and speciation. The heavy metal behavior in the unsaturated zone is treated with regard to the formation of colloidal suspensions of heavy metals and their precipitation-coprecipitation and absorption on the solid phase.

Acid rain creates one of the most significant pollution problems of the industrialized countries. The man-induced acidification of the system: surface water-unsaturated zone-groundwater, is described by Jacks et al. (Chap. 7), who emphasizes the situation in the Scandinavian countries. For a natural acid soil, the anthropogenic addition of sulfate and hydrogen ions may lead to a significant decrease of pH in the unsaturated zone and affects the groundwater quality. The pollution of the unsaturated zone is regarded in general as a local contamination process. Steinnes (Chap. 8) however, presents the case of long-distance atmospheric transport of heavy metals as a potential pollution hazard to the soil surface.

The control of the nitrate balance in the unsaturated zone by proper management of water and crops, considering the climate and soil limitations, is one of the solutions presented by Pratt and Jury (Chap. 4) for prevention of nitrate pollution. The effect of speciation upon transport and surface reactions with the solid phase of phosphorus and heavy metals is one of the main topics discussed in the papers of Avnimelech (Chap. 5) and Matthess (Chap. 6). Acidification of the unsaturated zone due to atmospheric pollutants – as exemplified for Sweden by Jacks et al. (Chap. 7) – affects the reactions between the inorganic chemicals and the solid phase surfaces.

The chapters presented in Part B suggest that the inorganic chemicals reach the unsaturated zone not as isolated compounds, but as part of a group of pollutants originating from various sources. The relationship between these various chemicals and their behavior in the unsaturated zone has not yet been in the focus of research. As a result, pollution control recommendations derived from existing research deal mainly with a single compound and not with a whole group of pollutants. Integral studies on interrelated behavior of the chemicals in the unsaturated zone will contribute toward preventing the deterioration of the system.

4. Pollution of the Unsaturated Zone with Nitrate

P. F. Pratt and W. A. Jury

4.1 Introduction

Previous to the late 1960's there was relatively little research on the side effects of the use of industrially fixed N on croplands. Because fertilizer N was inexpensive relative to other inputs no great effort was expanded on crop, fertilizer, and water management to increase the efficiency of N use. However, in the late 1960's, partly as a result of papers by Commoner (1968, 1970), the increasing use of fertilizer N became a subject for debate. Consequently, in the last 15 years a tremendous effort has been expended on N use efficiency, denitrification, leaching of NO_3^- from the root zone, and the rate of movement of water and NO_3^- through soil.

At about the same time that environmetalists were criticizing the increasing use of fertilizer N, there was also an increased awareness of the impact of animal and human wastes on water quality. This awareness stimulated research on the use of these wastes on croplands as well as on noncultivated areas. Thus, there was a renewed interest in the N contained in wastes, because of the potential for overloading the land with NO_3^- from these materials. The use of a small area for waste disposal can produce severe NO_3^- leaching problems in localized areas, but by far the most extensive problem of NO_3^- in the unsaturated zone is generated by vast areas of agricultural land under intensive crop production.

Increased research on the N cycle has led to many publications in technical journals and also many conference proceedings, special reports, and books dealing with the soil-water-plant system and the leaching of NO_3^-. Among these many publications are Panel on Nitrates (1978), SIDA-FAO (1972), IAEA (1980), Nielsen and MacDonald (1978), McKim (1978), Loehr (1974), Aldrich (1980), and Pratt (1979 b). In fact, the literature on N in the soil-plant-water system in the past 15 years is so vast that only a few papers and reports can be cited here to illustrate the behavior of N in the unsaturated zone.

4.2 Nitrate as a Pollutant

Nitrate in water can have adverse effects on humans, animals, and plants, although the concentrations at which these adverse effects begin to appear cover a large range. Concentrations of NO_3^--N greater than a few tenths mg l^{-1} are adequate to promote eutrophication in surface waters, if other conditions are not limiting (Committee on Water Quality Criterial 1973 a). For example, Sawyer (1947) found that the critical level for inorganic N was 0.3 mg l^{-1} at the time of spring overturn for nuisance algal blooms during the summer in Wisconsin lakes. The criteria for domestic water supplies is 10 mg l^{-1} primarily because of the need to keep drinking water at less than this level for the protection of human infants (U. S. Environmental Protection Agency 1976). The Committee on Water Quality (1973 b) set a criteria of 100 mg l^{-1} for farm animals. Criteria for irrigation water suggested by Ayers and Westcot (1976) are <5, 1.5 to 30, and >30 mg l^{-1}, respectively, for no problems, increasing problems, and severe problems in the production of sensitive crops or in the limitation of types of crops that can be produced economically.

The overriding concern for NO_3^- as a pollutant is the result of the correlation between NO_3^- concentration in drinking water and methemoglobinemia in human infants less than 4-months-old. During the first few months of life, human infants are susceptible to NO_2^--induced methemoglobinemia. At this age, the infant's gastric pH is sufficiently high (5 to 7) that the growth of NO_3^--reducing bacteria is not inhibited: Consequently, NO_3^- can be reduced to NO_2^- in the stomach before it is absorbed into the blood stream. In older children and adults, the NO_3^- is not reduced in the stomach because of lower pH values and it is absorbed into the blood before it reaches the reducing conditions of the intestines. The NO_3^- absorbed does not produce methemoglobin and is excreted in the urine. The NO_2^- absorbed into the blood of infants reacts with hemoglobin to produce methemoglobin, which does not transport oxygen. When more than 5% of the hemoglobin is converted to methemoglobin in infants, O_2 deficiency (cyanosis) can be observed. If conversion exceeds 50% death is likely.

Other effects of NO_3^- in drinking water on human health are possible. A linkage to carcinogenic N-nitroso-compounds has been proposed, but an assessment of this is not possible because of lack of evidence (Panel on Nitrates 1978).

4.3 Sources

The overall N cycle consists of (1) conversion of N_2 in the atmosphere into fixed forms, either as inorganic ions or as organic compounds, and (2) the return of N_2 in the atmosphere by many pathways and after a variable number of cycles through biological systems. Thus, the ultimate source of NO_3^- is atmospheric N_2, which is converted to fixed forms by biological or industrial fixation. In both processes, the N is fixed in the NH_4^+ form, after which it is converted into many other forms by biological and industrial processes.

Fixed N comes to or is incorporated into the land surface in many forms and materials. Fixation by rhizobium-legume combinations or by free-living organisms add N directly to the crop or the soil in organic forms. Animal manures, sewage sludge, crop residues, and roots also add organic materials, which then mineralize at various rates to give NH_4^+ and then NO_3^-. Net mineralization of the organic N accumulated in soils over decades or centuries also represents a production of NO_3^-. Industrial and municipal wastewaters, when spread on land or stored in holding ponds, also contribute NO_4^- to the land surface. Minor amounts of N reach the land surface in rainfall and in some areas NH_3 adsorption by soil and absorption by plant leaves can contribute small amounts of N.

The main direct sources of N are animal manures, commercial fertilizers, and fixation by the rhizobium-legume symbiosis. The potential for mismanagement of the fixation by legumes to create a hazard for NO_3^- leaching from the root zones is small, since fixation is proportional to crop production and in most cases the N removed in harvested crops is approximately that which was fixed. With fertilizers, animal manures, and human waste there is a potential for excessive applications. Where large populations of animals or people are concentrated in small areas, the costs of transportation of wastes prohibit wide distribution and large amounts are then applied per unit area near the source of production. Excess applications of fertilizer N can result from lack of knowledge of crop needs and also from unpredictable crop losses from drought and/or pest damage.

Estimates of the quantities of N fixed biologically range from 169 to 269 million metric tons (Mt) per year on a worldwide basis (Burns and Hardy 1975, Pratt et al. 1977, Soderlund and Svensson 1976). Estimates for fertilizer N use on a world basis show an increase from about 24 Mt in 1962–1968 to about 48 Mt in 1980–1981 (White 1980). Irrespective of the uncertainty of the amounts of N fixed by biological processes, there is no doubt that the fraction of total fixation by industrial processes has increased dramatically during the last 3 decades and that this fraction will likely continue to increase into the indefinite future (Winteringham 1980). Because of the potential for mismanagement of an increasing amount of industrially fixed N and the consequences of losses into air and water, efforts must be made to improve the efficiency of use of this N and the products that result from its use.

4.4 Behavior

The N cycle has been presented in many different forms and degrees of detail involving transformations in soils, waters, air, plants, and animals. However, the simplified biological N cycle presented by the Panel on Nitrates (1978) illustrates the behavior of N in the soil (Fig. 1). Organic N, the reservoir of N in the soil, mineralizes by ammonification and then by nitrification. The conversion of NO_2^- to NO_3^- is usually much faster than conversion of NH_4^+ to NO_2^- so that NO_2^- exists in very small concentrations and NO_3^- is the final stable product of the mineralization process in well aerated soils. Any of the mineral forms (NH_4^+, NO_2^-, and NO_3^-) can be assimilated by plants and microorganisms and return to the soil as organic N. The denitrification-fixation reactions represent a loss to the atmosphere and a return to the soil system.

A steady-state conceptional model of inputs and outputs of N into a soil-plant system is illustrated in Fig. 2. In this case, the soil-plant system is assumed to maintain a constant amount of total N so that outputs equal inputs. Of course, inputs and outputs are not continuous so that small increases and decreases occur even if, over a period of few years, the system is at steady state.

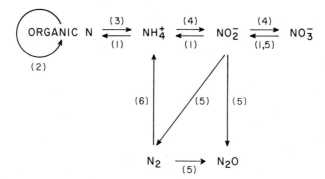

Fig. 1. Simplified biological N cycle showing major reactions. The numbers in parentheses indicate processes: (*1*) conversion to organic forms, i.e., assimilation; (*2*) transformations within the organic pool: (*3*) ammonification; (*4*) nitrification; (*5*) denitrification; (*6*) fixation (Panel on Nitrates 1978)

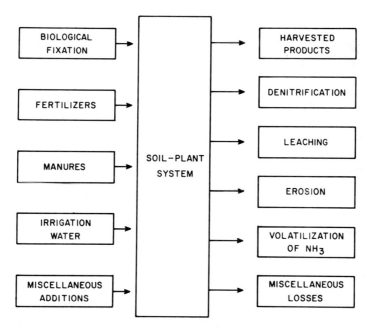

Fig. 2. Nitrogen inputs and outputs from the soil-plant system

   The major inputs are biological fixation by legumes and the application of fertilizers and animal wastes. In irrigated areas, there is the possibility of substantial inputs of N in irrigation waters. Other sources are minor except in small areas that might receive sewage sluges and/or wastewaters, in which cases large amounts per unit area are likely.

   The major outputs are removal in harvested products, leaching of NO_3^-, and denitrification. When management is adequate to protect the soil from erosion and to prevent volatilization of NH_3 from fertilizers and manures, the main losses of N from the soil system are by these three processes. For protection of ground-water from NO_3^- accumulations, the removal in harvested products and/or deni-trification must be maximized. Because denitrification is an economic loss, the ideal management of croplands must maximize removal in harvested products.

Tanji et al. (1977, 1979) expanded this conceptual model, originally presented by Fried et al. (1976) to include transient changes in N within the soil system and fluxes of water through the soil profile (the root zone). When tested with 2 years of data from corn field trials at Davis, California, computed results compared fa-vorably with (1) N in the grain plus stover, (2) the residual inorganic soil N after harvest, and (3) the NO_3^- concentration in the drainage water below the root zone.

Assuming losses by erosion and by volatilization of NH_3 are negligible, the mineral N added to or produced in soils can be assimilated by absorption by plant roots or by microorganisms, converted to N_2O and/or N_2 by denitrification or leached from the root zone in percolating water. The distribution of mineral N among these three processes depends on the many variables that control them. Absorption by plants is dependent on the species and variety and on the many factors that influence growth. Denitrification is increased by decrease in O_2 sup-ply, increase in available C in the soil, and by increase in temperature within the temperature range normally found in soils. Denitrification is usually complete in flooded soils. In well-aerated, well-drained sands, denitrification is very slow. But denitrification does occur in poorly aerated microvolumes of soils that are well aerated from a macroscopic point of view.

As a first approximation, NO_3^- in the soil system, including the unsaturated zone beneath the soil root zone, moves with and at the same rates as the move-ment of the percolating water. However, the relative rate of flux of NO_3^- with re-spect to that of water is dependent on the chemical properties of the soil material. The relative NO_3^- flux in positively charged soil materials that adsorb NO_3^- is less than unity, whereas in negatively charged soil materials that repel NO_3^-, it is greater than unity. In soils with little negative or positive charges, the relative flux is near unity.

A number of investigators have found that acid kaolinitic and amorphous soils adsorb NO_3^- and that this adsorption is greatest in volcanic ash derived soils having largely allophanic minerals (Rich and Thomas 1960, Thomas 1960, Singh and Kanehiro 1969, Kinjo and Pratt 1971, Schalscha et al. 1974). Kinjo and Pratt (1971) and Leon and Pratt (1974) found that 3 or 4 pore volumes of effluent were required to move peak concentrations of NO_3^- to the bottom of columns of allo-phanic subsoil materials and that about 6 or 7 pore volumes were required to re-move essentially all of the NO_3^-.

On the other extreme, the flux of NO_3^- in soils that contain dominantly smec-tite clays (montmorillonites and micaceous minerals) is faster than that for water because of anion exclusion from a fraction of the soil water (Berg and Thomas 1959, Dyer 1965, Thomas and Swaboda 1970, Smith 1972, Bresler 1973a, Bresler and Laufer 1974, Appelt et al. 1975, Tullock et al. 1975). Tullock et al. (1975) cor-related the relative rate of flux of Cl^- and water through laboratory columns with the cation-exchange capacity (CEC) of soils. From all comparative studies, the movement of Cl^- and NO_3^- behave identically in such studies so that data for Cl^-

can be used to predict the behavior of NO_3^-. Relative flux for NO_3^- and/or Cl^- in such column studies range from 1.0 to 2.0 for soils containing smectite clays, indicating that the effects of exclusion volume can be demonstrated in laboratory columns. However, the practical effects of both negative adsorption (exclusion volume) and positve adsorption on the movement of NO_3^- in field situations have not been demonstrated.

4.5 Movement

All dissolved chemicals, including the species of N, move through the soil by convection and diffusion. The former process refers to mass flow of solute with moving water and the latter process to movement within fluid by molecular collisions. Under most conditions, diffusion is small compared with convection. In addition, there is a spreading of solute due to multidimensional convection paths in soil, which is known as hydrodynamic dispersion. This effect is formally treated in the same way as molecular diffusion and the two influences are not separated.

Beginning with the classic experiments of Nielsen and Biggar (1961, 1962) and Biggar and Nielsen (1962), soil physicists described the movement of dissolved chemicals through soil using the so-called convection-dispersion equation.

$$\frac{\partial C}{\partial t} = D \frac{\partial^2 C}{\partial z^2} - V \frac{\partial C}{\partial z} + \Phi, \tag{1}$$

where C is solution concentration, D is the effective diffusion-dispersion coefficient, V is the pore water velocity, and Φ is a general reaction term. Application of this theory to N movement has been extensively reviewed by Gardner (1965).

Nitrogen reactions during movement through soil were studied by Cho (1971) and in a series of papers by Misra et al. (1974 a–c). For simultaneous oxidation of NH_4^+ to NO_3^- and reduction of NO_3^- to gaseous N_2, Misra et al. used a series of first-order rate coefficients describing the $NH_4^+ \rightarrow NO_2^-$, $NO_2^- \rightarrow NO_3^-$, $NO_3^- \rightarrow N_2$ reactions and solved for the rate coefficients by observing laboratory column breakthrough data. Recent work on denitrification has emphasized the intermediate N_2O state as well (Bremner and Blackmer 1978). A mathematical approach similar to that of Misra et al. (1974 a–c) was used to study the transformations of urea N in soil columns by Wagenet et al. (1977).

Numerous laboratory studies have proved successful in describing leaching processes for dissolved chemicals so that Eq. (1) is valid beyond question where applied to inert solutes. However, when soil reactions are also involved, such as adsorption or biological-chemical transformations, the interaction between flow rate and adsorption kinetics is not well understood even in the laboratory. Under equivalent flow conditions, however, the first-order rate coefficients obtained in the above mentioned laboratory experiments and others have proven to be useful for estimating the fate of NO_3^- moving through soil columns.

Under field conditions, nonuniform water flow has created additional problems both in measurement and modeling of chemical movement. The first attempt to apply the convection-dispersion Eq. (1) under large field conditions was conducted by Biggar and Nielsen (1976) who measured a wide distribution of dispersion coefficients and water velocities when Eq. (1) was applied to NO_3^- movement at different locations in the field. Similarly, Jury et al. (1976) found that a one-dimensional model of NO_3^- movement was not sufficient to describe NO_3^- leaching under a potato crop and Cameron et al. (1979) found that spatial variations due to differential rates of leaching caused by runoff and transport variability created large differences between predictions using Eq. (1) and measurements. However, Watts and Hanks (1978) found a good agreement with a model predicting NO_3^- movement and uptake by corn using Eq. (1) together with a water model and simple reaction terms similar to those discussed above.

Although only a limited number of field studies have been conducted, it is becoming apparent that nonuniform leaching patterns arising from variation in soil physical properties create substantial difficulties for those who would model the movement of dissolved chemicals such as NO_3^- through soil. At least near the surface, these variations arise primarily from variations in water movement and therefore will be resolved only when a better model for water velocity variations is found. In recent years, several such approaches have emerged. Dagan and Bresler (1979) used scaling theory together with a statistical model water input variability to describe chemical leaching across a field. Jury (1982) and Jury et al. (1982) proposed a stochastic transfer function model for calibrating the water flow variability from a single measurement. The calibration allows a later prediction of chemical movement to be made in the region of calibration.

The management implications of such variability are largely unexplored at this time. A recent discussion (Jury 1983) of differences in prediction between stochastic and deterministic models such as Eq. (1) showed large differences in predicted leaching behavior with substantial management implications for pollution. Under spatially variable field conditions, if an understanding is required of the extreme behavior of a field, such as rapid leaching due to cracks, worm holes, etc., then several of our existing management notions may have to be revised.

In spite of the absence of a useful general theory to predict movement of water an NO_3^- flux through the unsaturated zone, calculations based on the piston flow model can provide estimates of the approximate magnitude of flux rates and residence times for deep unsaturated zones in valleys filled with alluvium. The residence time is calculated from

$$t = \frac{S\theta}{D},$$

where t is time in years, S is depth in cm, θ is the volumetric water content of the soil materials through which the water is percolating, and D is the drainage volume in cm yr^{-1}. Assuming no cracks or channels through which the water flows, Pratt et al. (1978) estimated flux rates of 0.5 to 3.0 m yr^{-1} or residence times of 10 to 60 years for a 30-m depth, based on data for nearly 100 irrigated sites where deep soil cores were analyzed.

The amounts of NO_3^- that leach from soil profiles (root zone) depends on the amount of water (drainage volume) that moves through the profile, the NO_3^- input, added to and produced in the profile, and the effectiveness of competing reactions, i. e., assimilation by roots and by microorganisms and dissimilation by denitrification. Because of large ranges in NO_3^- input, assimilation and dissimilation in various climate-soil-crop-management systems, a unique solution to leaching of NO_3^- is impossible. However, the collective competing reactions can be expressed as an effective first-order reaction rate coefficient combined with water flux to develop a conceptually useful model.

An approximate relationship between drainage volume and NO_3^- leached was developed by analyzing a steady-state system under continuous irrigation. The assumptions are: (1) uniform NO_3^- concentration (C_I) in the irrigation water; (2) uniform water uptake in the root zone over the depths $O < Z < L$, where L is the depth to the bottom of the root zone; (3) an effective first-order reaction k (yr^{-1}) for the sum of all reactions that compete with leaching (e. g., denitrification, plant uptake); and (4) steady water flow with uniform water content and constant evapotranspiration, ET.

With these assumptions, the steady state drainage concentration is approximately

$$C(L, t) = CONST = \frac{C_I}{LF} e^{-k\tau(LF)}, \tag{2}$$

where

$$\tau(LF) = \int_0^L \frac{\theta dz}{J_w(z)} = \frac{L\theta}{ET} \ln(1/LF), \tag{3}$$

is the residence time for NO_3^- in the root zone and $J_w(z)$ is the steady state water flux at z.

The mass of NO_3^- leached in 1 year is calculated by

$$M = \int_0^t C(L, t) J_w(L) dt = \frac{C_I}{LF} e^{-k\tau} J_w(L) t. \tag{4}$$

The expression in Eq. (3) may be simplified by noting that the drainage water flux $J_w(L) OET \cdot LF/(1-LF)$ and using Eq. (2). Thus,

$$M = \frac{C_I ET \cdot t}{(1-LF)} \cdot (LF)^K, \tag{5}$$

where $K = kL/ET$.

If we set $t = 1$ year, we can plot annual drainage volume, $C = J_w(L)t$ against M for different values of K. Note that since annual cumulative ET is the same for all cases that LF and D are related by

$$LF = \frac{D}{ET+D}. \tag{6}$$

It is useful to notice that M is proportional to C_I and to ET and that one obtains the same curve for a given value of

$$K = kL/ET. \tag{7}$$

Relationships between NO_3^--N leached and drainage volume for various assumed values of k, using $L = 100$ cm, $ET = 80$ cm yr^{-1}, and $C_I = 20$ mg l^{-1} are

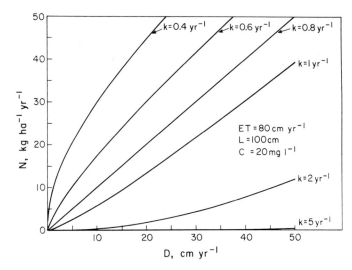

Fig. 3. Relationship between NO_3^- leached and drainage volume, D, calculated for a steady-state model for various first-order reaction rates for competing reactions. ET was 80 cm and C_i was 20 mg NO_3^--N l^{-1}

presented in Fig. 3. These curves reflect not only the effects of k values, but also the effects of the residence time in the root zone. The longer the residence time, the greater the amount of assimilation and dissimilation of NO_3^-. For all curves, a linear relationship between NO_3^- leached and D could be used without a large departure from the calculated lines when the D value is greater than 5 cm yr^{-1}. Thus, considering the scatter of data points that would be obtained from field measurements, a linear relationship would be expected for any given set of soil, climate, crop, and N and water management conditions where D is 5 cm yr^{-1} or larger. If the effective k value is > 1, the intercept for the linear relationship would be negative and for k values < 0.8, it would be positive. For values of 0.8 to 1.0, the intercept would likely be near zero.

The relationships expressed in Fig. 3 are calculated for highly specific conditions and should be considered only as qualitative indications of the effects of water flux and rates of competitive reactions on leaching of NO_3^-. However, they do suggest that field measurements might produce linear relationships between NO_3^- leached and drainage volume and that negative intercepts might be found.

Data for the number of studies for which the relationships between NO_3^- leached from the root zone and drainage volume were possible are presented in Table 1. In each case, there was no indication of curvilinearity in the data. The linearity of the relationship and the range in intercepts indicate that the relationships presented in Fig. 3 have some value in qualitatively predicting the real world. The correlation between the regression coefficient, the slope of the linear relationship between M and D, and N fertilizer N inputs showed a significant coefficient of 0.92, indicating that the effects of drainage volume are greater when more NO_3^- is in the soil system.

Table 1. Regression and correlation between $NO_3^- - N$ leached, M, and drainage volume, D, for various studies

Reference	Conditions			Regression equation	Correlation coeffcient[a]
	Fertilizer N kg ha^{-1} yr^{-1}	Crop	Drainage		
Letey et al. (1979)	Variable mean = 430	Various	Tile	$M = -4.5 + 2.7 \, D$	0.89
Rible et al. (1979)	Variable mean = 250	Various	Free	$M = 11.7 + 3.0 \, D$	0.77
Nielsen et al. (1979)	0	Corn	Free	$M = -2.5 + 0.90 \, D$	0.90
	90	Corn	Free	$M = -8.8 + 1.31 \, D$	0.98
	180	Corn	Free	$M = -16.1 + 1.89 \, D$	0.97
	360	Corn	Free	$M = -30.0 + 4.15 \, D$	0.93
Robbins and Carter (1980)	0	Various	Free	$M = 6.0 + 0.56 \, D$	0.92
	Crops following alfalfa plus those that received N	Various	Free	$M = -45.0 + 3.2 \, D$	0.71
Pratt et al.[b] (1976)	0	Barley-sudangrass	Free	$M = -48.0 + 1.2 \, D$	–
	400	Barley-sudangrass	Free	$M = 0.00 + 5.2 \, D$	–
	750	Barley-sudangrass	Free	$M = 97.0 + 5.1 \, D$	–

[a] All coefficients are significant at the 0.05 level except the value 0.71, which approaches significance at this level

[b] Not sufficient data points were determined to calculate correlation coefficients

4.6 Possible Solutions

Reductions in the amounts of NO_3^- in the unsaturated zone will result from improved efficiency of use of N by plants and by increased denitrification. In some areas, management can be changed to increase denitrification. Raveh and Avnimelech (1973) reduced the leaching of NO_3^-, presumably because of denitrification, as a result of increasing the saturated zone into the zone of microbiological activity (root zone) in organic soils of the Hula Valley. Gilliam et al. (1979) reduced the NO_3^- leaching from poorly drained mineral soils by raising the water table. In areas of waste disposal, denitrification can be increased by decreasing aeration and by applications of easily decomposed organic materials.

However, in most croplands denitrification should be reduced along with leaching losses. Both represent losses of a costly resource. Thus, in the vast majority of cropped fields, the solution is to manage for increased efficiency of use of available N in the soil or to obtain high yields with smaller fertilizer N inputs.

The many direct and indirect factors involved in NO_3^- leaching from the root zone of cropped soils are illustrated in Fig. 4. For a given land area with its soil, climate, and surrounding economic conditions, the farmer has a limited number

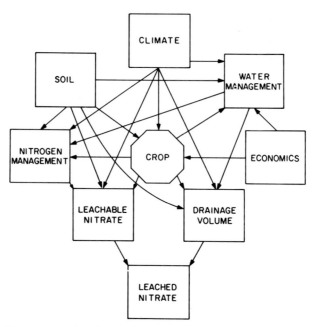

Fig. 4. Relationships among various factors that control NO_3^- leaching from the root zone in irrigated
lands. *Arrows* indicate dominant effects (Pratt 1979)

of options. He must select crops that will grow under the set of environmental
conditions and he must select crops that will more than pay the production costs.
After crop selection, he can manage the crop, N inputs, and water.

An ideal system for management to maximize N use by the crop and minimize
NO_3^- leaching has several requirements. The first is a crop that has a root system
that is efficient in absorption of NO_3^-. The second is that the system is so efficient
that there are no increases in leachable NO_3^- with increases in available N inputs
up to the point of maximum yield. The third is that the N needs of the crop for
maximum yield or for maximum economic return can be predicted so that the application of excess N can be avoided. A fourth requirement is that water movement through the soil is small during periods when large amounts of NO_3^- are
available in the root zone. This ideal system can serve as a goal and also as a model against which research and practice can be compared.

The efficiencies of crops in utilizing available N in the soil, as measured by the
fraction of input N that is removed in harvested material, cover a wide range. Removal in harvests are typically 25%–35%, 35%–50%, and 50%–70%, respectively, for fruit and nut crops, vegetable crops, and grain and forage crops, but
these ranges vary widely depending on other factors. Miller and Smith (1976) estimated removals of 6% for plums to nearly 100% for cereal hay for a large irrigated area in southern California. Many researchers have demonstrated that
there are only small amounts of NO_3^- leached from alfalfa fields (Letey et al. 1979,
Robbins and Carter 1980). This crop is also known as an excellent scavenger of
NO_3^- in soil profiles (Mathers et al. 1975, Muir et al. 1976). Hills et al. (1978) dem-

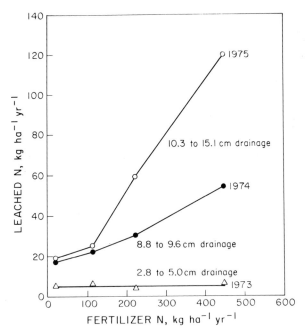

Fig. 5. Relationships among leached NO_3^--N, drainage volume, and fertilizer N. (Data from Gast et al. 1978)

onstrated that sugar beets were efficient users of available N in soils when fertilized with only enough N to produce maximum sugar yields.

The data of Gast et al. (1978), presented in Fig. 5, illustrate the effects of fertilizer inputs and drainage volume on the NO_3^- leached from tile lines. This 3-year experiment started with a soil containing only small amounts of available N so that the effects of repeated N inputs along with different drainage volumes were measured. The maximum yield for the corn crop was obtained at a N input of 112 kg ha^{-1} yr^{-1}. At low drainage volumes in 1973, there were no increases in NO_3^- leached with increase in N inputs. At higher drainage volumes, the leached NO_3^- increased slightly with increase in N inputs up to the point of maximum yield. Also, residual soil NO_3^- showed only very small increases up to the N input for maximum yield. Data from Nielsen et al. (1979) were used to calculate the relationships presented in Fig. 6, showing essentially the same effects of N inputs and drainage volume as shown in Fig. 5. In this case, maximum yield of corn was obtained at 190 kg N ha^{-1} yr^{-1}. The effects of application of excess N on leaching of NO_3^- from the root zone have been illustrated by the research of many studies, including that of Vaisman et al. (1982), which showed a large reduction in percent removal of N by a harvested forage crop as the N input increased. At maximum yield, the removal was 75% of a N input of 400 kg ha^{-1} yr^{-1}, whereas at an input of 850 kg N ha^{-1} yr^{-1}, the removal by the crop was only 31%. The amount of N added minus removal in harvested forage averaged 15, 118, 328, and 581 kg ha^{-1} yr^{-1}, respectively, for average inputs of 154, 399, 628, and 848 kg N ha^{-1}

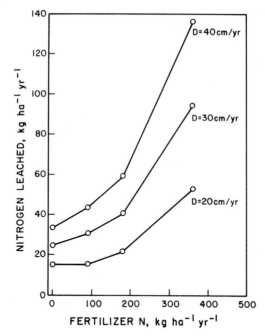

Fig. 6. Relationships between N leached and fertilizer N input for three drainage volumes. (Calculated from data by Nielsen et al. 1979)

yr^{-1}. Thus, the critical factors in the reduction of NO_3^- percolating through the unsaturated zone are crop, N, and water management. Crop management can help by use of varieties of high yield potential within the climate and soil limitations, by protection from weeds, insects, and diseases und by proper timing of planting, cultivating, and harvesting. Water needs to be managed for minimum leaching consistent with needs to leach salts from irrigated lands and to dispose of excess water to prevent water logging during periods of excess rain. Nitrogen management should focus on (1) timing on fertilizer inputs to coincide with crop needs; (2) accurate estimates of quantities of N needed to obtain maximum yields; (3) placement of fertilizer within the root zone; and (4) selection of sources to fit the soil and crop. Recommendations for crop N and water management are so specific for each agricultural production area that no general detailed recommendations can be made.

A general solution to increased N use efficiency has been sought through the use of chemicals that inhibit conversion of NH_4^+ to NO_3^-. Because NH_4^+ is adsorbed into the cation-exchange complex of soils, it does not move readily with water and is not subject to denitrification. Because it is available to plants, effective nitrification inhibition would reduce losses and increase use by plants. Unfortunately, the nitrification inhibitors produced to date, 1982, have been found to be effective in some situations and not others (Meisinger et al. 1980) and do not provide a generally useful solution. Local research must be performed to evaluate the effectiveness of nitrification inhibitors.

References

Aldrich SR (1980) Nitrogen in relation to food, environment and energy. Agric Exp St, College of Agricultur, Univ of Illinois, Urbana-Champaign Special Publ 61

Appelt H, Holtzclaw K, Pratt PF (1975) Effect of anion exclusion on the movement of chloride through soils. Soil Sci Soc Am Proc 39:264–267

Ayers RS, Westcot DW (1976) Water quality for agriculture. Irrigation and Drainage Paper no. 29. Food agric Organizat US (Rome, Italy) 00100

Berg W, Thomas GW (1959) Anion exclusion patterns from soils and soil clays. Soil Sci Soc Am Proc 23:348–350

Biggar JW, Nielsen DR (1962) Miscible Displacement. II. Behavior of tracers. Soil Sci Soc Am Proc 26:125–218

Biggar JW, Nielsen DR (1976) Spatial variability of the leaching characteristics of a field soil. Water Resour Res 12:78–84

Bremner JM, Blackmer AM (1978) Nitrous oxide emission from soils during nitrification of fertilizer nitrogen. Science 199:295–296

Bresler E (1973 a) Anion exclusion and coupling effects of nonsteady transport through unsaturated soils. I. Theory. Soil Sci Soc Am Proc 37:663–559

Bresler E (1973 b) Simultaneous transport of solutes and water under transient unsaturated flow conditions. Water Res 1:975–986

Bresler E, Laufer A (1974) Anion exclusion and coupling effects in nonsteady transport through unsaturated soils. II. Laboratory and numerical experiments. Soil Sci Soc Am Proc 38:213–218

Burns RG, Hardy RWF (1975) Nitrogen fixation in bacteria and higher plants. Springer, Berlin Heidelberg New York

Cameron DR, Kowalenks CG, Campbell CA (1979) Factors affecting nitrate nitrogen and chloride leaching variability in a field plot. Soil Sci Soc Am J 43:455–460

Cho CM (1971) Convective transport of ammonium with nitrification in soil. Can J Soil Sci 51:339–350

Committee on Water Quality (1973 a) Recreation and aesthetics. In: Water Quality Criteria 1972. NAS-NES. EPA-R3-73-033-March 1973. US Gov Print Office, Washington DC Stock number 5501–00520

Committee on Water Quality (1973 b) Agricultural uses of water. In: Water Quality Criteria 1972. NAS/NAE. EPA-R3-73-033-March 1973

Commoner B (1968) Balance of nature. In: Konkle WW (ed) Providing Quality Environment in Our Communities. Graduate School Press, USDA Washington DC

Commoner B (1970) Threats to the integrity of the nitrogen cycle: Nitrogen compounds in soil, water, atmosphere and precipitation. In: Singer SF (ed) Global Effects of Environmental Pollution. Springer, Berlin Heidelberg New York, pp 70–95

Dagan G, Bresler E (1979) Solute dispersion in unsaturated heterogeneous soil at field scale. I. Theory Soil Sci Soc Am J 43:461–446

Dyer KL (1965) Unsaturated flow phenomena in Panoche sandy clay loam as indicated by leaching of chloride and nitrate ions. Soil Sci Soc Am Proc 29:121–126

Fried M, Tanji KK, Van De Pol RM (1976) Simplified long term concept for evaluating leaching of nitrogen from agricultural land. J Environ Qual 5:197–200

Gardner WR (1965) Movement of nitrogen in soil. Agron Monog 10:550–572

Gast RG, Nelson WW, Randall GW (1978) Nitrate accumulation in soils and loss in tile drainage following nitrogen applications to continuous corn. J Environ Qual 7:258–261

Gilliam JW, Skaggs RW, Weed SB (1979) Drainge control to diminish nitrate loss from agricultural fields. J Environ Qual 8:173–242

Hills FJ, Broadbent FE, Fried M (1978) Timing and rate of fertilizer nitrogen for sugarbeets related to nitrogen uptake and pollution potential. J Environ Qual 7:368–372

IAEA (1980) Soil Nitrogen as Fertilizer or Pollutant. Proceeding Conference in Piracicaba, SP Brazil. July 1978. International Atomic Energy Agency, Vienna

Jury WA (1982) Simulation of solute transport using a transfer function. Water Resour Res 18:363–368

Jury WA (1983) Chemical transport modeling-current approaches and unresolved problems. In: Nelson DW (ed) Chemical mobility and reactivity in soil system SSSA Spec Public 11

Jury WA, Gardner WR, Saffigna PG, Tanner CB (1976) Model for predicting simultaneous movement of nitrate and water through a loamy sand. Soil Sci 122:36–43

Jury WA, Stolzy LH, Shouse P (1982) A field test of the transfer function model for predicting solute movement. Water Resour Res 18:369–374

Kinjo T, Pratt PF (1971) Nitrate adsorption. I. In some acid soils of Mexico and South America. Soil Sci Soc Am Proc 35:722–725

Leon LA, Pratt PF (1974) Efectos agronomicos de la retencion y lixiviacion de nitratos en los andepts de Colombia. Turrialba 24:408–413

Letey J, Blair JW, Devitt Dale, Lund LJ, Nash P (1979) Nitrate in effluent from specific tile-drained fields. In: Nitrates in Effluents from Irrigated Lands. NSF Report, May 1979. Available from NTIS. US Dept Commerce, Springfield VA 22161 USA

Loehr RC (1974) Agricultural waste management, problems, processes and approaches. Academic Press, London New York

Mathers AC, Steward BA, Blair B (1975) Nitrate nitrogen removal from soil profiles by alfalfa. J Environ Qual 4:403–405

McKim HCL (1978) State of knowledge in land treatment of wastewater Vols 1, 2. McKim HL Coordinator. Int Symp August 1978. Hanover, New Hampshire US Army Corps Eng

Meisinger JJ, Randall GW, Vitosh ML (1980) Nitrifications inhibitors-potentials and limitations. ASA Spec Pub Number 38. ASA, 677 S Segoe Rd, Madison, Wisconsin 53711

Miller RJ, Smith RB (1976) Nitrogen balance in the southern San Joaquin Valley. J Environ Qual 5:274–278

Misra C, Nielsen DR, Biggar JW (1974 b) Nitrogen transformations in soil during leaching. I. Theoretical considerations. Soil Sci Soc Am Proc 38:289–293

Misra C, Nielsen DR, Biggar JW (1974 b) Nitrogen transformations in soil during leaching. II. Steady state nitrification and nitrate reduction. Soil Sci Soc Am Proc 38:294–299

Misra C, Nielsen DR, Biggar JW (1974 c) Nitrogen transformations in soil during leaching. III. Nitrate reduction in soil columns. Soil Sci Soc Am Proc 38:300–304

Muir J, Boyce JS, Seim EC, Mosher PN, Deibert EJ, Olson RA (1976) Influence of crop management practices on nutrient movement below the root zone in Nebraska soils. J Environ Qual 5:255–259

Nielsen DR, Biggar JW (1961) Miscible displacement in soils. I. Experimental information. Soil Sci Soc Am Proc 25:1–5

Nielsen DR, Biggar JW (1962) Miscible displacement in soils. III. Theoretical considerations. Soil Sci Soc Am Proc 26:216–221

Nielsen DR, MacDonald JG (eds) (1978) Nitrogen in the environment Vols 1, 2. Academic Press, New York, London

Nielsen DR, Simmons CW, Biggar JW (1979) Flux of nitrate from a spatially variable field soil. In: Nitrate in Effluents from Irrigated Lands. NSF Report May 1979. Available from NTIS, US Dept Commerce, Springfield VA 22161

Panel on Nitrates (1978) An environmental assessment. Environmental Studies Board, Nat Res Council, Nat Acad Sci, 2101 Constitution Avenue, NW, Washington DC 20418

Pratt PF (1979 a) Integration, discussion and conclusions. In: Nitrate in Effluents from Irrigated Lands. NSF Report May 1979. Available from NTIS, US Dept Commerce, Springfield VA 22161

Pratt PF (1979 b) Nitrate in effluents from irrigated lands, In: Pratt PF (ed) NSF Report May 1979. Available from NTIS, US Dept Commerce, Springfield VA 22161

Pratt PF, Davis S, Sharpless RG (1976) A four-year field trial with animal manures. Hilgardia 44:99–125

Pratt PF, Barber JC, Corrin ML, Goering J, Hauck RD, Johnson HS, Klute A, Knowles R, Nelson DW, Pickett RC, Stephens ER (1977) Effect of increased nitrogen fixation on stratospheric ozone. Also CAST Report No. 53 January 1976, Climatic Change 1:109–135

Pratt PF, Lund LJ, Rible JM (1978) An approach to measuring leaching of nitrate from freely drained irrigated field. In: Nielson DR, MacDonald JG (eds) Nitrogen in the Environmental Vol 1 Academic Press, London New York, pp 22–256

Raveh A, Avnimelech Y (1973) Minimizing nitrate seepage from the Hula Valley into Lake Kinneret (Sea of Galilee). I. Enhancement of nitrate reduction by sprinkling and flooding. J Environ Qual 2:455–458

Rible JM, Pratt PF, Lund LJ, Holtzclaw KM (1979) Nitrates in the unsaturated zone of freely drained fields. In: Nitrates in effluents from irrigated lands. NSF Report May 1979. Available from NTIS, US Dept Commerce, Springfield VA 22161

Rich CI, Thomas GW (1960) The clay fraction of soils. Adv Agron 12:1–39

Robbins CW, Carter DL (1980) Nitrate-nitrogen leached below the root zone during and following alfalfa. J Environ Qual 9:447–450

Sawyer CN (1947) Fertilization of lakes by agricultural and urban drainage. FN Engl Water Works Assoc 61:109–127

Schalscha EB, Pratt PF, Domecq TC (1974) Nitrate adsorption by some volcanic-ash soils of southern Chile. Soil Sci Soc Am Proc 38:44–45

SIDA-FAO (1972) Report of Expert Consultation convened at FAO, Rome. Swedish Intern Develop Authority and FAO, Soils Bull 16

Singh BR, Kanehiro Y (1969) Adsorption of nitrate in amorphous and kaolinitic Hawaiian soils. Soil Sci Soc Am Proc 33:681–683

Smith SJ (1972) Relative rate of chloride movement in leaching of surface soil. Soil Sci 114:259–263

Soderlund R, Svensson BH (1976) The global nitrogen cycle. In Svenssen BH, Soderlund R (eds) Nitrogen, Phosphorus and Sulfur-Global Cycles. SCOPE Report 7, Ecol Bull (Stockholm) 22:23–73

Tanji KK, Fried M, Van De Pol RM (1977) A steady-state conceptual model for estimating nitrogen emissions from croppled lands. J Environ Qual 6:155–159

Tanji KK, Broadbent FE, Mehran M, Fried M (1979) An extended version of a conceptual model for evaluating annual nitrogen leaching losses from croplands. J Environ Qual 8:114–120

Thomas GW (1960) Effects of electrolyte inhibition upon cation-exchange behavior of soils. Soil Sci Soc Am Proc 24:329–332

Thomas GW, Swoboda AR (1970) Anion exclusion effects on chloride movement in soils. Soil Sci 110:163–166

Tullock RJ, Coleman NT, Pratt PF (1975) Rate of chloride and water movement in southern California soils. J Environ Qual 4:127–131

US Environmental Protection Agency (1976) Quality Criteria for Water. Office of Water and Hazardous Materials. US Environ Protect Agency, Washington DC 20460

Vaisman IJ; Shalhevet J, Kipnis T, Feigen A (1982) Water regime and nitrogen fertilization for Rhodes grass irrigated with municipal waste water on sand dune soil. J Environ Qual 11:230–232

Wagenet RJ, Biggar JW, Nielsen DR (1977) Tracing the transformations of urea fertilizer during leaching. Soil Sci Soc Am J 41:896–902

Watts DG, Hanks RJ (1978) A soil-water-nitrogen model for irrigated corn on sandy soils. Soil Sci Soc Am Proc J 42:492–499

White WC (1980) Personnel communication. The Fertilizer Institute, 1015 18th Street, NW Washington DC 20036

Winteringham EPW (1980) Food and agriculture in relation to energy, environment and resources. Intern Atomic Energy Agency Vienna 1980 Atomic Energy Review 18:223–245

5. Behavior of Phosphates in the Unsaturated Zone*

Y. AVNIMELECH

5.1 Introduction

Phosphorus is an element dealt with as a pollutant due to its effect on algal growth and thus on the eutriphication of surface water bodies. Phosphorus is considered to be a major factor controlling algal growth in many lakes and impoundments (Wetzel 1975). Very low phosphorus concentrations lead to excessive algal growth. When total phosphorus concentration exceeds 30 μ Pl^{-1}, the lake tends to be eutrophic (Vollenweider 1970). This concentration range is way below that found in fertilized soils. A different presentation of the problem can be given by comparing quantities of phosphorus leading to eutriphication in a water body to those applied to agricultural crops. According to Vollenweider (1970), a lake (average depth 10 m) receiving more than 0.2 g P m^{-2} y^{-1} tends to be eutrophic. The average amounts of fertilizer P applied to arable soils in developed countries are in the range of 1–2 g P m^{-2} y^{-1}. Assuming a contributing watershed/lake surface ratio of 10, it may be shown that a transfer of 1% of the land-applied P to the lake is large enough to induce eutriphication. The differences in orders of magnitude between concentrations and amounts of phosphorus in cultivated soils and limits for the equivalent entities demanded for high quality surface water pose a problem. From the farmer's viewpoint, a few percent leaching of the applied phosphorus with the drained water is hardly measurable and economically insignificant. Yet, as far as water quality is concerned, such leaching can be deleterious. An important route of phosphorus to surface water bodies is the erosion of soil particles containing phosphorus. The mechanisms of this process and the questions related to the availability of the particular phosphorus in surface water and its significance in water quality is beyond the scope of this presentation. The topic to be discussed here is the leaching of soluble phosphorus through the unsaturated zone to any outlet leading to a surface water body.

Soil scientists considered phosphorus to be practically immobile in most soils. The interest in the leaching of phosphorus has started only recently, in light of the above-mentioned water quality considerations. Different approaches are being used to quantify the phosphorus leaching process. A number of models are based on the assumption that a local equilibrium exists at any given point or layer in the soil profile. Phosphorus movement is presented by describing a sequence

* This work is based, in part, on the M. Sc. research work of Moshe Ronen. Moshe was a dedicated scientist having an original and clear way of thinking. Moshe Ronen was killed in a car accident at the very start of his scientific career. This work is dedicated to his memory

of equilibrium states along the soil profile. Two sets of equilibrium conditions are being used, namely, adsorption and solubility equilibria. A different approach is based on the use of a kinetic model independent of the establishment of any equilibrium state.

These different approaches will be critically reviewed.

5.2 Equilibrium Models Based on Adsorption Isotherms

Adsorption isotherms are commonly used to describe phosphorus immobilization in soils (e. g., Black 1970). The most commonly used is the Langmuir isotherm.

$$\frac{\Delta P}{M} = \frac{K\,bC}{1+KC} \tag{1}$$

or its linearized form

$$\frac{C/\Delta P}{M} = \frac{1}{Kb} + \frac{1}{b}C, \tag{2}$$

where C is the phosphate concentration in the solution, ΔP the amount of adsorbed phosphorus, M, K, and b are constants. The constant b is assumed to represent the adsorption saturation capacity and K to be related to the free energy of adsorption (Black 1970). The Langmuir isotherm was developed for a reversible, homogeneous monolayered adsorption of gases on solids. Phosphorus immobilization does not follow any of the above-mentioned requirements. It was shown that the Langmuir isotherm has a limited capacity to describe properly the interaction between soluble phosphorus and soil for either wide concentration range or different time duration (Ryden and Pratt 1980). Nevertheless, the Langmuir adsorption isotherm provides an approximation as to the interaction between phosphorus and soil. This approximation is frequently used for describing P mobility and availability in soil.

Hajek (1969) assumed a linear adsorption isotherm (such an assumption is valid at low solute concentration). The linear adsorption isotherm implies a constant distribution coefficient, Kd, defined as

$$Kd = \frac{\bar{P}}{P}, \tag{3}$$

where P and $\bar{P}$ are the solution and adsorbed phosphorus concentrations, respectively. Phosphorus breakthrough curves in soil columns could be calculated by considering the soil column to be divided into several segments and by calculating successive equilibrations during the simulated flow of a phosphatic solution. The calculated results were in good agreement with experimental measurements. Avnimelech et al. (1975) used a similar approach using the nonlinear Langmuir isotherm.

$$P_a = \frac{b}{1+K \cdot P_s}, \tag{4}$$

where P_a is the adsorbed P and P_s is the soluble P (both in mg P/g soil), while K and b are the experimentally obtained constants for the Langmuir isotherm. The mass balance equation for phosphorus in each soil segment and for each subsequent leaching unit is

$$P_s(i-1,j-1)+P_a(i,j-1)=P_s(i,j)+P_a(i,j), \qquad (5)$$

where i is a running index counting the soil segments and j is the index counting the sequential number of leaching units. Each leaching unit has a volume equal to the volumetric water capacity in a single soil segment. The left-hand side of Eq. (5) represents the amount of phosphorus getting into a given soil segment prior to the equilibration, i.e., the adsorbed P present in the segment plus the soluble phosphorus leached in from the segment above. The phophorus distribution in the soil segment after the equilibration (obtained by Eq. 4) is given by the right-hand side of Eq. (5). Calculated curves were found to be good approximations for phosphorus leaching curves obtained in small columns and in an experimental plot.

The approaches outlined above imply two basic assumptions. The first is that the adsorption equilibrium is obtained locally during the flow of water through the soil. The second assumption is that physical dispersive mechanisms can be neglected. Both computed and experimental breakthrough curves demonstrate a sharp phosphorus front, indicating that dispersion is mild. Dispersion is known to be lowered with decreasing flow rate (Bear 1979). Phosphorus flow rate in the soil is very low, since a relatively low dispersion coefficient is expected. Another reason for the possibly negligible effect of dispersion is the nonlinearity of the adsorption curve. Consider a front of phosphorus in the soil (Fig. 1). The P concentration downstream is lower than that upstream. Due to the curvilinearity of the adsorption isotherm, a higher percentage of P is adsorbed at the low concentration side as compared to the high concentration side of the front. As a result, the effective diffusion or other movement of phosphorus below the front is slower than that above the front and the front is thus maintained as a sharp front.

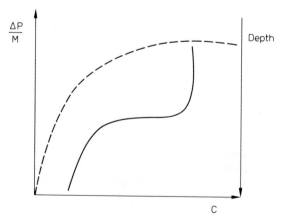

Fig. 1. Phosphorus adsorption isotherm (*broken line*) and phosphorus breakthrough curve (*solid line*). (*C* concentration of soluble phosphorus, $\Delta P/M$ phosphorus adsorption)

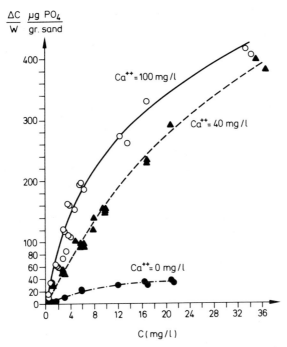

Fig. 2. Phosphorus adsorption on sandy soil at different calcium concentrations (Ronen 1971)

Ronen (1971) tested the adsorption approach as a means to predict phosphorus leaching to the aquifer through sand in the Dan Region wastewater reclamation project. Using batch equilibration techniques, he found that phosphorus immobilization could be attributed to adsorption only at a low concentration range, while solubility reactions seem to be dominant at high concentrations. The adsorption isotherms and the derived Langmuir parameters have shown a marked dependence on the calcium concentration in the solution (Fig. 2). This dependence is not implied in the simplistic Langmuir model. Another important point was demonstrated by comparing the breakthrough curves obtained in column flow experiments with these in adsorption experiments. A typical breakthrough curve (Fig. 3) consisted of three parts – a first, during which no phosphorus penetrated the sand column, a second, during which a steep increase in effluent P concentration took place, and a third flat part, the saturation part of the curve. The amount of phosphorus retained in the column until the start of phosphorus breakthrough (Point A in Fig. 3) was calculated for a number of leaching experiments. The average amount of P retained from solutions containing 40 mg Ca l^{-1} was 6 μg P g^{-1} sand and 12 μg P g^{-1} P when calcium concentrations were 100 mg/l^{-1}. These amounts are very low compared to the maximal P adsorption capacity for equivalent calcium concentrations (578 and 675 μg P, respectively; Fig. 2). Yet, the maximal capacity is relevant only for high P concentrations (above 40 mg P l^{-1}). The actual P concentrations in the column experiments at the point of breakthrough were 0.35 and 0.18 mg P l^{-1} for the 40 and 100 mg Ca

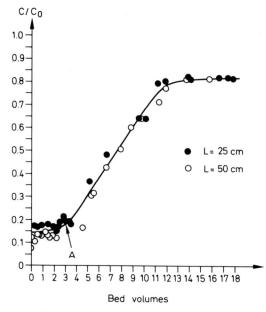

Fig. 3. Breakthrough of phosphorus through dune sand columns (Ronen 1971)

1^{-1}) solutions, respectively. The relevant adsorption capacity is, obviously, the one that corresponds to the actual P concentration. These adsorption capacities were in the range of 7–10 μg P g^{-1} sand, similar to the amounts retained in the column until the breakthrough point. This demonstration is often applicable to actual field conditions. Phosphorus may move in the soil without delay if the adsorption capacity as relevant to the actual solution concentration is saturated. The actual effective P adsorption capacity of soil is usually well below the maximal adsorption capacity of the soil.

5.3 Solubility Equilibrium

Ronen's experiments (Ronen 1971), mentioned above, indicating the marked effect of calcium concentration on phosphorus adsorption may be used to demonstrate the fact that phosphate adsorption is an empirical term, i.e., the disappearance of phosphate from the solution, and does not implicate any mechanism. Moreover, it was demonstrated that phosphorus interaction with the soil cannot be dealt with as an interaction of a single component with the soil, since other components (Ca^{2+} in this case) are also involved.

The dominant process controlling phosphorus levels in the soil solution is its chemical interactions and the formation of sparingly soluble salts. Phosphorus solubility in soils is limited mainly by the limited solubility of calcium, iron, and aluminum phosphates. Crystals of such salts are formed and have been identified when a concentrated phosphatic solution is interacting with the soil e.g., at the vicinity of fertilizer granules, (Huffman 1968, Black 1970). Interaction of phosphorus with Ca, Fe, or Al ions on the surface of the soil particles is probably the dominant process when dilute phosphatic solutions are interacting with soils (Beek and Riemsdijk 1979).

Solubility equilibria can be given schematically as in Eq. (5)

$$n\ A_{(l)} + m\ B_{(l)} \rightleftarrows C(s) \tag{6}$$

and are defined by the equality of the sum of chemical potentials of the reactants with the free energy of formation of the solid (Eq. 6).

$$n\mu^0_A + nRT\ \ln(A) + m\mu^0_B + mRT\ \ln(B) = F^0_C \tag{7}$$

or

$$n\ \ln(A) + m\ \ln(B) = \frac{F^0_C - n\mu^0_A - m\mu^0_B}{RT} = \text{constant}.$$

The commonly used solubility product [Eq. (8)] is equivalent to Eq. (7).

$$(A)^n + (B)^m = Ksp. \tag{8}$$

(Values in parenthesis represent activities, μ^0 standard chemical potentials, and F^0 free energy of formation.)

Any definition of phosphate behavior in soils by the relationships given above can lead to a rigorous quantitative evaluation of phosphorus mobility through the soil column.

The characterization of phosphorus in soil by its potential and its solubility behavior was studied by a number of investigators (e. g., Aslyng 1954, Larsen and Court 1961, Cole and Olsen 1959).

The technique used for establishing the solubility equilibrium controlling the system involves a series of equilibrations, calculation of phosphatic species activities in the solution, and then plotting the two reactant activities on a potential diagram, i.e., log activity of one species is plotted against that of the second species.

The species used are $Ca(OH)_2$ versus H_3PO_4 or $Ca(H_2PO_4)_2$ versus $Ca(OH)_2$. The two options are in essence equivalent. If a system is controlled through a solubility equilibrium, the data points in the potential diagram should fall on a straight line, the slope of which is equal to m/n [m/n of Eq. (5) is equivalent in this case to the Ca/P ratio of the reaction] and its location is defined by the solubility of the solid phase (a detailed discussion of this technique is given by Lindsay 1979). Most of the work related to equilibrium solubility of phosphates in soils was based on the comparison of solubility functions obtained in the soil system to equivalent functions obtained for well-known and defined phosphates (Ryden and Pratt 1980). This approach often led to results inconsistent with expected behavior of phosphates. Thus, for example, it was found that phosphate solubility in calcareous soils or aquatic sediments is determined by a $CaHPO_4$ stochiometry, having a solubility product different than those known for dicalcium phosphates (Cole and Olsen 1959, Avnimelech 1975).

The ability to use simple potential diagrams and to compare solubility behavior in soils and in pure systems is limited due to a few reasons. The first is the effect of particle size and crystal imperfections on the solubility behavior of phosphates

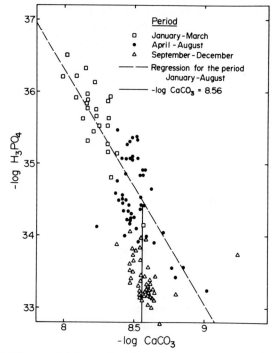

Fig. 4. $CaCO_3$–H_3PO_4 potential diagram for Lake Kinneret bottom water. (Data for the period 1975–1977)

(Weir et al. 1971). A more basic problem is that the soil system is a complicated multicomponent system, compared to the limited number of components in reference laboratory systems.

Calcium phosphate solubility is usually studied in the three component system $Ca(OH)_2$–H_3PO_4–H_2O. A $Ca(OH)_2$–H_3PO_4 potential diagram is thus appropriate to these systems. A major additional component, H_2CO_3, is found in calcareous soils and is known to interact with phosphate ions and with calcium phosphates (Olsen 1953). This additional component was included by Avnimelech (1980) in the analysis of phosphate solubility behavior in calcareous systems. A laboratory study has revealed that a calcium carbonate phosphate $Ca_3(HCO_3)_3$ PO_4 is formed and controls the solubility when dilute phosphatic solutions interact with $CaCO_3$.

The use of a $CaCO_3$–H_3PO_4 potential diagram allowed the characterization of phosphate solubility in Lake Kinneret water (Fig. 4) as well as in soils and drainage water in the calcareous clay soils of the Izrael Valley (Fig. 5). The obtained results indicated that phosphorus concentrations in the soil solution and in drainage water are controlled by the solubility product

$$(Ca)^3 \cdot (HCO_3)^3 \cdot (PO_4) = 10^{-28} \tag{9}$$

and that its concentration can be predicted using this equation. Preliminary results indicate that equilibrium is achieved in soil within a period of a few days.

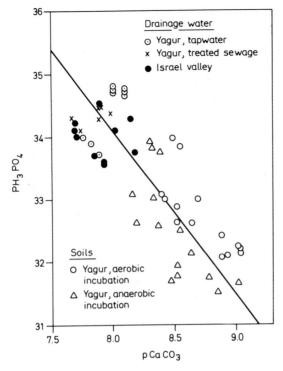

Fig. 5. $CaCO_3$–H_3PO_4 potential diagram for drainage water and soil extracts. (Yagur drainage water was sampled from plots irrigated with tap water or reclaimed sewage water. Soils were incubated prior to extraction for 14 days)

The establishment of a rigorous framework for the description of phosphorus solubility allows the prediction of phosphorus movement through soils down to the drainage outlets or to the aquifer.

5.4 The Kinetic Approach

The kinetics of phosphorus mobilization and immobilization in soils have been studied by many investigators. These works have also been used to describe and predict phosphorus leaching.

The different kinetic models are based on a series of assumptions and expressions. Some models assume a release of phosphorus to an infinite medium (Amer et al. 1955, Evans and Jurinak 1976), while others (e. g., Shah et al. 1975) assume a finite volume of solution. Several investigators assume first-order kinetics of P release from the solid (Amer et al. 1955, Li et al. 1972, Overman et al. 1980) and in some cases a series of first-order reactions.

Others have used second-order kinetics (Kuo and Lotse 1972, Griffin and Jurinak 1973), while some works assume a Freundlich equation having an intercept as a time variable (Kuo and Lotse 1974). A power form equation is also used (Chien and Clayton 1980). Engfield et al. (1976) tested five kinetic models for orthophosphate reactions in soils: linear first-order, first-order Freundlich adsorption, diffusion-limited Langmuir, and Freundlich adsorption and an empirical function. All of these fitted the data ($r > 0.8$).

Different mechanisms are also considered. Sharpley et al. (1981) assumed a diffusion-controlled process having an effective diffusion coefficient (including geometry term) as a time variable and a capacity term defined as the amount of phosphorus desorbed during a 2-h standard laboratory test. Overman et al. (1980) assumed that phosphorus is distributed in four phases: solution phase, adsorbed phase, transformed phase, and fixed phase. The reaction between the first three phases is reversible, while the transfer to the fixed phase is irreversible. The last point warrants discussion.

Many slow reactions may be neglected in short-term laboratory experiments or when fast processes are considered. Phosphate migration to the aquifer is a very slow or an extremely lengthy process. Slow processes cannot be neglected accordingly. The irreversibility of phosphorus fixation is an approximation for short-term processes only, otherwise the only form of phosphorus existing would be fixed phosphorus having zero rate of release.

An additional mechanism was proposed by Shah et al. (1975) and Novak et al. (1975). In their model, the source-sink term is a function of $(Y - Y^*)$, where Y is the concentration of phosphorus in the liquid phase and Y^* is the equilibrium concentration of phosphorus with the existing level of phosphorus in the solid phase. This approach is similar to models used in dissolution kinetic studies (e. g., Nancollas et al. 1979), where the rate of dissolution is considered to be a function of the driving force, the potential gradient between an equilibrium state, and an existing level in the solution. The equilibrium term used by Shah, Novak, and coauthors is the one obtainable from phosphorus adsorption isotherms and not the thermodynamic equilibrium term.

The large number of proposed mechanisms and models is an indication of the confusion involved. It seems that this confusion stems mainly from the trial to study the leaching of phosphorus per se. Phosphorus (P) does not exist in the soil. The phosphatic species present in the soil solution are the different orthophosphate ions ($H_2PO_4^-$, $HPO_4^=$, $PO_4^\equiv$) as well as organic and polyphosphates. The relative concentrations of these species are functions of the activities of other components in the system, mainly that of the hydrogen ion. The reactions of any of these species in the soil (i. e., adsorption or precipitation reactions) is controlled by the presence of other ions, such as Ca^{2+}, Fe^{3+}, Al^{3+}, and others. Negligence of these factors leads to an inaccurate description of the system.

5.5 Conclusions

A small sample of the existing literature describing or related to migration of phosphorus in the unsaturated zone has been reviewed here. One optimistic conclusion that can be drawn is that we have a number of quite good approximations related to the prediction of phosphorus leaching. The agreement found between calculated breakthrough curves, based upon different kinetic or equilibrium models and between actual laboratory or short-range field data is rather good.

The problem we are faced with is how accurate these models are and, moreover, how far can one extrapolate these models to longer time or geometric ranges.

A basic simplistic assumption involved in most existing models and theories related to phosphorus migration is that phosphorus moves and reacts with the soil as a single component. This is obviously a wrong assumption. Phosphorus can never move alone. Moreover, all known interactions of phosphorus with the soil involve reactions with additional components: adsorption as an exchangeable anion involves other anions; interaction with clay minerals involves hydrogen ions, silica, and other components; and chemisorption or solubility-precipitation reactions are never reactions of phosphorus alone.

Any breakthrough in our ability to predict, understand, or control phosphorus leaching has to be based on the development of models that take the actual chemical reactions occuring in the soil into account. Such models have to be multicomponent interactive models. Our understanding of the different processes involved and our computing ability may allow this task.

References

Amer F, Bouldin DR, Black CA, Duke FR (1955) Characterization of soil phosphorus by anion exchange resin and P^{32} equilibration. Plant Soil 6:391–408

Aslyng HC (1954) The lime and phosphate potentials of soils: The solubility and availability of phosphates. Royal Vet Agric College Yearbook, Copenhagen Danemark

Avnimelech Y (1975) Phosphate equilibrium in fish ponds. Verb Int Verein Limnol 19:2305–2308

Avnimelech Y (1980) Calcium-Carbonate-Phosphate surface complex in calcareous systems. Nature 290:255–257

Avnimelech Y, Lansman S, Lacher M (1975) Leaching of phosphorus through a soil column. Mekorot Water Co Res Rep, Nazareth (Hebrew)

Bear J (1979) Hydraulics of ground water. McGraw Hill, New York, 242 p

Beek J, van Riemsdijk WH (1979) Interaction of orthophosphate ion with soil. In: Bolt GH (ed) Soil Chemistry, B Physico-Chemical models. Elsevier North-Holland Amsterdam Oxford New York

Black CA (1970) Behaviour of soil and fertilizer phosphorus in relation to water pollution. In: Willrich TL, Smith GE (eds) Agricultural practices and water quality. Iowa State Univ Press, Ames Iowa

Chien SH, Clayton WR (1980) Application of Elovich equation to the kinetics of phosphate release and sorption in soils. Soil Sci Soc Am J 43:265–268

Cole CV, Olsen SR (1959) Phosphorus solubility in calcareous soils. I. Dicalcium phosphate activities in equilibrium solutions. Proc Soil Sci Soc Am 23:116–118

Enfield CG, Harlin CC (Jr), Bledsoe BE (1976) Comparison of five kinetic models for orthophosphate reactions in mineral soils. Soil Sci Soc Am J 40:243–249

Evans RL, Jurinak JJ (1976) Kinetics of phosphate release from a desert soil. Soil Sci 121:205–211

Griffin RA, Jurinak JJ (1973) Test of a new model for the kinetics of adsorption desorption processes. Soil Sci Soc Am Proc 37:869–872

Hajek BF (1969) Chemical interactions of wastewater in a soil environment. J Water Pollut Control Fed 41:1775–1786

Huffman EO (1968) The reactions of fertilizer phosphorus with soils. Overlook Agr 5:202–207

Kuo S, Lotse EG (1972) Kinetics of phosphate adsorption by calcium carbonate and Ca-kaolinite. Soil Sci Soc Am Proc 36:725–729

Kuo S, Lotse EG (1974) Kinetics of phosphate adsorption and desorption by hematite and gibbsite. Soil Sci 116:400–406

Larsen S, Court MN (1961) Soil phosphate solubility. Nature 189:164–165

Li WC, Armstrong DE, Williams JD, Harris RF, Syers JK (1972) Rate and extent of inorganic phosphate exchange in lake sediments. Soil Sci Soc Am Proc 36:279–284

Lindsay WL (1979) Chemical equilibria in soils. Wiley-Interscience, New York

Nancollas GH, Amjad Z, Koutsoukos P (1979) Calcium phosphates-speciation, solubility and kinetic considerations. In: Jenne EA (ed) Chemical modelling in aqueous systems. ACS Symp Ser 93:475–497

Novak LT, Adriano DC, Coulman GA, Shah DB (1975) Phosphorus movement in soils: Theoretical aspects. J Environ Qual 4:93–99

Olsen SR (1953) Inorganic phosphorus in alkaline and calcareous soils. Academic Press, New York Agronomy 4:89–122

Overman AR, McMahon BR, Chu RL (1980) Velocity dependence of phosphorus transport in a packed-bed reactor. J Water Pollut Control Fed 52:2471–2476

Ronen M (1971) Sorption of phosphorus on dune sand. M Sc Thesis, Technion Israel Instit Technol, Haifa Israel

Ryden JC, Pratt FE (1980) Phosphorus removal from wastewater applied to land. Hilgardia 48:1–36

Shah DB, Coulman GA, Novak LT, Ellis BG (1975) A mathematical model for phosphorus movement in soils. J Environ Qual 4:87–92

Sharpley AN, Ahuja LR, Yamamoto M, Menzel RG (1981) The kinetics of phosphorus release from soil. Soil Sci Soc Am J 45:493–496

Vollenweider RA (1970) Scientific fundamentals of the eutriphication of lakes and flowing water, with particular reference to nitrogen and phosphorus as factors of eutriphication. OECD, Paris

Weir DR, Chien SH, Black CA (1971) Solubility of Hydroxyapatite. Soil Sci 111:107–112

Wetzel RG (1975) Limnology, Saunders, Philadelphia London Toronto

6. Unsaturated Zone Pollution by Heavy Metals

G. Matthess

6.1 Introduction

Heavy metals are usually defined as metals with densities larger than $5 \, g \, cm^{-3}$. This group includes 70 elements with the atomic numbers 23–32, 40–51, 57–84, and 87–106. The heavy metals with the atomic numbers 61 and 93–106 are artifical elements.

The element technetium (atomic number 43) was found in a sample of molybdenum, which had been bombarded with neutrons in the Berkeley Cyclotron. The search for this element in terrestrial materials was without success so far, but it has been reported from the spectra of stars (Correns 1969).

Heavy metals in the unsaturated zone can be discussed from two aspects. The first one is their abundance in the unsaturated zone, the second one is their importance for human health, either in a beneficial or a noxious sense. As far as the abundance of heavy metals in the unsaturated zone is concerned, only very little information is available except for those elements which influence the agricultural use of soils and the quality of crops or which are considered to be dangerous for the underlying groundwater.

Thus, health effects are involved in both cases, and have aroused public interest in some heavy elements, especially cadmium and mercury, since the recent cases of human poisoning in highly industrialized environments in Japan. For the most important heavy metals, maximum permissible concentrations in soils, crops, and drinking water are fixed.

Actually the sparsity of data and the hygienic importance of selected heavy metals decreases the number of interesting elements to some 14 species.

6.2 Abundance

The importance of any heavy metal in the unsaturated zone is a result of its abundance in the earth crust and its geochemical mobility. Table 1 gives an idea of the natural abundances, showing the concentrations of the elements in magmatites as their primary source. Their mobility may be derived from the proportional decrease or increase in the products of weathering, in the resistates (sandstones), in the hydrolysates (shales), in the precipitates (carbonate and sulfate rocks), in the evaporites (salt deposits), and lastly in seawater. The very low con-

Table 1. Abundance of heavy elements in rocks (Horn and Adams 1966, Hem 1970) and seawater (ppm) (Turekian 1969)

Element	Magmatites	Sandstones	Shales	Precipitates	Evaporites	Seawater
Cd	0.192	0.0199	0.0476	–	–	1.1×10^{-4}
Co	23	0.328	8.06	0.123	1.6	3.9×10^{-4}
Cr	198	120	423	7.08	10.6	2×10^{-4}
Cu	97.4	15.4	44.7	4.44	2	9.1×10^{-7}
Fe	42,200	18,600	38,800	8,190	265	3.4×10^{-3}
Hg	0.328	0.0574	0.272	0.0456	–	1.5×10^{-4}
Mn	937	392	573	842	4.4	4×10^{-4}
Ni	93.8	2.57	29.4	12.8	1.4	6.6×10^{-3}
Pb	15.6	13.5	80	16.5	0.9	3×10^{-5}
Ra	–	–	–	–	–	1×10^{-10}
Ru	–	–	–	–	–	7×10^{-7}
Tl	1.1	1.5	1.6	0.065	–	–
U	2.75	1.01	4.49	2.2	0.2	3.3×10^{-3}
Zn	80	16.3	130	15.6	0.6	5×10^{-3}

centrations of heavy elements in seawater as compared to their abundance in primary rocks show their poor geochemical mobility resulting from low solubilities, from adsorption, and ion exchange effects on minerals or organic materials, e. g., clay or humic substances, from precipitation and coprecipitation, and from the fixation in biological material or in newly formed mineral substances.

A special group are the radioactive heavy metals. The most important radionuclides belong to the radioactive series of the uranium-radium series with 14 daughter substances and the thorium series with 10 daughter substances.

6.3 Contamination of Groundwater by Heavy Metals

Heavy metals are used in various ways, as raw materials for numerous industrial products or as catalysts in chemical processes. Some of them are constituents of pesticides or fertilizers, which are distributed over large areas in connection with industrial, agricultural, or hygienic activities. Principially, all heavy metals may act as contaminants in gaseous, liquid, or solid wastes. Appreciable amounts of some heavy metals are set free by the combustion of fossil fuels.

Radioactive heavy metals may occur as fission products in connection with the processing and smelting of uranium ores, the production and reprocessing of nuclear fuel and explosives, the disposal of nuclear wastes, and the escape of volatile radionuclides at nuclear power plants and of the various radionuclides used for medical or technical purposes. The use and disposal of radioactive material is controlled by appropriate laws, which may be considered as a model for environmental protection against any dangerous element or compound.

Gaseous wastes, which are apt to propagate a contamination within very short time intervals over wide areas, usually contain some heavy metals in small quantities, e. g., lead from traffic exhaust (Golwer 1973) and from sulfide ore smelters,

Table 2. Heavy metal immission [g (ha^{-1} y^{-1})] in precipitation in West Germany and Sweden (Scheffer and Schachtschabel 1982)

Element	Black Forest Open land	Solling (Forest Area)		Göttingen Township Open land	Uppsala/ Sweden
		Beech tree	Fir		
Pb	110	340	532	40	29.2
Cd	4.5	16	20	3.5	0.74
Cr	–	12	22	–	–
Cu	18	150	230	52	–
Ni	34	34	39	–	–
Hg	–	–	–	0.2	–
Zn	210	2,720	2,700	324	–

and the fallout-radionuclide ruthenium-106 (Aurand et al. 1971). The immission of heavy metals depends on the distance to the emittants and on the plant cover: open land or forest areas (Table 2).

Waste water, especially of industrial origin, contains heavy metals in higher or lower concentrations. They are enriched during sewage treatment in the sludges, which are used as fertilizers (Table 3).

Solid waste dumps and residues of mining, ore processing and smelting operations are commonly sources of higher local concentrations of heavy metals in the groundwater (Matthess 1972, 1974, Schöttler 1972).

Table 3. Heavy metal contents in municipal sewage sludges in West Germany (mg kg^{-1} dry residue) (Scheffer and Schachtschabel 1982)

Element	Maximum concentration	Average concentration
Pb	1,500	290
Cd	150	21
Cr	3,000	–
Cu	2,600	390
Ni	200	131
Hg	55	4.8
Zn	15,750	2,140

6.4 Solubility

The solubility of the heavy metals is generally controlled by the most abundant anions in natural water. These are hydroxide, hydrogen carbonate, carbonate, sulfate, chloride, nitrate, and – in a reducing environment – sulfide. Therefore, the mobility of the heavy metals depends on the solubility of their hydroxides, carbonates, sulfates, chlorides, and sulfides. Some of them are listed in Table 4 in comparison with the poorly soluble barium sulfate.

Table 4. Solubility products of different heavy metals and of
BaSO$_4$

			(at 25 °C)
BaSO$_4$	[Ba^{2+}]	]SO$_4^{2-}$]	1.08×10^{-10}
CuCO$_3$	[Cu^{2+}]	[CO$_3^{2-}$]	1.37×10^{-10}
FeCO$_3$	[Fe^{2+}]	[CO$_3^{2-}$]	2.11×10^{-11}
RaSO$_4$	[Ra^{2+}]	[SO$_4^{2-}$]	4.25×10^{-11} (20 °C)
ZnCO$_3$	[Zn^{2+}]	[CO$_3^{2-}$]	$6 \quad \times 10^{-11}$
PbCO$_3$	[Pb^{2+}]	[CO$_3^{2-}$]	1.5×10^{-13}
Ni(OH)^2	[Ni^{2+}]	[OH$^-$]^2	1.6×10^{-14}
Cd CO$_3$	[Cd^{2+}]	[CO$_3^{2-}$]	2.5×10^{-14}
Fe(OH)$_2$	[Fe^{2+}]	[OH$^-$]^2	1.65×10^{-15}
Mn(OH)$_2$	[Mn^{2+}]	[OH$^-$]^2	7.1×10^{-15}
Hg$_2$CO$_3$	[Hg$^+$]^2	[CO$_3^{2-}$]	$9 \quad \times 10^{-17}$
Hg Cl	[Hg$^+$]	[Cl$^-$]	$2 \quad \times 10^{-18}$
Ni S	[Ni^{2+}]	[S^{2-}]	$2 \quad \times 10^{-21}$
Zn S,β	[Zn^{2+}]	[S^{2-}]	1.1×10^{-24}
Pb S	[Pb^{2+}]	[S^{2-}]	3.4×10^{-28} (18 °C)
Cd S	[Cd^{2+}]	[S^{2-}]	3.6×10^{-29}
Fe(OH)$_3$	[Fe^{3+}]	[OH$^-$]^3	$4 \quad \times 10^{-38}$
Cu$_2$S	[Cu$^+$]^2	[S^{2-}]	$2 \quad \times 10^{-47}$ (18 °C)

The concept of the solubility product K_{SP} is derived from the law of mass action.

In natural waters, which are nonideal solutions, the ions present are not infinitely dilute. Thus, they influence other ion species through the electric fields surrounding them, so that the latter appear to be present in smaller amounts. This factor is taken into consideration by using the so-called activities instead of the measured concentrations.

The activity of a dissolved substance is a relative quantity, which only in very dilute solutions is numerically almost equal to its molar concentration. The activity of a certain molecular or ion is designated by the symbol a in conjunction with the chemical symbol for the molecule or ion, thus, a Ca^{2+}, or the chemical symbol is enclosed in square brackets thus, [Ca^{2+}].

The activities of dissolved substances and of the solvent can be determined by measurement of the solubility, the dissociation constants, the vapor pressure, the freezing point, and boiling point, as well as by measurement of the electromotive force of galvanic cells. However, the activities are more easily calculated from the analytic values by multiplying by the activity coefficient f_i:

$$a_i = f_i \cdot m_i. \tag{1}$$

Since the activity is dimensionless by definition (Garrels and Christ 1965), the activity coefficient has formally the dimension l/mole. For dilute solutions, the activity coefficients of individual ions can be obtained to sufficient accuracy from Eq. (2) which is based on the Debye-Hückel theory of interionic interaction:

$$-\log f_i = \frac{A z_i^2 \sqrt{I}}{1 + å_i B \sqrt{I}}. \tag{2}$$

A and B are pressure and temperature dependent empirical constants specific to the solvent, z_i is the valence of the ion in the solution specified, and $å_i$ is a term dependent on the effective diameter of the ion in the solution. For the ionic strength I, the empirical equation for dilute solutions holds:

$$I = \frac{1}{2} \sum_i C_i \cdot z_i^2 . \qquad (3)$$

The molar ionic strength is calculated in terms of molar concentration C_i [in mol l^{-1}(liter-molarity), in mol kg^{-1}solvent (kilomolarity)] and valence z_i.

The law of mass action describes the equilibrium

$$v_A A + v_B B \rightleftharpoons v_C C + v_D D,$$

in terms of a constant, which is dependent on the given temperature and pressure:

$$K = \frac{[a_C]^{v_C} \cdot [a_D]^{v_D}}{[a_A]^{v_A} \cdot [a_B]^{v_B}} . \qquad (4)$$

In this thermodynamic version of the law of mass action, the activities of the species A, B, C, and D and the stoichiometric coefficients v_A, v_B, v_C, and v_D are used.

For the dissociation reaction:

$$AB \rightleftharpoons v_A A^{Z_A^-} + v_B B^{Z_B^+} ,$$

the activity of the solid phase AB is unit by definition so that Eq. (4) can be simplified to the solubility product in its usual form [Eq. (5)]

$$K_{SP} = K_{AB}[AB]_{sat} = [A^{Z_A^-}]^{v_A} \cdot [B^{Z_B^+}]^{v_B} . \qquad (5)$$

The Gibbs free energy change for a chemical reaction under standard conditions (1 bar, 25 °C) can be calculated from the ΔG^0-values of the various substances involved.

$$\Delta G_r^0 = \sum \Delta G_{reaction\ products}^0 - \sum \Delta G_{reactants}^0 . \qquad (6)$$

Reactions in which the free energy in the reaction products is less than that in the original substances will probably proceed spontaneously unless retarded. As the solid phase surface energy is added to the Gibbs free energy, it follows that fine-grained material is less stable and thus more soluble than coarse-grained. This is independent of the kinetic effects of smaller grain-sizes, which result in faster equilibration (Langmuir 1971).

The standard free energy change is related to the equilibrium constants, thus:

$$\Delta G^0 = - RT \ln K, \qquad (7)$$

where R is the gas constant 8.315 J $\cdot °C^{-1}$ mol^{-1} and T is the absolute temperature in K.

These equations can be used for the calculation of the equilibrium constants of the reactions. The necessary thermodynamic data are listed, e. g., in Garrels and Christ (1965). For the calculation of aquatic equilibrium systems, computer programs are available, e.g., WATEQF (Plummer et al. 1976) and GEOCHEM (Sposito and Mattigod 1980). They adjust the ion concentrations measured in water samples with the help of iterative approaches to a chemical thermodynamic equilibrium model, taking into account the in situ temperature, pH, and Eh values and the thermodynamic constants of minerals and mineral phases in the system. As a result, the activities and the quantities of the anorganic species are listed.

The tendency toward dissolution or precipitation can be estimated by comparing the actual ion activity product of the dissolved constituents of a solid species with its solubility product, yielding the saturation index, SI,

$$SI = \log (IAP/K_{SP}), \tag{8}$$

where IAP is the ion activity product $\{A\}^x \{B\}^y$ and K_{SP} is the solubility product of the solid species $[A_x B_y]$. If SI is less than 0 the water is undersaturated with respect to $[A_x B_y]$ and a net dissolution of the solid should occur; at $SI = 0$, the water is at equilibrium with the solid; when SI is greater than 0, the water is supersaturated with the solid's constituent ions and net precipitation should occur. However, kinetic delays may prevent dissolution or precipitation (Edmunds 1977, Matthess 1982).

The solubility of some heavy metals, for example, iron, manganese, copper, and uranium is very strongly influenced by the pH and Eh of the groundwater. Important equilibrium systems include the processes dependent on the redox potential and pH value. These systems can be described with sufficient accuracy by means of equilibrium models based on the principle of the law of mass action and the laws of chemical thermodynamics (Garrels and Christ 1965, Stumm and Morgan 1981).

Oxidation can be broadly defined as loss of electrons and reduction as gain of electrons. The reduction-oxidation (redox) potential serves as a measure of the relative state of the oxidation or reduction in an aqueous system. In a solution which contains various oxidation states of an element the redox potential is measured as an electrical potential between an inert metal electrode and a standard reference electrode, both immersed in the solution.

The redox potential is usually denoted by the symbol E, or, if referred to the standard hydrogen electrode of zero volts, by Eh or E_H. The standard reduction potential E^0 holds for a redox pair at activities of unity and standard temperature and pressure.

The redox potential is related to the standard free energy change of a reaction, its equilibrium constants, the amounts of the reacting substances present in the equilibrium, and the standard potential, as shown in Eqs. (7), (9), and (10). Equation (10) is generally known as the Nernst equation.

$$E^0 = -\frac{\Delta G^0}{nF}, \tag{9}$$

$$Eh = E^0 + \frac{RT}{nF} \ln \frac{a_{Ox}}{a_{red}} \tag{10}$$

ΔG^0 = standard free energy of formation of the chemical reaction (in kJ mol^{-1})

R = universal gas constant (8.315 J $°C^{-1}$ mol^{-1})

T = absolute temperature (K)

n = number of electrons, represented by a multiple of e in the redox equation (difference of electrons between the oxidized and reduced substances)

F = Faraday's constant (unit of charge on 1 mole of electrons, 96.564 kJ V^{-1} or 96.487 coulomb mol^{-1})

E^0 = standard potential (V) ($a_{ox} = a_{red} = 1$)

Eh = redox potential (V)

a_{ox} and a_{red} = activities of the oxidized or reduced substances in the chemical system under consideration.

The redox potential can be determined directly by voltage measurement for the voltage between the above mentioned inert electrode (mostly a platinum electrode, less often gold) and the reference electrode (mostly a calomel electrode) (Garrels and Christ 1965). The same holds for pH measurements, which are performed with a glass electrode and a calomel electrode as reference electrode.

The entrance of atmospheric oxygen into systems of low initial Eh causes a rapid upward shift of Eh values during water pumping from wells or at emergence from springs. However, when atmospheric oxygen can be excluded, very accurate measurements can be achieved. Instruments have been specially designed for Eh measurements of groundwater.

Many hydrogeologically important oxidation-reduction reactions are defined by Eh, by Eh and pH, or by Eh, pH, and the activity of the individual elements in the solution. These controls of solubility can be represented on stability field diagrams.

The stability field diagram is a useful model with pH plotted along the abscissa and Eh along the ordinate. The limits of the stability field of the ion species or solids at chemical equilibrium can be calculated for given concentrations of dissolved ions with the help of thermodynamic equilibrium relationships. The diagrams are not suitable for predicting reaction rates.

An Eh-pH diagram is a theoretical model which contains only that information which has been used in its construction. It is always an oversimplification, because not all the components in the system are known or else cannot be taken into account. Such "mixed potentials" are mostly not amenable to quantitative interpretation (Stumm and Morgan 1981).

Eh-pH diagrams as published by Hem (1970) and Garrels and Christ (1965) are useful qualitative and comprehensive guides toward understanding the relations between the solid and dissolved species even if they cannot be used as quantitative descriptions. One of the reasons for this restriction is the presence of various species and the forming of soluble complexes with dissolved organic materials. In a similar way, the mobility of manganese, cadmium, copper, mercury, uranium, and many other heavy elements is affected by the Eh and pH of groundwater.

6.5 Complexation

In concentrated aqueous solutions (for ionic strengths above 0.1, but in many cases also for much lower values) complex ions appear in the solution, which can be treated as thermodynamic entities and are in dynamic equilibrium with the free ions. Complexes are combinations of cations, which are called the central atoms, with molecules or anions containing free pairs of electrons (bases), which are referred to as ligands. Bases containing more than one ligand atom, e. g., oxalate and citrate, are referred to as multidentate complex formers. Complex formation with multidentate ligands is called chelation and the complexes are called chelates (Stumm and Morgan 1981). As a measure for the stability of a complex, a stability constant can be defined on the base of the law of mass action [Eq. (11)]. For the association of a complex M_aL_b from d moles of a metal M and b moles of a ligand holds

$$K = \frac{[M_dL_b]}{[M]^d[L]^b}.$$ (11)

Complexation by inorganic or organic complex forming substances may change the solubility in pore solutions of the unsaturated zone, e. g., fulvic acids, tartaric acid, citric acid, and salicilic acid tend to increase solubility, whereas humid acids form chelates of low solubility. For the numerous complex forming substances in the soil, only a general decrease of complex stability of the complex can be stated following the line: $Fe^{3+} > Al^{3+} > Cu^{2+} > Pb^{2+} > Fe^{2+} > Ni^{2+} > Cd^{2+} > Zn^{2+} > Mn^{2+}$. Humic acid complexes of Cu, Pb, Cd, and Zn have log K-values of 8.65, 8.35, 6.25, and 5.72, whereas fulvic acid complexes of Cu, Pb, and Zn log K-values of 4.0, 4.0, and 3.6 (Scheffer and Schachtschabel 1982).

Water insoluble complexes, e. g., insoluble humic acids, may irreversibly fix cations, which will occupy two-thirds of the positions of the total binding capacity of about 200–600 mEq metal ion/100 g humic acid (Förstner and Müller 1974).

The solubility of an element in natural waters can be determined from the activities of all the dissolved species in equilibrium with the stable solids and from the ratios of the activities of the dissolved species to their concentrations. The difference between the calculated solubility and the measured value gives the possibility of estimating the degree of the formation of complexes or of assessing the completeness of the analyses.

6.6 Colloidal Solutions of Heavy Metals

Colloidal solutions contain electrically charged particles in the size range 10 nm to μm. The charge originates through three principal causes: individual molecules on the surface of a particle can dissociate; thus, H^+ or OH^- ions can be separated off and the colloids are left with positive or negative charges, respectively. Surface charge may be caused by lattice imperfection at the particle surface and by isomorphous replacement within the lattice. Furthermore, positive or negative ions can be adsorbed from the liquid and these will charge the particles accordingly (Stumm and Morgan 1981).

In natural waters (i. e., pH 5–9), colloids of ferric hydroxide are positively charged, whereas colloids of MnO_2 are generally negatively charged.

The occurrence of highly charged ions generally leads to the breakdown of the colloidal state and causes flocculation of the colloids. In the opposite sense, the effect of charged substances is known as peptization.

Emulsions and suspensions contain larger, more widely dispersed particles, which occur within the water as a few large molecules. Thus, emulsions and suspensions are transitional to colloidal dispersions.

6.7 Precipitation and Coprecipitation

Processes which change the chemical properties of a pore solution, e. g., its temperature, Eh, pH, or the admixture of other pore solutions with different dissolved constituents may give rise to the precipitation of some compounds. In the precipitates some foreign ions are commonly trapped in or substituted within the structure of the newly formed substances. This process, the so called coprecipitation, is very effective in removing trace elements, e. g., copper, lead, zinc, tungsten, and vanadium, when ferric hydroxide and manganese hydroxide are precipitated. Radium is coprecipitated when barite is formed.

The significant feature of coprecipitation processes is that the new solid phase is more stable and insoluble in the final solution than the original solid phase. Thus, the propensity of an ion to be absorbed into a mineral lattice can be expressed by solubility arguments (Jackson et al. 1980). Stumm and Morgan (1981) proposed for the solid solution of B in A

$$A(s) + B(aq) = B(s) + A(aq),$$

where (s) denotes the solid phase and (aq) the aqueous one, a coprecipitation index, which can be calculated from the quotient of the solubility products of the two mineral species:

$$CI = K_{SP}(A)/K_{SP}(B). \tag{12}$$

Consequently, in a solution of cadmium ions B(aq) undersaturated with respect to CdS ($-\log K_{SP} = 28.44$), but saturated with respect to ZnS ($-\log K_{SP} = 23.96$), it is not unreasonable to expect the coprecipitation of (Zn, Cd)S ($CI \sim 10^{4.5}$). The coprecipitation index merely indicates the likelihood of a reaction occurring ($CI > 1$), but does not guarantee that such will be the case.

6.8 Adsorption and Ion Exchange

The underground materials, both inorganic and organic, have surfaces with small unbalanced electrical charges, which attract ions from the groundwater. The forces, which bind these ions, range from Van der Waals forces to chemical adsorption by valence bonds. Valence bond ions are eventually adsorbed into the internal structures of the minerals. This adsorption mechanism is important for the ion exchange where equivalent quantities of bound ions are displaced by other ions from the solution. The most effective adsorbing substances in the ground are clay minerals, zeolites, hydroxides of iron and manganese, humic substances, plant roots, microbial slimes, and microorganisms.

Heavy metals are subject to adsorption on underground particles. This may be described for diluted suspensions by the Freundlich isotherm, which defines the equilibrium between the concentration of the suspended (C_s) and adsorbed (C_a) species (13):

$$C_a = kC_s^n, \tag{13}$$

where k and n are assumed to be specific constants for the investigated rock and heavy metal. The Freundlich isotherm shows that the adsorption of heavy metals can be reversible. Another possible description may be the Langmuir isotherm (14), which possibly is the better mathematical definition of the nature of the adsorption processes:

$$C_a = \frac{K \cdot b \cdot C_s}{1 + K \cdot C_s}, \tag{14}$$

where K is a constant relating to the bonding energy and b the adsorption maximum when the adsorbent is completely saturated. The continuous adsorption-desorption reactions cause a retardation of the heavy metal with respect to the surrounding water, which is described by the retardation factor R_d, the quotient of water velocity v_w to the transport velocity of heavy metals v_m (15). The retardation factor can be calculated if the distribution coefficient K_d of the heavy metals is known (15):

$$R_d = \frac{v_w}{v_m} = 1 + \frac{\varrho_b}{n} \cdot K_d \tag{15}$$

with the bulk density of the aquifer material ϱ_b, its porosity n, the mean water velocity v_w, and the mean transport velocity of the heavy metals v_m. The empirical distribution coefficient K_d ($ml\,g^{-1}$) defines the affinity of the soil material for a certain contaminant and can be used as a useful geochemical measure for the mobility of a substance in aqueous systems. K_d is in diluted solutions or suspensions equal to the coefficient of the Freundlich or Langmuir isotherm. K_d values can be calculated by laboratory and field experiments using Eq. (15). The K_d value may be affected by reactions other than adsorption, such as precipitation, isomorphous substitution, redox processes, complex ion formation, and acid base buffering (Jackson et al. 1980).

The heavy metals Pb, Cd, Cu, Ni, and Hg are adsorbed on unspecific and on specific bonding positions of soil components. The unspecific adsorption due to electrostatic forces of grain surfaces has relatively small bonding forces. The specific adsorption especially by hydroxocomplexes of the heavy metals [$Me(OH)^+$, $Me(OH)_2^0$] on the surface of Fe-, Al-, and Mn oxides have stronger bonding forces. Therefore, the specific adsorption on oxide surfaces increases with increasing tendency to form hydroxocomplexes and hydroxides in the line $Cd < Ni < Zn \ll Cu \leq Pb$ (Scheffer and Schachtschabel 1982).

The adsorption of heavy metals in the soil is mainly controlled by the presence of humic substances. This is evident, e. g., by considering the heavy metal load of the soil near highways. In the humic soil layers, heavy metals are enriched depending on the traffic intensity and age of the highway. The concentrations, for which maximum values of 700 mg Pb kg^{-1}, 1,700 mg Zn kg^{-1}, 40 mg Cd kg^{-1}, and 700 mg Cu kg^{-1} have been reported (Golwer 1973, Krämer 1976), decrease with increasing depth and increasing distance from the highway.

The organic carbon contents of the different soil horizons and the average organic carbon contents of the different soil types vary in wide ranges between 0% and more than 10%. The content of organic substances, in which organic carbon content ranges between 45% and 55%, varies between almost 100% in the soil litter and less than 1% in deeper soil horizons.

The stoichiometric adsorption-desorption process of bound and dissolved ion is called ion exchange. Direction, extent, and velocity of an ion exchange process depends upon the relative concentration and the properties of the involved ions and of the foreign ions. The exchange processes are reversible and can be described by the law of mass action. The intensity of the bond is different for the various ions and exchange substances. The intensity increases with the valence state and within one valence state with the atomic number and probably with the ion radius.

The ion exchange process with the competing ions A^+ and B^+ and the cation exchanger R^-

$$A^+ + B^+R^- \rightleftharpoons A^+R^- + B^+$$

can be described by a selectivity quotient:

$$K_B^A = \frac{(A^+R^-)(B^+)}{(A^+)(B^+R^-)} = \frac{(A^+R^-)/(A^+)}{(B^+R^-)/(B^+)}, \tag{16}$$

where (A^+) and (B^+) are the concentrations of these ions in solution and (A^+R^-) and (B^+R^-) are their concentrations in the adsorbed phase. The selectivity quotient may be considered as a mass action equilibrium constant if (1) the activities of all four forms may be calculated , (2) the adsorption reaction is reversible, and (3) secondary reactions, which are not readily reversible, such as ion fixation due to lattice collapse, do not occur and so prevent the attainment of true equilibrium. The estimation of the activities of the adsorbed phases, A^+R^- and B^+R^-, has been discussed by Truesdell (1972).

The amount of exchangeable ions in mEq/100 g solids at pH 7 is commonly known as the ion exchange capacity Q.

The selectivity quotient is related to the distribution coefficient K_d, if the cation exchange capacity, Q, of the exchanger, R^-, and the total competing cation concentration in solution, C, are known and the system under consideration is at equilibrium (Jackson et al. 1980):

$$K_B^A = K_d^A \frac{[C-(A^+)]}{[Q-(A^+R^-)]}. \tag{17}$$

If the contaminant A^+ is present in amounts much less than B^+, (A^+) is much less than C and (A^+R^-) than Q, the relationship can be simplified to

$$K_d^A = \frac{K_B^A \cdot Q}{C}. \tag{18}$$

Therefore, the distribution coefficient is directly proportional to the cation exchange capacity and the selectivity quotient and inversely proportional to the total competing cation concentration. If (A^+) and (A^+R^-) are very small, (B^+)

is close to C, and therefore, K_d^A is proportional to $(B^+)^{-1}$. More complex cases, for example, involving monovalent-divalent exchange, may be developed using similar reasoning.

From this discussion follows that adsorbed heavy metals can be partially desorbed by other competing ions in the groundwater, if these ions occur in sufficiently large concentrations. The higher the ionic concentration, the faster the contamination will move through successive exchanges.

Selective adsorption of some ions are found: oxihydrates of iron favor zinc, copper, lead, mercury, chromium, molybdenum, tungsten, and vanadium, whereas the oxihydrates of manganese prefer copper, nickel, cobalt, chromium, molybdenum, and tungsten; and clay minerals zinc, copper, lead, and mercury (Krauskopf 1956).

The hydrogeological implications of the complexing of the transition metal ions (e. g., Fe, Mn, Co, Zn, Pb, Cu, Ni, Cr) by inorganic or organic ligands alter their adsorption or precipitation from solution. The adsorption of radioactive cobalt and zinc is significantly decreased in the presence of dissolved organic carbon (DOC) compounds, which are to be found in all natural waters. This decrease is due to the formation of organo-cobalt and organo-zinc complexes, whose sorption behavior may differ from that of the hydrated metal ions and whose increased solubility in solution reduces the effectiveness of precipitation (Jackson et al. 1980).

6.9 Radioactive Decay

Radionuclides decay with time following Eq. (19)

$$C_t = C_0 \cdot e^{-\lambda t} \tag{19}$$

with C_t = concentration at time t

C_0 = initial concentration

λ = decay constant = $\dfrac{\ln 2}{T_{1/2}} = \dfrac{0.693}{T_{1/2}}$

$T_{1/2}$ = half-life (Table 5)

t = time.

6.10 Biochemical Processes

Organic substances, including organic complex forming compounds, are degraded by microorganisms, which obtain carbon and hydrogen for their cell synthesis and energy for their metabolism. The microbial degradation follows Eq. (19). Microbial activities are also involved in many reactions of the sulfur, nitrogen, iron, and manganese geochemistry in such a way that they accelerate the reactions. Nevertheless, it has to be kept in mind that microorganisms do not affect the direction of any reaction that results from the thermodynamic constraints

Table 5. Radioactive heavy metals

Element	Mass number of radio isotope	Half-life	Radiation
Cerium	144	285 d	β^-,γ
Chromium	51	27.8 d	γ
Cobalt	57	267 d	γ
Plutonium	238	86,4 a	α,γ
	239	$2 \cdot 10^4$ a	α,γ
	240	6580 a	α
	242	$3.8 \cdot 10^5$ a	α
Radium	226	1620 a	α, γ
	228	6.7 a	β^-
Ruthenium	103	40 d	β^-,γ
	106	385 d	β^-,γ
Uranium	235	$7.1 \cdot 10^8$ a	α,γ
	238	$4.5 \cdot 10^9$ a	α,γ
Zinc	65	245 d	β^-,γ

of the system and can be deduced from the stability field diagrams (Matthess 1982).

Mercury ions can be transformed by microbial activities into water soluble monomethyl mercury (CH_3Hg^+) and partly into water insoluble dimethyl mercury [$(CH_3)_2Hg$]. Vice versa a microbial demethylization can produce Hg^{2+} and Hg^0 (Scheffer and Schachtschabel 1982).

6.11 Gas Exchange

Gas exchange in the ground air is controlled mainly by diffusion, furthermore, by thermal and barometric influences. In this context, the removal of the volatile mercury species Hg^0 and $(CH_3)_2Hg$ is to be mentioned.

6.12 Filtration

The filtration effect of the subsurface materials is a complex physical and chemical phenomenon. It includes the mechanical separation of that fraction of suspended particles that is too large to pass through the subsurface flow channels in pores and joints. Smaller particles, e. g., iron III hydroxide flakes or colloids are retained by sedimentation, diffusion, and interception (Yao et al. 1971), so that even particles <0.2 μm are removed from the seepage water in fine porous soils.

The sedimentation is very important for the accumulation of inorganic mineral suspension (density ≥ 2 g cm^{-3}).

For particles with diameters of less than 1 μm the diffusion is very important, becoming increasingly effective with decreasing particle size. The interception is the most effective process, although this process has its minimum at particle diameters of about 1 μm. Mainly Van der Waals (mass) forces and Coulomb (electrostatical) forces act as bonding forces. The effects of Van der Waals forces are restricted to very short distances ($r \leq$ few nm) and their power decreases rapidly with increasing distances ($\sim 1/r^7$), whereas the Coulomb forces decrease much less with increasing distances ($\sim 1/r^2$).

The solid particles of an aquifer are usually negatively charged. The ferric iron hydroxides are positively charged at pH 5–9, but MnO_2 colloids at the normal pH values in groundwater (pH 7) are negatively charged.

The positively charged ferric iron hydroxides are adsorbed by the soil materials. The negatively charged colloids are strongly adsorbed by anionic adsorbents and only slightly by cationic adsorbents. It is well-known from filtering processes that the negatively charged particles stay in suspension in sandfilters, as the repulsive electrostatical forces are stronger than the Van der Waals forces. The dissolved cations in water decrease the repulsive forces of the grain surfaces. The cations are adsorbed by the solid substance and decrease their charge deficiency. Under these conditions, the mass forces are more effective and an accumulation of particles can take place.

References

Aurand K, Matthess G, Wolter R (1971) Strontium-90, Ruthenium-106 und Caesium-137 in natürlichen Wässern. Notizbl Hess L-Amt Bodenforsch 99:313–333

Correns CW (1969) The discovery of the chemical elements. The history of geochemistry. Definitions of geochemistry. In: Wedepohl KH (ed) Handbook of Geochemistry, Vol. 1. Springer, Berlin Heidelberg New York, pp 1–11

Edmunds WM (1977) Groundwater Geochemistry – Controls and Processes, Groundwater Quality, Measurement, Prediction and Protection. Proc Conf Reading, England

Förstner U and Müller G (1974) Schwermetalle in Flüssen und Seen als Ausdruck der Umweltverschmutzung. Springer Berlin Heidelberg New York

Garrels RM, Christ CL (1965) Solutions, minerals and equilibria. Harper and Row, New York, Weathershill, Tokyo

Golwer A (1973) Beeinflussung des Grundwassers durch Straßen. Z Deutsch Geol Ges 124(2)39–50

Hem JD (1970) Study and interpretation of the chemical characteristics of natural water, 2nd edn. US Geol Surv Water Supply Pap 1473:363

Horn MK, Adams JAS (1966) Computer-derived geochemical balances and element abundance. Geochim Cosmochim Acta 30:279–297

Jackson RE, Merritt WF, Champ DR, Gulens J, Inch KJ (1980) The distribution coefficient as a geochemical measure of the mobility of contaminants in a ground water flow system. Panel Proc Ser STI/PUB/518:209–225 IAEA, Vienna

Krämer F (1976) Erste Untersuchungen zur Erstellung eines Bodenbelastungskatasters (Pb, Zn, Cd, Cu) im Raume Duisburg-Dinslaken. Schriftenr Landesamt Immissions Bodenschutz NW, 39:45–48

Krauskopf KB (1956) Factors controlling the concentration of thirteen rare metals in sea-water. Geochim Cosmochim Acta 9:1–32

Langmuir D (1971) Particle size effect on the reaction Goethite = Hematite + Water. Am J Sci 271:147–156

Matthess G (1972) Bleigehalte in Gestein, Boden und Grundwasser. In: Aurand K (ed) Blei und Umwelt. Berlin-Dahlem (Ver Wasser, Boden-Lufthygiene), pp 21–27

Matthess G (1974) Heavy metals as trace constituents in natural and polluted Groundwaters. Geol Mijnbouw 53:149–155

Matthess G (1982) The properties of groundwater. Wiley and Sons, New York, 406 p

Plummer LN, Jones BF, Truesdell AH (1976) WATEQF – a fortran IV version of WATEQ, a computer program for calculating chemical equilibrium of natural waters. US Geol Surv Water Res Invest 76/13:615

Scheffer F, Schachtschabel P (1982) Lehrbuch der Bodenkunde, 11th edn. Enke, Stuttgart pp 442

Schöttler U (1972) Hydrochemische Untersuchung von Sickerwässern unterhalb von Abfallablagerungen, ihr Verhalten im Untergrund und Methoden zur statistischen Darstellung. Thesis, Aachen

Sposito G, Mattigod SV (1980) GEOCHEM: a computer program for the calculation of chemical equilibria in soil solutions and other natural water systems. Kearney Foundat Soil Sci Univ. California, Riverside

Stumm W, Morgan JJ (1981) Aquatic chemistry. 2nd edn. Wiley and Sons, New York

Truesdell AH (1972) Ion exchange. In: Fairbridge RW (ed) Encyclopaedia of geochemistry and environmental sciences. Van Norstrand, New York

Turekian KK (1969) The oceans, streams, and atmosphere. In: Wedepohl KH (ed) Handbook of Geochemistry, Vol. 1. Springer, Berlin Heidelberg New York, pp 297–323

Yao KH, Habilian MT, O'Melia CR (1971) Water and wastewater filtration: concepts and applications. Environ Sci Technol 5:(11) 1105–1112

7. Effect of Acid Rain on Soil and Groundwater in Sweden

G. JACKS, G. KNUTSSON, L. MAXE, and A. FYLKNER

7.1 Introduction

The acidification of the environment is probably the greatest environmental problem in many industrialized parts of the world at present and in the near future. The acidification is different to other types of impact on the environment in several ways:

– it has an imperceptible course during a long initial phase;
– it affects the whole environment of a country, not only a region or a place as most other impacts on the environment;
– it has several causes and the most obvious one, the acid rain, must be prevented on an international level.

The acidification problem was first described in Sweden and Norway (Odén 1968), then in the eastern USA and Canada. The most striking effect in these countries has been the acidification of surface water and the following damage to the ecology in lakes and streams. But other signs of acidification are now appearing in many other countries, e. g., corrosion, decay of historical monuments, and death of trees. The point of this paper is if it can be proved that there are any effects of acid rain on soil and groundwater.

7.2 Acidification of Precipitation and Surface Water

7.2.1 Precipitation

Unaffected precipitation has small quantities of oxides of sulphur and nitrogen. The burning of fossil fuels has, however, given rise to an increasing content of such oxides in the precipitation. This anthropogenic contribution must have followed the growth of industries and the urbanization in western Europe and eastern North America. It is known that the emissions of sulfur dioxide slowly increased in Europe during the first half of the 20th century (OECD 1977, cited after Environment '82 Committee, 1982).

From the mid-1950's to the beginning of the 1960's, a rapid increase of the content of sulfate and nitrate in precipitation was documented in many parts of Europe by means of a network of precipitation sampling stations, established at that time. The pH of the precipitation decreased during the same period in southern Scandinavia from 5.5 to 4.5 (Fig. 1).

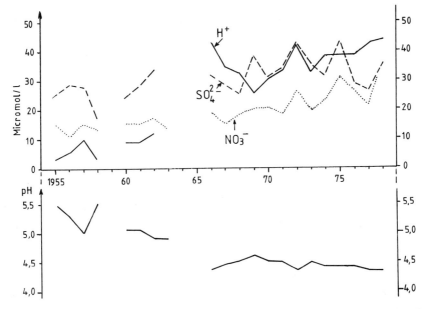

Fig. 1. The fluctuations of pH, sulfate-, nitrate-, and hydrogen ions in precipitation at Forshult, Värmland county. pH is lowered more than one pH unit since 1955 (Monitor 1981)

The distribution pattern of the acidification air pollutants is also well-documented. It shows that the largest part of the sulfur emissions comes from urban and industrial areas in central Europe and the British Isles (Fig. 2). The pollutants are transported by dominating winds to other parts of Europe, such as Scandinavia. About 75%–80% of the sulfur deposition over Sweden comes from other European countries (Environment '82 Committee, 1982).

7.2.2 Surface Water

Most lakes and streams in Sweden, Finland, and some parts of Norway are situated in areas where the bedrock and the soil mainly consist of acid crystalline rocks, mostly granite and gneiss. These rock types are rather "poor" and resistant to weathering, which means that the buffering capacity of the surface water in such catchment areas is generally low and the sensitivity to acidification rather high. The initial pH value of this type of surface water must, however, have been around 7. Due to the humid climate and the weathering-resistant rocks in Sweden, there has been natural acidification since the latest Ice Age and the pH value of lake water in western Sweden has slowly decreased to around 6 at the beginning of this century (Renberg and Hellberg 1982). This change is documented in lake sediments, where the microscopic shells of siliceous algae, diatoms, are preserved. The different species of diatoms can be classified by their pH preference/tolerance in five categories (from alkalibiontic forms at pH > 7 to acidobiontic forms at pH ≤ 5.5). So the fossil diatoms can be used as a biological pH meter. During the last decades, the change in the composition of diatoms shows

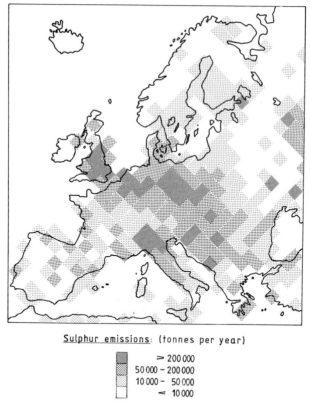

Sulphur emissions: (tonnes per year)

> 200 000
50 000 – 200 000
10 000 – 50 000
≺ 10 000

Fig. 2. Antropogenic emissions of sulfur to the atmosphere in Europe at the end of the 1970's. (After Environment '82 Committee, 1982)

that there has been a dramatic decrease from pH 6.0 to pH 4.5 in lake water in western Sweden (Renberg and Hellberg 1982; Fig. 3). This very rapid decrease has also been shown by comparing pH measurements 50 years ago in lakes from different parts of western Sweden with pH measurements in the same lakes in 1971 (Monitor 1981). Today, more than 18,000 lakes (>1 ha $= >10,000$ m^2) in Sweden are acidified (pH < 5.5 one time during the year) and about 4,000 of these lakes are very seriously acidified (pH < 5.0 during the whole year; Monitor 1981). Most of these lakes are situated in southwestern Sweden, where the amount of acidifying air pollutants is ten times higher than in northern Sweden.

7.3 Acidification of Soil and Groundwater

7.3.1 Terrestrial Water Flow, Especially in Moraine Terrain

Most of the water in rivers and lakes has previously been considered to be a result of surface runoff. Only the base flow in a river during dry seasons should originate from groundwater. The models of Tóth (1962) and Gustafsson (1966)

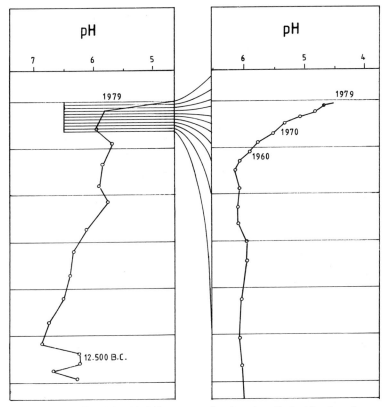

Fig. 3. The lowering of pH in lake Gårdsjön, western Sweden, since the last Ice Age, documented by fossil diatoms in the lake sediment. (After Environment '82 Committee, 1982)

concerning the flow pattern of groundwater in systems of different scales (local to regional) have given an increased understanding of terrestrial water flow. The terrain can be divided into recharge and discharge areas with varying extension owing to the fluctuations of the groundwater level. Lundin (1982) has shown that in moraine terrain, the permeability decreases very rapidly with increased depth. The higher permeability in the surface layers of tills is due to, e. g., frost action, tree roots activities, and washout processes in glacial time. This means that during the wet seasons most water flows in the permeable upper part of the soil, near the surface (Fig. 4). Current research using oxygen isotopes as tracers indicates that three-quarters of the spring flood has passed through the soil and can be designated as "subsurface" water or groundwater (Rodhe 1981). As most tills are very heterogenous, groundwater flow at greater depths is mostly concentrated to "veins" in beds and layers of sand and gravel within the more compact till (Knutsson 1971). So the chemical processes have to work in the uppermost part of the soil (about 0.5–1.0 m) and in restricted portions of the soil at greater depths. Another consequence is that if a lake has been acidified, some effects of acidification must also appear in the soil and the groundwater within the catchment area of the lake.

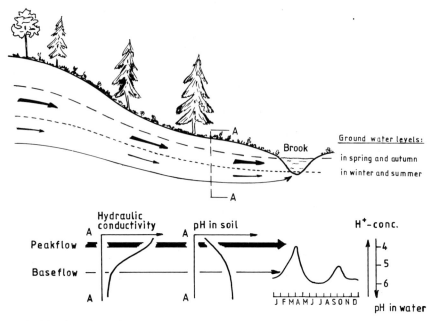

Fig. 4. Hydrogeology of a slope in till during different seasons. (Modified after Lundin 1982)

7.3.2 Acidification Processes

Soil acidification may be caused by several processes, both natural and man-made. The most important ones are the following:

– acid rain, in the form of wet and dry deposition,
– nutrient uptake by vegetation,
– oxidation of sulfur and nitrogen compounds in the soil,
– oxidation and hydrolysis of ferrous iron in the soil,
– soil respiration giving carbonic acid.

Many of the processes are reversible, the hydrogen ions secreted to the soil solution when plants take up cations (Fig. 5) are neutralized when the plant is decomposed by bacteria. Likewise the reduction of sulfate and nitrate is a sink for hydrogen ions. However, in general, we find that these processes tend to be more acidifying than neutralizing. We remove the nutrients along with the timber from the forests and the crops from the fields giving a corresponding net hydrogen ion addition to the soil. An attempt has been made to quantify the most important acid inputs to different ecosystems by Nilsson and Nilsson (1981) (Table 1).

The figures in Table 1 apply to the situation in southwestern Sweden. Thus, in comparison with the biological acidification, acid rain is of importance only in forest ecosystems, especially in the poorer ones.

The oxidation of sulfur and nitrogen compounds may comprise both natural "pools" and added fertilizers. Very acid groundwaters with sulfate contents exceeding 1,000 mg l^{-1} were found by Hultberg and Johansson (1981) in an area on

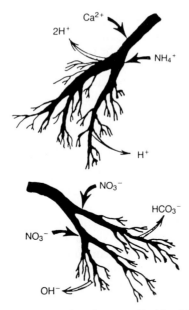

Fig. 5. Effects of root uptake of nutrients on soil acidity (Monitor 1981)

the Swedish west coast. This was essentially an effect of an unusually low ground-water level resulting from a couple of years of subnormal precipitation. In anaerobic environments, reduction may prevail. Actually, the acid precipitation is the release, over a few decades, of acid trapped in sulfate reduction during time spans of millions of years in the process of coal and oil formation.

The oxidation of ferrous iron and hydrolysis of formed ferric iron may be acidifying at least locally where anaerobic conditions turn to aerobic. This process contributes in the order of 0.1 kilo equivalent/ha $\cdot$ year H^+ in a podzolic soil.

Soil respiration is in this connection not considered to be an acidifying factor of importance. Carbonic acid is a weak acid with a pK_a of 6.5. It hardly dissociates at all below a pH of 5. Even before acidification, the ecosystems concerned had a pH close to this value. Furthermore, carbonic acid is a precursor of bicarbonate, the most important buffer in natural waters.

Table 1. Input of hydrogen ions by acid rain and by biological processes

Acidifying factor	Ecosystem		
	Agricultural land	Spruceforest on medium good soil	Pineforest on poor soil
	H^+ Kilo equivalents $ha^{-1} yr^{-1}$		
Acid rain	0.8	1.5	1.2
Nutrient uptake	5–10	0.2–0.7	0.1
Accumulation in litter		0.7–1.4	0.2

It has been argued that the lake acidification in southern Scandinavia has been caused by afforestation of previous grassland (Rosenqvist 1977). Thus, biological sources rather than acid rain should be the reason for water acidification. Forest fires release cations locked up in organic matter which is neutralizing. However, modern forestry has decreased the incidence of forest fires. This effect has also been proposed to contribute to the acidity of soils (Rosenqvist 1982), However, several studies have failed to show any relation between change of land use, such as afforestation of grassland and acidification of lakes (Johannessen and Wright 1980, Wright and Henriksen 1980).

7.3.3 Processes in Podzols

As was evident from the previous section, forest soils are vulnerable to acidification. Very shallow soils above the forest line in parts of southern Norway are even more sensitive (SNSF 1980). Medium to poor forest soils are formed into podzols. Thus, considerable research has been focused on the reaction of podzols to acid input. The buffering against acidification of the soil solution is effected mainly by two processes, ion exchange and weathering. Ion exchange evens out the composition of the soil solution in a short perspective, ranging up to a forest generation. This buffer has to be "recharged" by weathering, which constitutes the ultimate buffer (Fig. 6).

Podzols exhibit a pronounced stratification. The litter accumulates on the top of the soil and is slowly decomposed into humus, mainly by fungi. The pH of the raw humus layer is often below 4 in medium to poor podzols (Fig. 6). A bleached horizon (A_2), usually grayish in color due to intense leaching, follows below the raw humus layer. Below this layer, there is an enrichment zone, the B-horizon, reddish to brownish. Iron leached from the bleached horizon precipitates here in trivalent form as ferric hydroxide or goethite (Fig. 7). Primary minerals are degraded into secondary ones by organic acids; high molecular weight fulvo acids

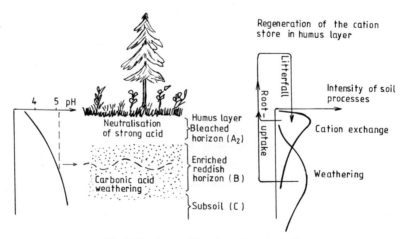

Fig. 6. Structure and function of a podzol soil

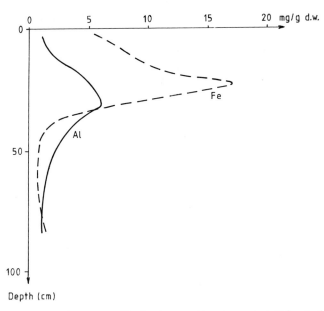

Fig. 7. Distribution of oxalate-extractable aluminum and iron in a podzol. (After Andersson 1977)

(Petersen 1980) and/or low molecular weight ones as oxalic acid (Graustein 1981). Iron and aluminum are carried with the soil solution in complexed form (Fig. 7). Likewise the concentration of heavy metals as lead in soil solutions is closely correlated with the content of dissolved organic matter (Tyler 1982). Gradually, when the organometal compounds meet a higher pH in the soil, they are metabolized or coagulated and the metals are precipitated more or less completely in the B-horizon (Fig. 7). Aluminum seems to be an important coagulant. If the soil is very acid, it is likely that there is an excess of dissolved aluminum, precipitating all the organic matter. Forest streams are usually brownish except in acidified areas, where the excess aluminum has precipitated all the organic matter in the soil (Environment '82 Committee, 1982).

In acid soil an inorganic component of aluminum is released instead of the organoaluminum complexes (Nilsson and Bergqvist 1983). Figure 8 indicates a considerable internal acid generation of biological origin in the portion between 5 to 15 cm.

Aluminum is a key element in the acidification process. It is believed to be the cause of ailing forests in West Germany (Ulrich 1982) and is the fish-killing agent in acid lakes (Environment '82 Committee, 1982). The solids controlling the dissolved concentrations of aluminum are not indisputably identified. In the bleached layer of podzols, vermiculite and montmorillonite are common (Gjems 1967). This is contradictory to the acid conditions. However, the till soils in Scandinavia are very young and the previously mentioned layer silicates may be in the process of loosing cations by a slow diffusion from interlayer positions (Graustein 1981). Kaolinite, common in acid environments, is rare in podzolic soils. On the basis of van Breemens work in acid sulfate soils (van Breemen 1973), it has been

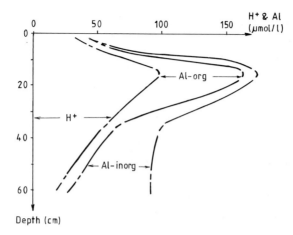

Fig. 8. Hydrogen ions and aluminum forms in a podzol profile extending down into the B-horizon. Compiled from lysimeter data. (After Nilsson and Bergqvist 1983)

postulated that basic aluminum sulfates may form in podzols (Eriksson 1981, Nilsson and Bergqvist 1983) and thus be a component in concentration control of dissolved aluminum. The compounds suggested in this connection are AlOH-SO_4 and basaluminite, $Al_4SO_4(OH)_{10} \cdot 5H_2O$ (Nilsson and Bergqvist 1983). A lysimeter investigation by Nilsson and Bergqvist (1983) points out that both compounds, together with gibbsite, may exhibit a check on the dissolved aluminum concentrations in soil solutions. Nordstrom (1982) suggests basic aluminum sulfates to be important in acid mine water environments, while in less sulfate-infested areas, gibbsite may check aluminum and adsorption may be important for sulfate.

7.3.4 Sulfate Behavior in Soils

Any acid soil may start to leak hydrogen ions if a mobile anion is added (SNSF 1980). This is the basis of the mobile anion concept introduced by Seip (1980). Thus, even if precipitation was not acid, but enriched in sulfate, it may effect acidification of natural waters. However, sulfate is not a perfectly mobile anion in the soil environment. In bogs, about 20%–30% is retained (Braekke 1981). Even in till areas, there is a retention, 10%–20% in a thoroughly investigated area on the Swedisch west coast (Hultberg pers. comm., 1982). Sulfate to chloride ratios in groundwater compared with that in deposition indicate the same condition (Jacks and Knutsson 1981) as does hydrochemical budgets for catchments in Sweden (Andersson and Eriksson 1978).

Tropical acid soils have pronounced sulfate adsorption capacities. The B-horizon of podzols contains extractable sulfate (Singh 1980). Increased acidity will promote the ability of sulfate adsorption. The sulfate released by extraction may be adsorbed on ferric and aluminum hydroxides (Singh 1980), but may also be present in the form of basic aluminum sulfates (Nordstrom 1982). Whatever may be the case, this represents a buffer against acidification, as on adsorption of sulfate, hydroxyl ions are generally released (Rajan 1978).

The sulfate retention in the B-horizon acts as a buffer, which means that if sulfur deposition is relaxed, soils will still for some time leak sulfate and accompanying cations.

7.3.5 Behavior of Soil Solutions as Affected by Soil Acidification

The acidification of soils has been studied in the lysimeter scale (Nilsson and Bergqvist 1983) and in minicatchments (SNSF 1980, Christophersen et al. 1982 b). Long records of analyses from wells and lakes provide indirect evidence of what happens in the soil. The main effects on the solute behavior are:

- increased washout of cations, mainly Ca^{2+} and Mg^{2+};
- decreased levels of bicarbonate;
- hydrogen ions start to accompany sulfate in percolating water when the store of exchangeable cations (Ca^{2+} and Mg^{2+}) is depleted;
- aluminum is mobilized in inorganic form in the lower portions of the soil profile;
- decreased concentration of dissolved organic matter in soil solutions and runoff water;
- heavy metals are released to the soil solution, the effect is most pronounced for zinc and cadmium;
- phosphate concentrations in soil solutions and runoff water will decrease due to precipitation of $AlPO_4$;
- selenium is likely to be immobilized in the B-horizon by adsorption, paralleling the behavior of sulfate.

The ecological effects of this has so far been most obvious in surface water, in lakes and running waters. Increased levels of aluminum have caused killing of fish. Decreased influx of phosphate to surface water has lowered primary production in the form of phyto- and zooplankton. In soils, the increased levels of aluminium are especially suspected to be harmful for spruce roots, according to findings by Ulrich (1982) in West Germany. In Sweden, forests have also been affected. Eventual harmful effects may have been compensated by the increased nitrogen deposition (Environment '82 Committee, 1982). Nitrogen is generally deficient in coniferous forests.

The decreased content of dissolved organic matter means a loss of metal complexing ligands and may cause increased metal toxicity in soil solutions, runoff water, and lake sediments (Environment '82 Committee, 1982).

Increased levels of cadmium have been recorded in the surface layers of lake sediments. So far, however, only a trickle of the cadmium found in acid soil solutions reaches the lakes (Fig. 9).

7.3.6 Acidification of Soil

A direct acidification of the soil has been difficult to prove. A decreased base saturation in the humus layer of podzols in central Sweden was recorded by

G. Jacks et al.

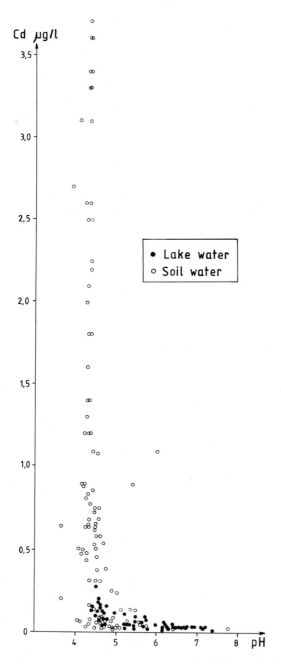

Fig. 9. Cadmium in soil solutions and lake waters in Sweden. (Data compiled from different sources by Nilsson, Swed. Environ. Board, 1983)

Troedsson (1980) when he compared samples from 1971–1973 with those from 1961–1963. Odén has observed an acidity gradient for soils of similar texture and origin from southwestern Sweden via central Sweden and up in the northern part of the country, along the deposition gradient (Environment '82 Committee, 1982). Soils around the town of Falun, where pyritic ore has been roasted for centuries, show a local acidification. A pronounced acidification has recently been revealed by Tamm (1983) in southwestern Sweden. Sites in till soils that were investigated 55 years ago have been resampled. They have shown lowering of the pH with 0.5–0.7 in the bleached layer. The decreases were less in the humus layer. Decreases were recorded even to a depth of 0.7 m. This indicates that weathering does not keep pace with the acidification.

In view of the considerable stores of exchangeable cations in most soil, water acidification by acid deposition has been disputed by some scientists (Rosenqvist 1977). However, as mentioned earlier, most of the water passes through very shallow portions of till soils and in "veins". Only a marginal displacement downward of the zone where carbonic acid weathering starts will leave the runoff water without alkalinity. The findings by Tamm (1983) indicate a considerable displacement of the soil acidity with increased depth.

7.3.7 Acid Groundwater

Natural state chemistry of groundwater in a region is, in principal, due to the climatic and hydrogeological conditions of the region. The biogeochemical processes in the unsaturated zone, as described above, are of great significance to the chemical composition of the groundwater. The geochemical processes in the groundwater zone as well as the flow pattern of the groundwater (Fig. 10) will give the final composition of the groundwater. The turnover rate of the groundwater, together with the weatherability of the ground, determines the degree of neutralization of the groundwater. The turnover time increases with the depth of the aquifer and along with the pH and the alkalinity (Fig. 10). As water passes from the soil into the bedrock, there is an abrupt increase in turnover time and in pH and alkalinity.

The soils of Sweden are very young in the geological sense. But as mentioned earlier in "poor" soils there has been a pronounced leaching of the uppermost layer of the mineral soil. In agreement with lake water, there must have been a slow decrease of pH in groundwater of shallow aquifers since the latest Ice Age in large regions of Sweden with acid rocks resistant to weathering.

A general conclusion, based on the above mentioned assumptions, the experiences, and the evaluation of hydrochemical data is that *natural acid groundwater can be found in many places in Sweden,* where some or all of the following natural conditions exist:

– the bedrock and the soil consist of minerals resistant to weathering,
– the quaternary deposits have a coarse grain size composition,
– the groundwater has a short turnover rate,
– the precipitation is high and the leaching of the soil is intensive.

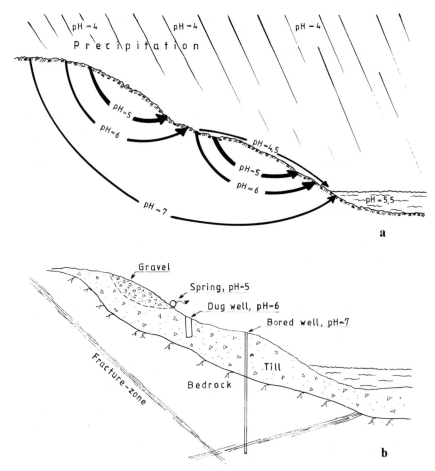

Fig. 10. a The principal flow pattern of groundwater with the resulting pH values (Jacks et al. 1981).
b A typical geological section in Sweden with different types of wells and their pH values (Jacks et al.
1981)

These geological and climatic conditions also mean that acid groundwater is
found mostly in regions dominated by coniferous forests and podzolic soils. One
typical example of acid groundwater is documented in the porphyry region of
southeastern Sweden. The soil cover is very thin and consists of gravelly till derived from quartz porphyry. The mean pH-value is 5.4 in groundwater from small
springs and shallow wells (Jacks and Knutsson 1981). Another example is reported from the west coast of Sweden with high precipitation. The groundwater of
the wells dug in sand has a mean pH-value of 5.1 whereas the groundwater of the
wells dug through a clay layer has a mean pH value of 6.1 (Abrahamson and
Tilosius 1979). The lowest pH value of the groundwater both in gravelly till and
sand was 4.0.

7.3.8 Changes of the Chemistry of Groundwater

The documentation of the changes of the chemistry of groundwater is not as good as for precipitation and surface water. The most uniform data concerning water chemistry belongs to the groundwater chemistry network of the Geological Survey of Sweden. It started in 1968 and consists of more than 100 sampling points, mostly springs, in 40 observation areas over the whole country. Samples are taken 2–6 times per year. Time series of hydrochemical data can also be obtained from municipal wells, sometimes for 20–25 years, but mostly for 10–15 years. Another possibility to obtain data with a time difference is to search for water analyses in old groundwater investigations and to repeat the samplings and analyses in the same wells and springs. All these types of documentation have been used in the research work by the authors. Hydrochemical data from about 400 municipal wells in six counties in different parts of Sweden was examined as well as water analysis from about 550 private wells and springs in six test areas (Fig. 11). Resampling was carried out in about 250 of these wells and springs (Jacks and Knutsson 1981, 1982, Jacks et al. 1981). Most figures were stored in the water data bank of the Geological Survey of Sweden, treated statistically, and plotted in diagrams. Furthermore, all reports from inventories of acid groundwater in Sweden and the very few papers concerning acidification of groundwater from other countries (Norway and Canada) were studied.

The evaluation of the hydrochemical data was accompanied by considerable difficulties. The composition of groundwater varies not only in place, but also by time. The annual variations are observable, especially in shallow aquifers. In addition, chemical fluctuations lasting several years may occur, which are dependent on the climatic changes. The most obvious relationship is that of the increasing content of sulfate after a series of dry years. This is apparently due to the oxidation of sulfides in the ground as the groundwater level is lowered. This was the case in the mid-1970's, especially in the county of Kopparberg (Fig. 11). In some areas, e.g., the county of Kronoberg, wet years are followed by increasing content of aggressive carbonic acid. This may be explained by the fact that the increasing content of soil moisture prevents the diffusion of carbon dioxide to the atmosphere. However, the increased content of aggressive carbonic acid in groundwater has not resulted in increased alkalinity, indicating a slow weathering of minerals in the ground. Changes of the chemistry of groundwater may also be due to technical modifications in or around a well. Furthermore, increased pumping of a well is generally accompanied by increased alkalinity as groundwater from greater depths starts to circulate.

Thus, considerable amount of different data has to be considered in order to evaluate the long-term changes of the chemistry of groundwater. The result of this evaluation mainly concerning municipal wells in sand and gravel can be summarized as follows:

– a real acidification, that is significantly lowered pH values, cannot be proved; but there is a tendency towards lowering of pH in water from shallow wells and springs in western and southeastern Sweden;

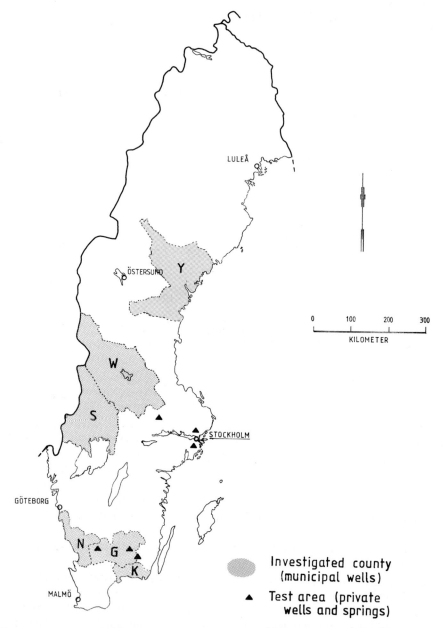

Fig. 11. Investigated counties and test areas in Sweden: *G* Kronoberg county, *K* Blekinge county, *N* Halland county, *S* Värmland county, *W* Kopparberg county, *Y* Västernorrland county

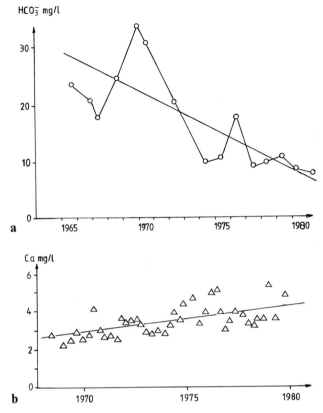

Fig. 12. a Decreased alkalinity in groundwater from Vittaryd, Kronoberg county (Jacks et al. 1981).
b Increased hardness in groundwater from Brattforsheden, Värmland county. (After Geological Survey of Sweden, Monitor 1981)

- the alkalinity decreases in groundwater from some regions in western and southeastern Sweden, where the bedrock consists of weathering resistant minerals (Fig. 12);
- the alkalinity increases in groundwater from regions where the bedrock consists of easily weathered minerals;
- the total hardness, that is the content of calcium and magnesium, increases in the groundwater of many wells in southeastern and western Sweden (Fig. 12);
- the content of sulphate increases in groundwater of more than 75% of the municipal wells. A temporary increase was observed in the county of Halland during the first part of the 1970's;
- the aggressive carbonic acid increases in groundwater, especially in southwestern Sweden; this is pronounced during periods of wet years;
- the content of aluminum is notable (1–1.5 mg l^{-1}) in some types of groundwater with pH less than 5. Increased contents of heavy metals have been found in the same type of groundwater, but the contents are still acceptable from toxicological considerations.

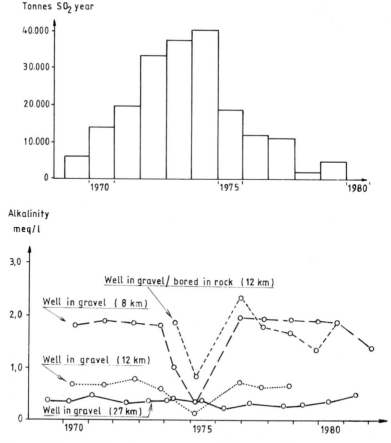

Fig. 13. Large emissions of sulfur oxides at Karlshamn, Blekinge county resulted in the mid-1970's in a temporary lowering of the alkalinity in some municipal wells near Karlshamn (in km). (Compiled from data of Jönsson 1980, Jacks and Knutsson (1982)

The acid deposition is highest in the southern and western parts of Sweden as most of it originates from western and central Europe (Fig. 2). However, large local emissions of sulfur oxides, e. g., from Stenungsund in western Sweden and Karlshamn in the county of Blekinge (Fig. 13) seem to have had a local effect on the chemistry of the groundwater. A decrease in alkalinity was observed in water from some municipal wells in the vicinity of Karlshamn during the mid-1970's. Since the emissions were reduced, the alkalinity has increased again (Fig. 13).

In private dug wells, groundwater is generally pumped out from very shallow aquifers. Therefore, groundwater in such wells is more vulnerable to acidification than groundwater from municipal wells. Unfortunately, groundwater in private wells has not been regularly sampled, so it is difficult to prove any changes of the water chemistry. However, the relationship between hardness and alkalinity can give interesting information, especially the change of the relationship from one sampling time to another. The changes are due to the fact that the hardness

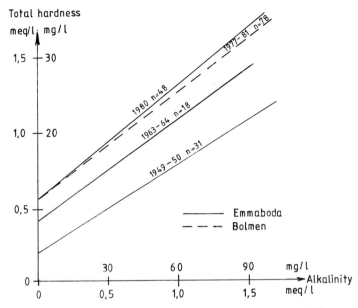

Fig. 14. The relationship between total hardness and alkalinity in water at three different times from private wells and springs near Emmaboda, southeastern Sweden and the same ratio in water from private wells near lake Bolmen, Kronoberg county (Jacks and Knutsson 1982)

has increased, indicating an elution of calcium and magnesium ions and or that the alkalinity has decreased. The inference is that dilute Ca-HCO_3-water has become more mineralized Ca-SO_4-HCO_3 waters. This is an effect of the acid precipitation and has been documented in some of the selected areas and in other areas too. A progressive impact of this kind is observable in southeastern Sweden for more than 30 years. Hardness relative to alkalinity has increased markedly, especially between 1949–1950 to 1963–1964 (Fig. 14), a period long before the acidification of lakes.

In conclusion it may be observed that the acid rain affects the groundwater of shallow aquifers in some regions of western and southeastern Sweden. The effects seems to be most pronounced in the county of Kronoberg (Fig. 11).

7.4 Modeling the Acidification

Several models describing the acidification processes have been brought forward. They range from very simple dose-response considerations to complex compartmentalized models. The former ones may be used for prediction, as for instance Henriksens plot (1980). By plotting lake water pH versus sulfur deposition of nonmarine origin, Dicksson and Bengtsson (1979) could show that very sensitive lakes in Sweden can stand no more than 0.5 g S m^{-2} year^{-1} in their catchments without being acidified. A similar approach that takes into account the partial pressure of carbon dioxide has been presented by Thompson (1982).

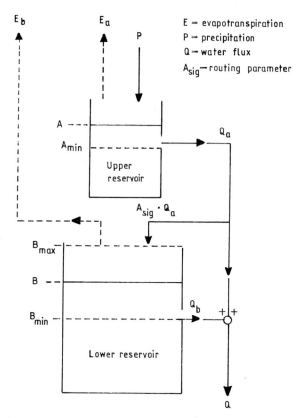

Fig. 15. Hydrological model used as a basis for a model of sulfate concentrations in streamwater (Christophersen et al. 1982a)

Among the more complex models, the soil chemistry model by Reuss (1980) may be mentioned. It takes into account ion exchanges and sulphate adsorption.

The stratified flow during different runoff conditions has been modeled by the so-called Birkenes model (Fig. 15; Christophersen et al. 1982a). The model includes as chemical processes cation exchange and sulphate adsorption. It is calibrated by means of daily values of precipitation, runoff, and chemical parameters. After the calibration, it is able to predict the quantity and chemistry of runoff from precipitation data as input.

Other models are the ILWAS model (Chen et al. 1982) and the Trickle down model (Schnoor et al. 1980).

The authors, together with Axelsson and Karlqvist at the Geological Survey of Sweden, have developed a qualitative model in order to classify the sensitivity to groundwater acidification in different parts of Sweden (Jacks and Knutsson 1982). Parameters as type of rock, soil texture, kind of solum, runoff, and relief were used with an areal resolution of 5 × 5 km squares. The parameters were given different weights and the result was calibrated by the alkalinity and pH values of groundwater in several test areas. The computer print was generalized to a sensitivity map with five classes. The map shows that the most sensitive areas are to be found in the western parts of southern Sweden, in the northwestern parts of central Sweden, and near the coastline of northern Sweden.

References

Abrahamson R, Tilosius M (1979) Undersökning av grundvatten med avseende på försurning i Stenungsunds, Kungälvs, Lilla Edets kommuner. Länsläkarorganisationen i Göteborgs och Bohuslän

Andersson A (1977) Heavy metals in Swedish agricultural soils: on their retention, distribution and amounts. Swed J Agric Res 7:7–20

Andersson U-M, Eriksson E (1978) Hydrochemical investigations in three representative basins in Sweden. Swed Nat Sci Counc IHP Rep 50:25

Braekke FH (1981) Hydrochemistry of high altitude catchments in South Norway – Dynamics in waterflow, and in release-fixation of sulphate, nitrate and hydronium. Norw For Res Inst Rep 36:(10) 21

Breemen Van N (1973) Dissolved aluminium in acid sulfate soils and in acid mine waters. Soil Sci Soc Am J 37:694–697

Chen CW, Dean JD, Gherini SA, Goldstein RA (1982) Acid rain model hydrologic module. J Environ Eng Div Am Soc Civil Eng 108:455–472

Christophersen N, Seip HM, Wright RF (1982 a) A Model for streamwater chemistry at Birkenes, Norway.Water Resourc Res 18:977–996

Christophersen N, Stuanes AO, Wright RF (1982 b) Runoff chemistry at a minicatchment watered with "unpolluted precipitation". Nordic Hydrol 13:115–128

Dicksson W, Bengtsson B (1979) Liming of lakes and watercourses. Swed Environ Board Rep Drottningholm Lab 8:67

Environment '82 Committee (1982) Acidification today and tomorrow. Swed Ministr Agric, Stockholm

Eriksson E (1981) Aluminium in ground water - possible solution equilibria. Nordic Hydrol 12:43–50

Gjems RE (1967) Studies on clay minerals and clay – mineral formation in soil profiles in Scandinavia. Medd Fra Det Norska Skogsforsøksvesen Vol XXI:81

Graustein WC (1981) The effect of forest vegetation on solute acquisition and chemical weathering: a study of the Tesuque watersheds near Santa Fe, New Mexico. Ph D Thesis, Yale Univ

Gustafsson Y (1966) The influence of topography on ground water formation. Ground water problems. Pergamon, New York London

Henriksen A (1980) Acidification of freshwaters – a large scale titration. In: Drabløs D, Tollan A (eds) Ecological impact of acid precipitation. SNFS, Oss-Ås, pp 68–74

Hultberg H, Johansson S (1981) Acid ground water. Nordic Hydrol 12:51–64

Jacks G, Knutsson G (1981) Känsligheten för grundvattenförsurning i olika delar av landet (förstudie). With english summary. KHM Tek Rap 11, Statens Vattenfallsverk, Vällingby

Jacks G, Knutsson G (1982) Känsligheten för grundvattenförsurning i olika delar av landet (huvudrapport). With english summary. KHM Tek Rap 49, Statens Vattenfallsverk, Vällingby

Jacks G, Knutsson G, Lundberg L, Maxe L (1981) Grundvatten och försurning i Kronobergs län – en undersökning av kommunala grundvattentäkter. With english summary. Inst Kulturtek KTH, Stockholm Meddelande 3:41

Johannessen M, Wright RF (1980) Regional surveys of lakes in southern Norway. Land-use changes in the catchments and fish population in the lakes. SNSF Proj TN53/80:28

Jönsson M (1980) Grundvattendata 1969–1979. Kommunala vattentäkter i Blekinge län. Examensarbete 1980:1, Högskolan i Kalmar

Knutsson G (1971) Studies of ground water flow in till soils. Geologiska föreningens i Stockholm Förhandlingar. Vol 93, Part 3

Lundin L (1982) Soil moisture and ground water in till soil and the significance of soil type for runoff. Uppsala Univ Dep Physic Geogr Rep 56:216

Monitor (1981) Försurning av mark och vatten, Statens Naturvårdsverk, Solna

Nilsson I, Bergqvist B (1983) Aluminium chemistry and acidification processes in a shallow podsol on the Swedish westcoast. Water, Air and Soil Poll 20:311–329

Nilsson I, Nilsson J (1981) Sources for soil acidification. Swed Environ Board SNV PM 1411: (in Swedish) 36

Nilsson J (1983) Miljökonsekvenser vid förbränning av ved och torv. Swed Environ Board (In preparation)

Nordstrom DK (1982) The effect of sulfate on aluminium concentrations in natural waters: some stability relations in the system Al_2O_3-SO_3-H_2O at 298 K. Geochim Cosmochim Acta 46:681–692

Odén S (1968) The acidification of air and precipitation and its consequences on the natural environment. Swed Nat Sci Res Counc Ecol Comm Bul 1:68 (in Swedish)

Petersen L (1980) Podsolization: mechanism and possible effects of acid precipitation. In: Hutchinson TC, Havas M (eds) Effects of acid precipitation on terrestrial ecosystems. Plenum, New York London, pp 223–227

Rajan SSS (1978) Sulfate adsorbed on hydrous alumina, ligands displayed and changes in surface charge. Soil Sci Soc Am J 42:39–44

Renberg I, Hellberg T (1982) The pH history of lakes in southwestern Sweden, as calculated from subfossil diatom flora of the sediment. Ambio 11:1

Reuss JO (1980) Simulations of soil nutrient losses resulting from rainfall acidity. Ecol Model 11:15–38

Rodhe A (1981) Vårfloden – smältvatten eller grundvatten? 6:e Nordiska Hydrologiska Konferensen i Vemdalen 1981. UNGI Rap 83

Rosenqvist I (1977) Acid soil – acid water. Ingenjørforlaget, Oslo (in Norwegian), 123 pp

Rosenqvist I (1982) Quantification of acidification. Seminar on acidification, Stockholm May 26th. Swed Acad Eng Sci

Schnoor IL, Carmichael GR, Van Schepen FA (1980) An integrated approach to acid rainfall assessment. In: Keith LH (ed) Proc of the Acid Rain Symp. Am Chem Soc Ann Arbour, Michigan

Seip HM (1980) Acidification of freshwaters – sources and mechanisms. In: Drabløs D, Tollan A (eds) Ecological impacts of acid precipitation. SNFS, Oslo-Ås, pp 358–365

Singh BR, Abrahamsen G, Stuanes A (1980) Effect of simulated acid rain on sulfate movement in acid forest soils. Soil Sci Soc Am J 44:75–80

SNSF (1980) Acid precipitation, effects of forest and fish: Final report. In: Overrein L, Seip HM, Tollan A (eds). Norges lantbruksvitenskaplige forskningsråd. Norges Teknisk-natur-vitenskaplige Forskningsråd Miljøverndepartementet 173

Tamm C-O (1983) Acidification goes underground. Miljöaktuellt 11:1 (in Swedish)

Thompson ME (1982) The cation denudation rate as quantitative index of sensitivity of eastern Canadian rivers to acidic precipitation. Water Air Soil Poll 18:215–226

Toth J (1982) A theoretical analysis of ground water flow in small drainage basin. Proc Hydrol Symp no 3, Groundwater NRC (Canada)

Troedsson T (1980) Ten years acidification of Swedish forest soils. In: Drabløs D, Tollan A (eds) Ecological impact of acid precipitation. SNFS, Oslo-Ås, p 184

Tyler G (1983) Does acidification increase metal availability and hereby inhibit decomposition and mineralisation processes in forest soils? In: Ecological effects of acid precipitation (Background papers 1982 Stockholm Conference on the Acidification of the Environment), National Swedish Environment Protection Board, Report 1636:245–256

Ulrich B (1983) An ecosystem oriented hypothesis on the effect of air pollution on forest ecosystems. In: Ecological effects of acid precipitation (Background papers 1982 Stockholm Conference on the Acidification of the Environment), National Swedish Environment Protection Board, Report 1636:221–231

Wright RF, Henriksen A (1980) Regional survey of lakes and streams in southwest Scotland, April 1979. SNSF Proj IR71/80:63

8. Heavy Metal Pollution of Natural Surface Soils Due to Long-Distance Atmospheric Transport

E. STEINNES

8.1 Introduction

A great part of the aerosols released to the atmosphere from high temperature anthropogenic processes are dispersed over large areas before being removed mainly by wet deposition. The significance of this long-distance atmospheric transport with regard to the chemical composition of surface soils has not been appreciably recognized in the past.

Norway forms a useful study area in this respect, because some parts of the country are considerably exposed to long-distance transported atmospheric pollutants, while other parts are not. Moreover, the regional distribution of heavy metal deposition in Norway is well-known from extensive measurements.

It may be appropriate to enter the present subject by stating the following facts:

1. The significance of *atmospheric supply* (natural and anthropogenic compounds) to the chemical composition of surface soils is far from being fully recognized in soil science (Låg 1968, Låg and Steinnes 1974, Allen and Steinnes 1979).
2. A very significant part of the aerosols released to the atmosphere from high temperature processes (fossil-fuel burning, metal smelters, automobile engines) are dispersed over *large areas*. The more volatile trace elements (e. g., Pb, Cd, As, etc.) are preferentially concentrated on the small particle fraction available for transport over very long distances.
3. Published material on heavy metal contamination of soil from airborne material deals almost exclusively with areas in the vicinity of some local source. *The long-distance atmospheric transport has been largely ignored* in this respect.

The purpose of the present paper is to show, on the basis of experience from investigations in Norway, that the long-distance atmospheric transport of pollutants may not be insignificant with respect to pollution of the unsaturated zone. The examples selected are from studies on heavy metals, but other air pollutants may show similar impacts.

8.2 Atmospheric Pollution in Norway

Air pollution research carried out during the last decade has convincingly demonstrated that the southernmost part of the Scandinavian peninsula is considerably affected by "acid precipitation" due to transport of polluted air masses from highly industrial areas, mainly in Central and Western Europe. For instance, the wet deposition of "excess sulfate" in southernmost Norway in 1980 was about ten times the level encountered in the northern part of the country (Overrein et al. 1980). A main reason for this difference is that the precipitation in southern Norway occurs predominantly with wind directions within the sector southeast-southwest, whereas westerly and northwesterly winds, carrying clean air from the Norwegian Sea, are mainly responsible for the precipitation falling in central and northern Norway.

The situation concerning heavy metal air pollution in Norway on a national basis has been extensively studied by means of air particulate samples (air concentrations), precipitation (wet deposition), and moss samples (relative deposition figures). On the basis of these and other studies (Hanssen et al. 1980, Rambæk and Steinnes 1980, Hanssen et al. 1981), the following major trends have been shown to be evident:

1. The deposition of many elements is significantly higher in the southern and southwestern part of the country than in areas further north. For some elements (Pb, As, Cd, etc.) this difference amounts to about a factor of 10.
2. A great part of this deposition is due to *long-distance atmospheric transport* from other parts of Europe.
3. For the above group of elements the long-distance transport is the predominant source of atmospheric deposition in Norway.

8.3 Heavy Metals in Norwegian Surface Soils

During the summer of 1977, sampling of natural surface soils was carried out all over Norway. The sampling network is shown in Fig. 1 and includes about 500 points regularly distributed over the country. Before discussing some results obtained on this material, it may be useful to review some properties of natural surface soils in Norway:

In more than 70% of the cases, the samples obtained were from podzol profiles, with the following typical properties:

pH (H_2O):   3.6–4.5
Organic matter: 50%–90%
Cation exchange capacity: 70–100 mEq/100 g
Base saturation: 10%–25%.

This material has been analyzed with respect to about 25 elements (HNO_3-soluble fraction). Some of these elements show a similar and rather characteristic geographic distribution in the soil, as illustrated in the case of Pb in Fig. 2. The distribution trend appears to be very similar to that of the atmospheric deposition of lead, indicating that the Pb content of surface soils is to a great extent dependent on long-distance atmospheric supply (Allen and Steinnes 1979).

Rather than showing similar plots for other elements of interest, we shall discuss the results in terms of regional mean values. For this purpose, the country

Fig. 1. Sampling network for the 1977 nation wide survey of natural surface soils in Norway

has been conveniently divided into 12 regions according to topographical and meteorological criteria. The regional mean values for Pb are shown in Figure 3. As evident from Figure 3, the Pb content of surface soils is about ten times higher in regions close to the south coast than in areas far north. The levels within each region were found to be quite consistent. Whereas in Region 1 only one sample out of a total of 55 showed a lead content below 40 ppm, 43 out of 44 samples

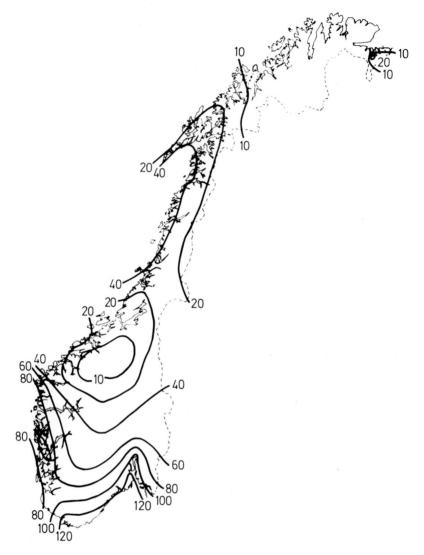

Fig. 2. Geographical distribution of lead in Norwegian natural surface soils (ppm)

analyzed from region 12 had Pb contents below 20 ppm. It seems very difficult to explain this trend on the basis of differences in bedrock geology.

In Figure 4, relative mean values for a number of heavy metals from the regions 1, 2, 3, 8, and 12 are plotted in the same diagram. It appears that the regional differences of the volatile group elements Cd, As, and Sb are very similar to those of Pb, whereas the concentration of the less volatile elements Cu and Zn do not show correspondingly large regional differences.

Further evidence on the importance of atmospheric input may be obtained from studies of soil profiles. Figure 5 shows the vertical distribution of Pb in two

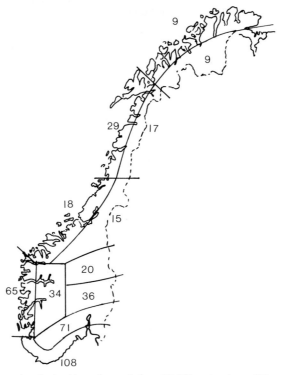

Fig. 3. Mean values for lead in surface soils from 12 different regions of Norway (ppm)

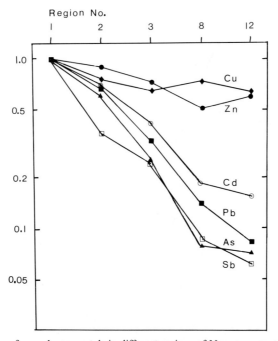

Fig. 4. Mean values of some heavy metals in different regions of Norway expressed relative to those of Region 1 (southern coast). Note the semilogarithmic scale

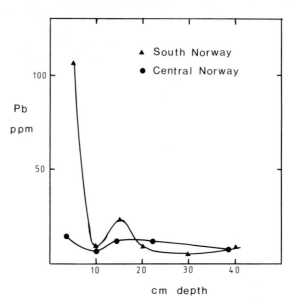

Fig. 5. Vertical distribution of lead in two typical podzol profiles from South Norway and Central Norway, respectively (ppm)

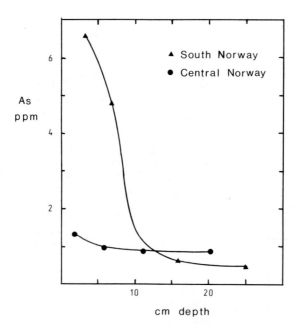

Fig. 6. Vertical distribution of arsenic in two typical brown earth profiles from South Norway and Central Norway, respectively (ppm)

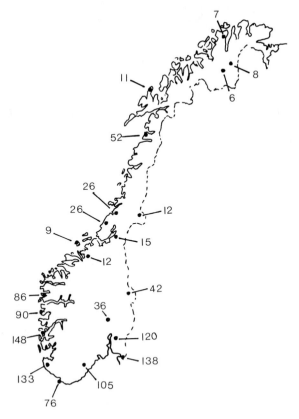

Fig. 7. Lead content of surface peat from 21 ombrotrophic bogs in different parts of Norway (ppm)

podzol profiles from regions 1 and 8, respectively, and in Figure 6 the As distribution in two brown earth profiles from the same two regions is shown. The large differences in depth distribution clearly demonstrate the importance of atmospheric input during recent times in southern Norway.

A final proof that the long-distance atmospheric transport really is the key factor in the respect discussed, is coming from results on peat profiles from 21 ombrotrophic bogs regularly distributed over the country (Hvatum et al. 1983). As evident from Pb values in surface peat (Fig. 7), the levels correspond very well to those obtained for the natural surface soils. Results for other trace elements (As, Sb, Cu, Zn, etc.) in the surface peat as well as depth profiles of the same elements in bogs from different parts of the country point in the same direction as the data referred to above.

8.4 Conclusions

The following conclusions may be drawn from the evidence presented in this paper.

1. Contribution from long-distance atmospheric transport can increase significantly the concentrations of certain trace elements in the surface layer of soils. This applies in particular to soils rich in organic matter.
2. Many concentration values regarded in the literature as "background" values may have been much more affected by atmospheric pollution than anticipated.

References

Allen RO, Steinnes E (1979) Contribution from long-range atmospheric transport to the heavy metal pollution of surface soil. Heavy Metals in the Environ. CEP Consultants, London, pp 271–274

Hanssen JE, Rambæk JP, Semb A, Steinnes E (1980) Atmospheric deposition of trace elements in Norway. In: Drabløs D, Tollan A (eds) Ecological Impact of Acid Precipitation. SNSF, Oslo-Ås, pp 116–117

Hanssen JE, Rambæk JP, Semb A, Steinnes E (1981) Atmospheric deposition of some heavy metals in Norway. Heavy Metals in the Environ. CEP Consultants, Amsterdam, pp 322–325

Hvatum OØ, Bølviken B, Steinnes E (1983) Heavy metals in Norwegian ombrotrophic bogs. In: Hallberg R (ed) Environmental Biogeochemistry. Ecol Bull (Stockholm) 35:351–356

Låg J (1968) Relationships between the chemical composition of the precipitation and the content of exchangeable ions in the humus layer of natural soils. Acta Agric Scand 18:148–152

Låg J, Steinnes E (1974) Soil selenium in relation to precipitation. Ambio 3:237–238

Overrein L, Seip HM, Tollan A (1980) Acid precipitation-effects on forest and fish. Res Rep FR 19/80 SNSF Proj, Oslo-Ås

Rambæk JP, Steinnes E (1980) Atmospheric deposition of heavy metals studied by analysis of moss samples using neutron activation analysis and atomic absorption spectrometry. Nucl Meth Environ Energ Res (CONF-800433), pp 175–180

C. Behavior of Organic Compounds

Introductory Comments

B. Y_{ARON}

Organic compounds reach the unsaturated zone as the result of natural processes or are added by man in his effort to control the ecosystem or to dispose of his wastes. Regardless of the source and the amount added, many of these compounds affect the chemical and physical properties of the unsaturated zone. The tremendous number of organic substances present in the unsaturated zone environment as a result of natural or antropogenic action, makes it impossible to present an exhaustive description of their behavior. It was up to the editors to select certain aspects of the problem and try to conclude what are the most important future trends in the research of the behaviour of organic compounds in the unsaturated zone.

Organic pollutants which come in contact with organic and inorganic solid phase of the unsaturated zone may move and interact with it, a process which leads to changes in the properties of the pollutants or of the unsaturated zone. The physico chemical processes involved consist of adsorption, formation of bonded residues, and surface conversion of the contaminants. Chemical changes in the system are occuring as a result of biological activity or of chemical reactions with the unsaturated zone media. The physical properties of the unsaturated zone could be affected when large quantities of organic wastes are added to the system or when specific organic polymers are applied in order to improve the solid phase structure.

The first chapter included in Part C deals with the natural organic compounds of the unsaturated zone. Hayes (Chap. 9) summarizes the existing knowledge on the chemical nature and reactivities of soil organic polymers. Starting with the presentation of the genesis of humic materials, the author presents the up-to-date results on the structure of these materials and concludes with data on the effect of the humic substances on the unsaturated zone properties. Humus could not be considered as a pollutant, but it plays an important role in controlling the behavior of both organic and inorganic contaminant. Complexation of metals and heavy metal cations by humic substances – for example – may affect the adsorption and transport of these substances in the unsaturated zone. The formation of "bound" residues in soils may also be caused by the interaction of contaminants with the natural organic polymers.

Out of numerous synthetic organic chemicals added by man to the unsaturated zone as agrochemicals or as waste materials, we chose to present pesticides as a case study. In the two chapters included in Part C, Calvet (Chap. 10) and Saltzman and Mingelgrin (Chap. 4) discuss the phenomena of adsorption, transport and chemical degradation in the unsaturated zones. Calvet's paper (Chap. 10) is devoted to the adsorption of chemicals from aqueous and organic solutions when the solid phase is saturated or partially hydrated. Since the transport of pesticides by convection follows the general pattern of solute transport, Calvet focuses his discussion mainly on the diffusion of these chemicals. The adsorption parameters used for modeling the transport of pesticides in unsaturated zones are generally measured at saturated conditions (batch experiments) leading to discrepancy between calculated and experimental results. Calvet points out the need to find new experimental procedures by which the adsorption of organic contaminants in unsaturated conditions can be measured experimentally. The persistence of a pesticide is governed by both biological and chemical degradation. Saltzman and Mingelgrin (Chap. 11) indicate that in the unsaturated zone – where biological activity is less developed – the chemical conversion becomes increasingly important in governing pesticide persistence. The biological processes involved in the pesticide degradation are mentioned in Lynch's paper (Chap. 13). Surface reactions of clay pesticide affected by changes in the hydration level are discussed and examples of pesticide conversion and formation of new metabolites under various environmental conditions and solid phase composition are presented (Chap. 11).

If pesticides – as other synthetic organic contaminants – are added to the unsaturated zone in minute amounts, organic solids originating from manure, sludge, and sewage effluent are added to the system in larger quantities. The last article included in Part C deals with both, the beneficial and hazardous aspects, of the organic solids reaching the unsaturated zone. Yaron et al. (Chap. 12) discuss the effect of suspended organic solids on the soil hydraulic properties, the transport of contaminants adsorbed on suspended organic solids in the unsaturated zone, and the effect of ammonification-denitrification processes of sludge on the pH and electrical conductivity of the soil solution. The results presented in this paper suggest possible irreversible effects in the unsaturated zone due to the addition of sewage effluent and sludge. These include changes in the hydraulic conductivity or dispersion of clays due to the saturation of the exchange complex with ammonium ions.

The papers presented in Part C suggest the importance of the bounding with organic ligands, the effect of surface reactions in unsaturated conditions on contaminant persistence, and the irreversible effects of organic solids on the unsaturated zone properties. Those suggestions are based mainly on laboratory investigations where unsaturated conditions were not always kept. Further large scale field studies are required to prove the application of laboratory findings to field conditions and to suggest ways and means of controling the behavior of organic compounds in the unsaturated zone.

9. Chemical Nature and Reactivities of Soil Organic Polymers

M. H. B. HAYES

9.1 Introduction

Soil organic polymers are components of humus substances. There is no definition or classification of humus which has general acceptance. Hayes and Swift (1978), on the basis of proposals by Kononova (1966, 1975), regarded as *humus*, the transformed products of plant and animal remains which bear no morphological resemblances to the structures from which they were derived.

The classification of humus substances adopted by Hayes and Swift will be used here. They considered that humified materials consist of both humic and nonhumic substances. Brown, amorphous, polymeric materials called *humic substances* can be differentiated on the basis of their solubilities in aqueous acids and bases. *Humic acids* are precipitated at pH 1 on acidification of aqueous alkaline extracts of soil organic matter; *fulvic acids* are soluble in aqueous acids and bases, and *humins* are insoluble in these solvents. Polysaccharides, polypeptides, and altered lignin products, produced by microorganisms when transforming soil organic residues or derived from altered plant and animal remains may be regarded as nonhumic components of humus substances.

This contribution will briefly review the genesis of humus materials and provide some indications of what is known about the structures of the humic and polysaccharide polymers, which are the most important components of humus. These structures will be interpreted in terms of the reactivities of the polymers in so far as their interactions with clays and soil applied organic chemicals are concerned.

9.2 The Genesis of Humus Materials

A portion of all the organic substances which enter the soil environment inevitably go to form humus materials. Jenkinson (1981, p 506) has pointed out that "in the long run no fraction of the organic matter in plants and animals can withstand decomposition to carbon dioxide and water. If this were not so, any completely resistant fraction would now cover the surface of the earth." He has presented data (ibid, p. 507), which indicates that the overall turnover time (TT) of soil organic matter may be written as:

$$TT = (\text{soil organic carbon})/(\text{net primary production} \qquad (1)$$
$$\text{of carbon on land per year})$$

and is of the order of 13 years. There are plenty of data which show, however, that some of the components of humus persist for a long time in the soil. For instance, Campbell et al. (1967) have calculated a mean residence time of 800–1500 years for nonhydrolizable humic substances in podzol and chernozem soils.

Bazilevich (1974) has estimated that the humus reserves in soils are of the order of 2.4×10^{12} tonnes and Kovda (1974) has suggested that 5.6×10^{10} tonnes of carbon are produced annually on land. Simple extrapolations from these data suggest that the humus reserves on land are 20–25 times greater than the amounts of organic materials photosynthesized annually. Although there is general acceptance about the extent of materials produced by photosynthesis, few agree about the extents of humus reserves; and the Bazilevich figure is generally considered to be high. Estimations from data by Jenkinson and others (Jenkinson 1981, p. 512) would suggest that humus reserves are at least ten times greater than the amounts of organic matter photosynthesized annually, but the data on which this estimate is made show considerable variations, depending on the soil, crop, climate, and other factors.

Humification is primarily a biological process. There is, however, some evidence to indicate that extracellular chemical reactions can also lead to the genesis of humus materials and some consideration will be given here to both types of processes.

9.2.1 Humification by Biological Processes

Humification of Plant Materials. Nearly a generation ago (Hayes 1960), this author investigated the fate of finely ground ^{14}C-labelled rye (*Secale cereale*) tissue (grown for 82 days as described by Stotzky 1956; see also Stotzky and Mortensen 1958) which was incorporated to a depth of 15 cm in columns of a humic histosol (containing in the 0–5 cm layer, 83% organic matter, 17% ash, and with a cation exchange capacity of 2,000 µEq g^{-1}) classified as a Rifle Peat from Celeryville, Ohio, USA. The columns, contained in 9 cm diameter and 28, 53, and 78 cm long glass or coated steel pipes, were obtained by driving the pipes into the soil profile. After a pre-amendment incubation period of 102 days at 28 ± 4 °C, in which aliquots of 50 cm^3 of water were added 19 times (to give a rainfall equivalent of 14.8 cm), carbon evolution was relatively constant. Then the plant tissue (3.2 g C per column, equivalent to ca. 5 tonnes C ha^{-1}) was incorporated and incubation was continued at the same temperature for a further 128 days. Water was added on eight further occasions (simulating an additional 6.25 cm of rainfall). A stream of water vapour saturated, CO_2-free air was passed over the columns and the CO_2 evolved was trapped and quantitatively measured. The ^{14}C label allowed a distinction to be made between CO_2 from biological oxidations of the rye tissue and of the indigenous soil organic materials; it also allowed estimates to be made of the amounts of organic matter from the amendment materials in the dried residues of the drainage from the columns.

Hayes and Mortensen (1963) have provided data for the organic matter lost as CO_2 and in the drainage waters from the soil columns. Figure 1 (from Hayes 1960) presents data for the rates of evolution of carbon during the first 300 h of

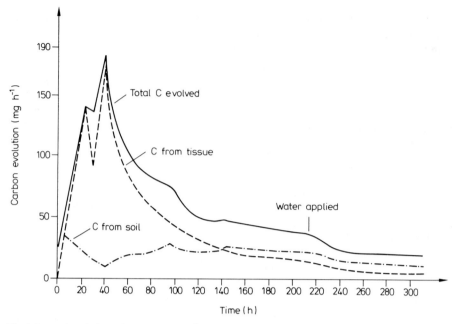

Fig. 1. Rate of evolution of carbon (mg h^{-1}) during the biological oxidation of a Rifle Peat soil amended with ^{14}C-labelled rye tissue (Hayes 1960)

incubation from the whole amended soil, from the histosol only, and from the tissue added in the case of a 78 cm column. This figure shows how the rate of evolution of carbon from the native soil fell between the 6th and 40th hours of incubation. (The simulated cultivation used to mix the tissue with the soil gave an enhanced evolution of carbon from the native soil during the first 6 h of incubation.) This depressive effect lasted for about 600 h and was followed by a small priming effect or enhanced loss of carbon from the soil. However, the priming was not significant, and at the end of the experiments there were no significant differences between the amounts of carbon lost from the soil in the amended columns and those lost from the unamended soils in the control experiments.

Reference to Fig. 1 shows that carbon evolution from the rye tissue was slow during the first 6 h of incubation. The slow initial utilization of the tissue by the microorganisms could indicate that multiplication of an appropriate population to utilize the plant material occured. However, the depression in the rates of CO_2 evolution from the soil during the 6–40 h post-incubation period suggests that the indigenous population preferentially utilized the tissue substrate and that the slow initial utilization of the tissue resulted from a need to imbibe moisture from the soil to a level at which the plant material could provide a substrate for the microorganisms.

Peak evolution of carbon took place in all instances between 40 and 50 h after addition of tissue; thereafter there was a slow decline in the production of CO_2, as is indicated in Fig. 1. The depression in the rate of carbon evolution from the tissue during the 20–30 h post-incubation period was consistent in all experiments and cannot be explained; similar trends were not evident for the evolution from the indigenous soil (whether amended or not) during the same interval.

Moisture had a profound influence on biological activity in the columns and on the rates of humification, as indexed by the rates and extents of CO_2 evolution. Addition of water equivalent to a rainfall of ca. 0.78 cm depressed to CO_2 evolved; however, the rate of evolution returned to normal after drainage had taken place. The optimum moisture content for the utilization of rye tissue by the microorganisms occured at P_w values of 160–175. P_w values were obtained from the relationship:

$$P_w = (M_w/M_s)100, \qquad (2)$$

where M_w is the mass of water and M_s is the mass of soil after drying at 105 °C for 24 h. Activity was essentially the same (though well below maximum) in the P_w range of 175–210, but biological activitiy was significantly depressed at values greater than 220. (It is of interest to note that the moisture content varied only from 61.5%–69% for the P_w range 160–220.)

After incubation for 128 days, 38%–45% of the rye tissue could be accounted for in the different soil columns and the rate of evolution of carbon from the tissue was of the order of 0.05 mg h^{-1}, whereas that from the soil was about 1 mg h^{-1}. About 1% of the carbon in the surface 15 cm of the column was derived from the rye and this material was being transformed about 20 times faster than the indigenous organic matter at that time.

Humic acid materials were isolated by extracting the H^+-exchanged soil seven times with 0.1 M NaOH and adjusting the supernatants to 2 with HCl. About 40% of the soil mass was isolated in this way as humic acids having ash and carbon contents of 8.5% and 55% (60% on a d.a.f. basis), respectively. About 25% of the tissue which remained in the soil after 128 days of incubation was present as humic acids and this suggested that about 11% of the carbon added had been transformed to humic acid structures.

The columns were frozen at the end of the 128 day incubation period. Subsequently, Chahal et al. (1966) took soil samples at different depths in the 23 and 53 cm columns, extracted these with solvents, such as 0.1 M sodium pyrophosphate at pH 7.0, with 0.5 M NaOH, and with 6 M HCl at 121 °C. Extracts were fractionated by means of gel filtration and continuous flow paper electrophoresis and various fractions were analysed for total organic matter, colour, phenols, total saccharides and uronic acids, nitrogen and radioactivity. Radioactivity was present in samples at 0–15, 15–30, and 30–45 cm depths and that in the 0–15 cm layer was primarily associated with the highest molecular weight materials, whereas that which migrated in the profile was associated with the lower molecular weight components.

Pyrophosphate, the mildest of the extractants used, isolated the most soluble and highly oxidized of the humus substances and the electrophoretic migration pattern of these was similar to that for the 6 M HCl hydrolysates. The more efficient alkaline solvent extracted considerable amounts of colourless polysaccharides, which could have been microbial and/or hemicellulose in origin. Phenolic components were invariably associated with the brown materials and the association of radioactivity with these indicated that the amendment tissues gave rise to humic substances. Such observations do not indicate the extents to which the labelled materials were scrambled during the synthesis processes or whether or not whole molecules from the plant tissues were incorporated in the humus structures.

Humification of Simple Chemicals. Several works have reported the formation of humus-like substances from the proliferation of fungi, bacteria, and even yeasts on simple organic chemicals. The most comprehensive work in this area was carried out by Martin and his colleagues (Martin and Haider 1976, 1977, Haider et al. 1977, and earlier references in these). They have shown that various di- and trihydroxybenzene structures, e.g. 1,3,5-trihydroxybenzene, 3,4,5-trihy-droxytoluene, 3,4,5-trihydroxybenzene-carboxylic acid (gallic acid) and its decar-boxylation product, can be isolated from fungal cultures which did not contain such aromatic compounds in the substrates. Several anthraquinone structures were also isolated. When the pH of the media reached 5–6, pronounced phenolase activity in fungal cells gave rise to browning, indicative of the polymerization of the phenol and anthraquinone structures. In the soil environment, phenols could arise from microbial metabolism or/and from the degradation of plant remains. These could be polymerised by extracellular or intracellular enzymes or even by chemical reactions.

9.2.2 Humification by Chemical Processes

The Maillard or the Browning Reaction is well recognized in food processing and it takes its name from Maillard, a French chemist who synthesized brown polymers by heating solutions of glucose and glycine and showed that these poly-mers had some properties comparable with humic acids. During the 1930's and 1940's, Enders and his colleagues in Germany described browning polymers from reactions of 2-ketopropanal and glycine and they showed that some of these had properties similar to those of soil humic acids. They postulated that 2-keto-propanal could be released from microorganisms in unfavourable growth con-ditions to form humic substances and they were able to show that the dione is present in most soils (for references see Hayes and Swift 1978, pp 244–246).

Flaig et al. (1975) have reviewed the extensive literature dealing with humus-like substances from phenolic precursors. The starting materials for such synthe-sis are extensively distributed in lignin residues and in microbial metabolic prod-ucts in soils. Polyhydroxybenzenes readily form quinones and polymerize under alkaline conditions to give humic-type products in the soil. There is ample evi-dence for phenoloxidase enzymes in soils and there is good reason to believe that phenolic substances released by degradation or biosynthesized by organisms could be polymerized to humic substances by chemical and enzymatic processes (refer to Hayes and Swift 1978, p 247).

9.3 Structures of Humus Substances

In order to describe the structure of any polymer, it is necessary to know what the repeating units are (the *primary structures*), the order in which these units are linked together (*secondary structure*) and the size of the molecule and its shape (*tertiary structure*). Structural studies of polymers, such as polysaccharides and proteins, are feasible because their synthesis is genetically controlled and in each molecular species the units are linked in invariant sequences through bonds, which are cleaved by enzymes, or by hydrolysis in acidic and basic media. Soil polysaccharides, but not humic substances, fall into this category.

Should humic substances arise from polymerization by random condensation reactions, it is likely that the primary structures are composed of single molecules randomly linked. Hence, any attempt to determine secondary structures would be meaningless. It is, however, very important to have an appreciation of the tertiary structures, because the reactivities of polymers are influenced by their shapes and sizes. These properties can, for example, influence the extent to which an adsorptive species can penetrate into and be trapped inside an adsorbent polymer matrix.

9.3.1 Extraction and Fractionation of Humus Substances

In order to carry out meaningful studies of the structures of humus substances, it is important to isolate the materials and to fractionate them into reasonably homogeneous components. Swincer et al. (1969), Finch et al. (1971), Greenland and Oades (1975); Hayes and Swift (1978) and Cheshire (1979) have reviewed procedures for the extraction and fractionation of soil polysaccharides. Neutralization with $BaCO_3$ of the dilute H_2SO_4 soil extract provided Finch et al. with two gross polysaccharide materials and these were further fractionated by elution through dextran gels (Sephadex) and ion exchange cellulose or gel preparations. Swincer et al. (1969) achieved a good separation of polysaccharides from the humic substances in the 0.5 M NaOH extract of soils pretreated with 1 M HCl or HF. The extracts were eluted through H^+-exchanged resins (e.g. H^+-Dowex 50) and the eluates were then passed through polyclar-AT [a poly(vinylpyrrolidone)product] or nylon-coated celite resins, which adsorbed most of the humic substances and allowed the polysaccharide materials through the columns. In this way, more than 50% of the soil carbohydrate could be isolated. Extraction was improved by ultrasonicating the medium during contact with the alkali. Subsequently, the residual soil was extracted at 60 °C using acetic anhydride containing 2.5% concentrated H_2SO_4. After dilution in water, the acetylated polysaccharides remaining were extracted in chloroform. Such sequential extraction processes removed nearly 80% of the total soil carbohydrate.

Hayes et al. (1975) made use of the properties of dextran gels to adsorb brown humic components of fulvic extracts and then employed anion exchange chromatography to bring about separations of the polysaccharides in the extracts on the basis of charge density differences. In this way, they isolated two polysaccharide substances composed of 70% of glucose units and these were thought to be amongst the purest polysaccharides isolated from soil at that time. Further purification might have been achieved by chromatography on ion-exchange media using borate buffers which form negatively charged borate complexes with adjacent cis hydroxyl groups in carbohydrate structures.

Hayes and Swift (1978, pp 182–190) have reviewd procedures used till their time of writing for the extraction of humic substances from soil and they have referred to principles involved in some of the extraction processes. They emphasized how humic polyelectrolytes are insoluble at the pH values of most fertile soils because of the contribution of di- and polyvalent cations to the neutralization of the negative charges originating largely from the carbonyl groups in the polymers. Such cations give a pseudo cross-linking effect, causing the polymer to shrink and to be difficult to hydrate. Effective solvation of humic molecules requires the replacement of these cations. For that reason, humic substances are H^+-ion exchanged, by treatment with dilute acid (treatment with some H^+-exchanged resins should, in principle, be effective also) before extraction. Low molecular weight, polar, and highly charged H^+-exchanged humic substances (fulvic acids) are soluble in water, but the less highly charged materials are not. The insoluble molecules are associated into moderately compact structures through inter- and intramolecular hydrogen bonding.

The reader is referred to Hayes and Swift (1978) for a discussion of the extraction and fractionation of humic substances. However, since that work was written, the International Humic Substances Society (IHSS) was formed (Malcolm 1981) and under its auspices two conferences have been organised to deal with genesis, extraction, fractionation, structures, and some interactions of humic substances. The first of these conferences (at Estes Park, Colorado, August 1983) carried extensive treatments of extraction and fractionation procedures. The contributions which dealt in considerable detail with procedures for aquatic as well as soil humic materials, are scheduled for publication (by Wiley, N.Y.) in 1984. For that reason, it is not appropriate to provide details here of extraction and fractionation procedures. Suffice it to say that the extraction procedure recommended by the IHSS involves pretreatment of the sample with 1 M HCl, neutralization (to pH 7) of the residue with 1 M NaOH, the addition (under N_2) of 0.1 M NaOH to a final extractant to soil ratio of 10:1 before shaking intermittently for 4 h and allowing to settle overnight. After appropriate centrifugation or filtration, the humic acids are isolated from the precipitate formed when the supernatant or filtrate is adjusted to pH 1 after acidification with 6 M HCl.

The work of Cameron et al. (1972) provides the most comprehensive study so far of the fractionation of soil humic acids. They isolated components which were reasonably homogeneous with respect to molecular weight by repeatedly refractionating components eluted in similar volumes of tris buffer [2-amino-2(hydroxymethyl)propane-1,3-diol] from gel columns. Pressure filtration through graded porosity membranes also provided fractions which were moderately homogeneous on the basis of molecular sizes. The 11 components isolated had molecular weight values ranging from 2×10^3 to 1.5×10^6.

Humic acids are also polydisperse with respect to charge. In theory, at least, separations on the basis of charge differences can be made by uses of electrophoretic and of ion-exchange chromatographic techniques. Stevenson et al. (1952) used moving boundary electrophoresis to show that some separation of components with different charge densities was possible. Continuous-flow (paper curtain) electrophoresis, as used by Chahal et al. (1966) and Waldron and Mortensen (1961), also indicates that humic substances contain materials with a spread of charges. However, no discrete fractions have yet been isolated which can be stated to be homogeneous in so far as charge density is concerned. Any appropriate approach to isolate fractions having such homogeneity should use materials which were previously fractionated on the basis of molecular size differences. The most meaningful studies in the future of humic structures will use materials which are homogeneous with regard to molecular weight and charge density properties.

9.3.2 Structural Information from Degradative Procedures

Classical procedures for the determination of structures of biological polymers degrade the macromolecules to identifiable components which can be related to structures within the polymer. Thus, hydrolysis has provided highly useful information where labile bonds, such as the glycosidic and peptide linkages of polysaccharides and proteins, hold the components together. It should therefore be relatively easy to determine fully the structures of soil polysaccharides when even modest supplies of homogenous, uncontaminated polymers are available. However, structural studies of humic substances are more difficult.

Studies of Soil Polysaccharide Hydrolysates. Sugars which have been identified in the hydrolysate digests of soil polysaccharides include the hexoses glucose, galactose, and mannose, the pentoses arabinose, ribose, and xylose, and the six hexose derivatives, fucose or 6-deoxy-L-galactose, and rhamnose or 6-deoxy-L-mannose (where the -OH groups on the C-6 carbons are replaced by H), glucosamine and galactosamine or 2-amino-2-deoxy-D-glucose or galactose, where the -OH groups on C-2 are replaced by NH_2, and glucuronic and galacturonic acids (where -CH_2OH groups on C-6 are replaced by COOH). It is highly likely that the amino groups in the sugar amines in the polysaccharides are acetylated (present as -$NHCOCH_3$); however, these acetyl groups would be released during hydrolysis. Other sugars detected in small amounts (Cheshire 1979) include the pentose and sorbose, the methylated sugars 4-0-methyl-D-galactose, 2-0-methyl-L-rhamnose, 2-0-methylxylose and 3-0-methylxylose, 2-0-methylarabinose, the sugar alcohols inositol and mannitol, and the amino sugar muramic acid (3-0-carboxyethyl-D-glucosamine).

Hydrolysates from some polysaccharide materials isolated from soils have been found to contain amino acids, possibly from glycoproteins, sulphur, possibly from sulphated polysaccharides from animal and microbial tissues, inorganic and organic (possibly inositols) phosphorus, and components of humic substances. Such materials can be regarded as contaminants in many instances because the contents of such extraneous materials can be decreased by careful fractionation procedures (Hayes and Swift 1978, Cheshire 1979). It is possible that stable humic-polysaccharide complexes can exist in soil through, for example, phenolic glycoside linkages.

As yet, no soil polysaccharide has been isolated which is sufficiently pure to warrant detailed studies of secondary and tertiary structures. Cheshire and his colleagues (see Cheshire 1979) have contributed most to concepts relevant to secondary structures and their data indicate that glycosidic linkages between the hexose units in soil polysaccharides are predominantly $(1{\rightarrow}3)$ and $(1{\rightarrow}4)$, but that $(1{\rightarrow}6)$ and $(1{\rightarrow}2)$ linkages are also present and that xylose units are linked $(1{\rightarrow}4)$.

Information from Humic Digest Products. Humic acids, when hydrolysed with 6 M HCl, can release up to 50% of their masses as volatile and as water soluble components. Most of the volatile losses arise from decarboxylation and the soluble components consist largely of sugars, amino acids, phenols, and metals. However, the "core" of humic polymers resists degradation in aqueous acids and bases and would appear to be composed of units linked by carbon to carbon bonds and by ether-type functional groups which are difficult to cleave. A high energy input is needed to cleave the core linkages in humic polymers and so the digest products identified are often derivatives of the actual structures in the polymer. Hayes and Swift (1978) and Maximov et al. (1977) concluded independently that the uses of highly energetic degradative procedures are likely to alter the structures of the polymer components even before they are released into the digests. Further alterations can be expected in the released molecules during their residence in reactive conditions prevailing in the digests. These problems can be overcome to some extent by appropriate uses of studies with model compounds. Such studies (Hayes and Swift 1978, Hayes 1984) would investigate the types, extents, and rates of alterations of organic structures of the types which might be present in humic materials when these molecules are subjected to the reactive conditions present in the degradation digests. The results would allow some predictions to be made of the parent compounds which gave rise to the structures identified. Haworth and his colleagues (see Atherton et al. 1967, Cheshire et al. 1967, 1968) concluded from the digest products identified from the *zinc dust distillation* at 500–550 °C of 6 M HCl boiled humic acids that the polymer cores are composed of fused aromatic structures. The digest compounds ranged from naphthalene to coronene. They observed that substantial amounts of sugars, amino acids, and some phenols and metals were released by the HCl hydrolysis and they deduced that polysaccharides, peptides, phenolic structures, and complexed metals were attached to the polycyclic aromatic core. However, the fact that polycyclic aromatics were obtained as artefacts when hydroxybenzene carboxylic acids, furfural, and polymers from quinones were subjected to similar digest conditions (Cheshire et al. 1968) weakened this theory somewhat.

Schnitzer and his colleagues have provided a substantial amount of data for products and yields obtained when humic and fulvic acids were oxidized with alkaline solutions of potassium permanganate (see references in Schnitzer and Khan 1972, Schnitzer 1978). By use of gas-liquid chromatography mass spectrometry (g.l.c.m.s.) and microinfrared spectroscopy techniques, they identified over 30 compounds in their methylated digest products. Compounds which could be assigned to structure types 3, 5 to 13, and 17 and 18 (Fig. 2) were identified. It is unlikely, however, that these compounds were components of the polymers; they were more likely to be oxidized products of some of the primary structures or they could have been artefacts formed in the reaction. The mechanisms of permanganate oxidations have been extensively studied and it is tempting to infer that the benzene polycarboxylic acids identified were the oxidation products of

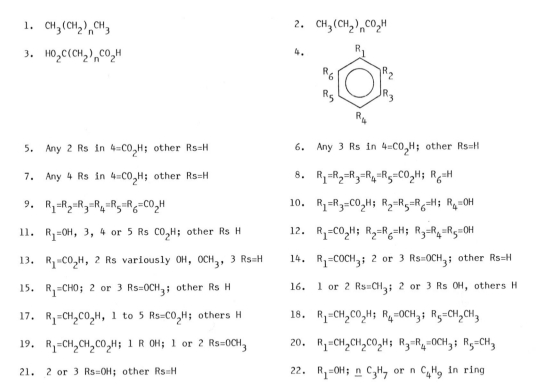

1. $CH_3(CH_2)_nCH_3$

2. $CH_3(CH_2)_nCO_2H$

3. $HO_2C(CH_2)_nCO_2H$

4.

5. Any 2 Rs in 4=CO_2H; other Rs=H

6. Any 3 Rs in 4=CO_2H; other Rs=H

7. Any 4 Rs in 4=CO_2H; other Rs=H

8. $R_1=R_2=R_3=R_4=R_5=CO_2H$; $R_6=H$

9. $R_1=R_2=R_3=R_4=R_5=R_6=CO_2H$

10. $R_1=R_3=CO_2H$; $R_2=R_5=R_6=H$; $R_4=OH$

11. $R_1=OH$, 3, 4 or 5 Rs CO_2H; other Rs H

12. $R_1=CO_2H$; $R_2=R_6=H$; $R_3=R_4=R_5=OH$

13. $R_1=CO_2H$, 2 Rs variously OH, OCH_3, 3 Rs=H

14. $R_1=COCH_3$; 2 or 3 Rs=OCH_3; other Rs=H

15. $R_1=CHO$; 2 or 3 Rs=OCH_3; other Rs H

16. 1 or 2 Rs=CH_3; 2 or 3 Rs OH, others H

17. $R_1=CH_2CO_2H$, 1 to 5 Rs=CO_2H; others H

18. $R_1=CH_2CO_2H$; $R_4=OCH_3$; $R_5=CH_2CH_3$

19. $R_1=CH_2CH_2CO_2H$; 1 R OH; 1 or 2 Rs=OCH_3

20. $R_1=CH_2CH_2CO_2H$; $R_3=R_4=OCH_3$; $R_5=CH_3$

21. 2 or 3 Rs=OH; other Rs=H

22. $R_1=OH$; $\underline{n}$ C_3H_7 or n C_4H_9 in ring

Fig. 2. Types of organic structures released into the digests of degradation reactions of humic substances

fused aromatic structures. However, aliphatic substituents on the aromatic nuclei would be oxidized in permanganate solutions to benzenecarboxylic acid structures.

The origins of the dicarboxylic acids of type 3 (Fig. 2), where $n=2$ to 8 and of the tricarboxylic acids identified could be ascribed to aliphatic hydrocarbon structures containing olefinic linkages or oxidizable groups (such as alcohols, aldehydes) in the hydrocarbon chains. Structure types 17 and 18 would suggest that at least some of these hydrocarbon structures were present as substituents on aromatic structures.

Schnitzer and his colleagues have extensively researched *alkaline cupric oxide* oxidative reactions for the degradation of humic substances (see Schnitzer 1974, 1978, Griffith and Schnitzer 1976) and they have identified more than 60 products in the digests. About 25 of these compounds were also contained in the alkaline permanganate digests. Products contained in both digests included several aliphatic di- and tricarboxylic acids and all of the benzenedi- and polycarboxylic acids. Straight chain aliphatic acids, of type 2, Fig. 2, where $n = 10$ to 22, branched chain carboxylic acids, straight chain aliphatic hydrocarbons, compound type 1, Fig. 2, where $n = 12$ to 38, as well as aromatic keto and aldehyde structures (types

14 and 15) were found in the alkaline cupric oxide and not in the permanganate digests.

Hayes and Swift (1978) have discussed the possible origins of many of the cupric oxide-humic substances digest products. They have shown how the hydrocarbons could have been released from long chain unsaturated hydrocarbon structures or even from saturated hydrocarbons containing carbonyl groups, and they agreed with the interpretations of Schnitzer and his colleagues, who suggested that the aliphatic hydrocarbons and the fatty acids could have come from microbially synthesized hydrocarbons adsorbed by the humic substances and from the saponification of phenolic esters, respectively.

The fact that all of the benzenecarboxylic acids found in permanganate digests were present also in those where alkaline cupric oxide was used, is of especial interest. Alkaline cupric oxide digestion would not degrade fused aromatic structures to benzenecarboxylic acids. Thus, the likely origins for these acids were in oxidizable aliphatic substituents attached to the aromatic structures. These same substituents would have oxidized in permanganate to benzenecarboxylic acids. It is possible, however, that the carboxyl substituents could have arisen from the carbonylation of aromatic structures in the alkaline media and it will be necessary to carry out carefully controlled model studies in order to investigate this possibility.

Burges et al. (1964) introduced the sodium amalgam procedure for the degradation of humic materials. The technique has been successfully used since then by Piper and Posner (1972) for the degradation of humic structures and by Martin et al. (1974) for the degradation of fungal and model phenolic polymers, as well as soil humic acids.

Compounds identified in the amalgam digests include those of types 12 to 16, 19, 20, and 21. It should be pointed out that methylation was necessary where all of the types of compounds in Fig. 2 were identified by g.l.c.m.s. For that reason, many of the methoxy substituents could have been present in the polymers as phenolic hydroxides. Although the acids were identified as the methyl esters, it is safe to assume that such esters were not present to any significant extent in the polymers; thus, all such esters are written as acids. Most of the compounds which were identified in the amalgam digests were not methylated and several polyhydroxybenzene structures were positively identified in these digests.

Sodium amalgam provides a mild, reductive degradative procedure and model studies have shown that degradation in the digests of structures of the types identified was small. It is reasonable to assume, therefore, that most of the structures in the amalgam digests are among the primary structures of humic substances. Thus, it is probable, on the basis of deductions from model studies, that many of the compounds identified were held through ether linkages in the polymer, or, less likely, as biphenyl-type structures where activating (CH_3- or -OH) substituents were *ortho* and/or *para* to the linking bond (see Hayes and Swift 1978, Hayes 1984).

Functional group analysis has confirmed that aldehyde and keto substituents are contained in humic substances and the digest products from amalgam degradation indicates that some of these groups are associated with phenolic structures. This procedure also shows that methyl groups are attached directly to the aromatic structures and the propanoic acid substituents in compounds of types 19 and 20 provide evidence for longer chain aliphatic substituents on these structures.

Only one carboxyl was present in any of the aromatic structures identified in the sodium amalgam digests. Carboxyl groups would not be generated as artefacts by this procedure and this confirms that direct attachment of carboxyl groups to aromatic structures occurs in humic polymers. It further suggests that fused aromatic compounds are not major components of the primary structures.

The author's laboratory (Swift 1968, Craggs 1972, O'Callaghan 1980) has been involved with the degradation of humic polymers using saturated (10%) *sodium sulphide* solutions and *phenol* and *para toluenesulphonic acid* (PTS) or *boron trifluoride* catalysts (Colclough 1980). Hayes and Swift (1978) have outlined the mechanisms of degradation involved where such systems are used and they have outlined the development of these procedures for studies of humic substances.

O'Callaghan (1980) has used g.l.c.m.s. techniques to identify more than 30 digest products from reaction of humic acids with 10% aqueous Na_2S solutions at 250 °C under autoclave conditions. Type 3 compounds (Fig. 2), where $n = 2$ to 5, were much in evidence in the digests, as were n alkanols $[CH_3(CH_2)_nOH$, where $n = 1$ to 4]. Structure types 5, 12, 14, 16, 18, 21, and 22 were among the aromatic structures identified and several of these structures had one to four carbon aliphatic substituents on the aromatic nuclei (cf. structure types 16, 20, 22). On the basis of information from model studies (O'Callaghan 1980), it can be concluded that such structures would arise where quinone methide intermediates are formed (see Hayes and Swift 1978, Hayes 1984).

Phenol-p-toluenesulphonic acid was first used for the depolymerization of soil organic colloids by Jackson et al. (1972). This procedure is thought to help identify interaromatic linkages in the humic polymers. Colclough (1980) has isolated up to 60% of the starting material as ether soluble components of the digest and many of these components still await identification by g.l.c.m.s. From the results obtained so far, it would appear that substantial amounts of aliphatic structures link aromatic components in the humic polymers. It is probable that these linkages carry hydroxyl, carboxyl, and possibly carbonyl functional groups. Long chain aliphatic hydrocarbons were abundant in Colclough's digests. As yet their origins cannot be fully explained, but it is likely that they arose from microbial synthesis and were adsorbed by the humic acids.

9.3.3 Structural Information from Nondegradative Procedures

Hayes and Swift (1978) have reviewed applications of spectroscopic procedures for studies of the structure of humic substances. Of the various procedures which are available, nuclear magnetic resonance spectroscopy (n.m.r.) would appear to give the most information. Grant (1977) analyzed soil extracts by proton (^1H) and carbon-13 (^{13}C) n.m.r. and he was able to show various environments of $CH_2(CO)$, CH_2-NH-, carbohydrate H-C-O, and only minor amounts of aromatic components.

Substantial data relevant to predictions of structures in solid coal (Barron and Wilson 1981, Havens et al. 1982, Maciel et al. 1982), and solid soil organic matter and humic substances (Barron et al. 1980, Hatcher et al. 1981, Wilson 1981, Wilson et al. 1981, Worobey and Webster 1981, Schnitzer 1982) have been obtained using recent developments in ^{13}C-n.m.r. instrumentation where facilities for cross polarization and magic angle spinning (CPMAS) have allowed meaningful spectra of solid samples to be obtained. Hayes (1984) has summarised the data which have been obtained. The evidence which is available suggests that in some instances as much as 60% of the primary structures of humic substances may be aromatic, but it would appear, for the most part, less than 35% of the structures are in fact aromatic. This is not surprising when it is considered that nearly 50% of the mass of these substances can be lost on hydrolysis and for the most part, the hydrolysates are composed of aliphatic structures.

9.3.4 Sizes and Shapes of Humus Substances

Hayes and Swift (1978) have reviewed applications of physical techniques for determinations of the sizes and shapes of humic materials. They have stressed the need to work with polymers which are relatively homogeneous with respect to size, and they have stressed the desirability of isolating structures which are as homogeneous as possible with respect to charge. The importance of adding electrolyte (such as 0.1 to 0.2 M KCl) to suppress the charges in the polymers is emphasized. Such additions are essential in order to avoid anomalous results from highly expanded polymer molecules.

Cameron et al. (1972) used equilibrium ultracentrifugation to determine the molecular sizes and shapes of humic acids which were carefully fractionated (as mentioned earlier) into 11 reasonably homogeneous components by use of techniques of gel filtration and of pressure filtration through graded porosity membranes. It is very probable that many of the components, which were not soluble in the tris buffer, had molecular weight values higher than the 1.5×10^6 measured.

Determinations of frictional ratios, f/f_o, where f is the frictional coefficient of the molecule under investigation, and f_o is that for a condensed sphere occupying the same volume, provide useful indications of the shapes of polymers in solution. The f/f_o value for a tightly coiled globular protein is close to 1, whereas the values for highly solvated and expanded molecules are substantially greater.

When Cameron et al. plotted frictional ratio data against molecular weight values, they observed a linear relationship for samples with mol. wts. up to 2 to 3×10^5. Linearity was not maintained for the higher molecular weight values. They interpreted these data to indicate that humic molecules assume random coil solution conformations. This would give rise to molecules which are randomly coiled with respect to time and space, and which are rougly spherical and with a Gaussian distribution of mass, greatest at the centre and decreasing to zero at the outer limits. The break from linearity in the f/f_o versus molecular weight plot at the higher molecular weight values could be attributed to increased branching or possibly to cross linking which would give rise to more compact structures.

9.4 Practical Applications of Information from Structural Studies of Humus Substances

There is considerable evidence indicating that soil polysaccharides stabilize soil aggregates, but it will not be possible to understand the mechanisms involved until more is known about the polymer structures, (Cheshire 1979, Theng 1979, Hayes 1980, Burchill et al. 1981). It is generally considered that polysaccharides of plant, as well as of microbial origins, are involved in the stabilization, although mucopolysaccharides from the earthworm gut could also be important. Polymers with substantial amounts of xylose and arabinose are likely to have plant origins, while those containing galactose, mannose, rhamnose, fucose, glucosamine, and galactosamine would be likely to be synthesized by microorganisms.

Hayes (1980) and Burchill et al. (1981) have discussed some concepts of bridging neighbouring components in order to stabilize soil aggregates. They have suggested that the β-(1→6)-linked dextran (Polytran) used by Olness and Clapp (1975) and having a mol. wt. of about 2×20^6 could have linear dimensions up to 7 μm. Such molecules could effectively bridge domains into microaggregate structures. Polysaccharides having β-linkages in the structures have more linear or helical structures than those with α-linkages and so the former can make more points of contact with adsorbent species. The extents to which polysaccharides which bind soil particles have β-linkages is unknown and we await structural investigations of purified components which bind to clays and other inorganic colloidal constituents.

The evidence which is accumulating from degradative and nondegradative investigations suggests that single ring and not fused aromatic compounds are contained in humic structures. Many of the aromatic substances are di- or even polysubstituted. These substituents include hydroxyl, methoxyl, and hydrocarbon structures. It would appear likely that aliphatic hydrocarbon and aromatic and aliphatic ether linkages connect the components of the polymers. There is good evidence to suggest that peptides and carbohydrate structures are linked to the humic polymers and that these contribute significantly to the overall aliphatic composition.

The data which indicate that humic substances are polyelectrolytes having random coil solution conformations have important practical considerations. In the soil environment, H^+, Ca^{2+}, Mg^{2+}, Al^{3+}, Fe^{2+}/Fe^{3+} can be expected to be the predominating exchangeable cations on the humic structures. H^+-exchanged humic acids are insoluble in water because the carboxyl groups are largely undissociated and hence lack the impetus to solvate which is provided by charged groups. In addition, inter- and intramolecular hydrogen bonding will pull together the strands in the random coil and exclude water. Divalent and polyvalent cations bridge two or more charged groups within the polymer matrix and cause a shrinkage of structure and loss of water from the matrix. Such cations can form bridges between the negative charges on the polymers and those on the inorganic colloids, and this mechanism is important in stabilizing soil aggregates (Theng 1979, Hayes and Swift 1981).

As humic substances dry, further shrinkage of the polymers occur and the closer proximities of the polymer strands promotes van der Waals interactions between them and especially between the hydrophobic structures. Thus, humic substances, when dried, are slow to rewet.

It is well-recognized that humic substances have high affinities for aromatic organic chemicals, and aromatic biocides are generally inactivated in soils containing high organic mater contents. The binding mechanisms could involve charge transfer reactions between the chemicals and the adsorbents, and the capacity of the humic substances for binding is governed by the extents to which the adsorptives can penetrate to the interior of the polymer adsorbents. Optimim recovery of the chemicals is achieved when the polymers are swollen by the extractants.

Turchenek and Oades (1979) have provided extensive data which indicate the extent to which humus materials and elements, such as Al, Ca, Fe, Mg, P, Si, and Ti are present in the different density fractions of soils. Their data show that there are definite differences between the humic materials associated with different size fractions of soil clays. It would appear that the more aliphatic humic materials are associated with the heavier fine clays, whereas the more aromatic polymers are preferentially associated with the lighter, coarser soil clays. The differences in affinities for the different clays might be explained in terms of shapes, polarities, charge densities, etc.

Soil clays are often contaminated with oxide and hydroxide structures (Schwertmann and Taylor 1981) and these in turn can be associated with organic polymers to form what might be described as the soil "conglomerate colloid" structures. There is much that needs to be known about these associations and about the organic components of the conglomerates. The observations of Turchenek and Oades has focused attention on possible differences in organic structures associated with the various soil clay materials. CPMAS n.m.r. spectroscopy should help resolve the differences between the various structures.

References

Atherton NM, Cranwell PA, Floyd AJ, Haworth RD (1967) Humic acid. I. ESR spectra of humic acids. Tetrahedron 23:1653–1667

Barron PF, Wilson MA (1981) Humic soil and coal structure study with magic-angle spinning ^{13}C CP-NMR. Nature 289:275–276

Barron PF, Wilson MA, Stephens JF, Cornell BA, Tate KR (1980) Cross-polarization ^{13}C NMR spectroscopy of whole soils. Nature 286:585–587

Bazilevich NI (1974) Soil-forming role of substance and energy exchange in the soil-plant system. Trans 10th Int Congr Soil Sci (Moscow) 6:17–27

Burchill S, Hayes MHB, Greenland DJ (1981) Adsorption. In: Greenland DJ, Hayes MHB (eds) The Chemistry of Soil Processes. Wiley and Sons, Chichester, pp 221–400

Burges NA, Hurst HM, Walkden B (1964) The phenolic constituents of humic acid and their relation to the lignin of plant cover. Geochim Cosmochim Acta 28:1547–1554

Cameron RS, Thornton BK, Swift RS, Posner AM (1972) Molecular weight and shape of humic acid from sedimentation and diffusion measurements on fractionated extracts. J Soil Sci 23:394–408

Campbell CA, Paul EA, Rennie DA, McCallum KJ (1967) Applicabilities of the carbon-dating method of analysis to soil humus studies. Soil Sci 104:217–224

Chahal KS, Mortensen JL, Himes FL (1966) Decomposition products of carbon-14 labelled rye tissue in a peat profile. Soil Sci Soc Am Proc 30:217–220

Cheshire MV (1979) Nature and Origin of Carbohydrates in Soils. Academic Press, London New York

Cheshire NV, Cranwell PA, Falshaw CP, Floyd AJ, Haworth RD (1967) Humic acid. II. Structure of humic acids. Tetrahedron 23:1669–1682

Cheshire NV, Cranwell PA, Haworth RD (1968) Humic acid. III. Tetrahedron 24:5155–5167

Colclough P (1980) Degradation of humic acids with phenol and p-toluenesulphonic acid. Internal Report, Chemistry Dept, Univ Birmingham

Craggs JD (1972) Sodium sulphide reactions with humic acid and model compounds. Ph D Thesis, Univ Birmingham

Finch P, Hayes MHB, Stacey M (1971) The biochemistry of soil polysaccharides. In: McLaren AD, Skujins JJ (eds) Soil Biochemistry Vol 2. Dekker, New York, pp 254–319

Flaig W, Beutelspacher H, Rietz E (1975) Chemical composition and physical properties of humic substances. In: Gieseking JE (ed) Soil Components Vol 1. Springer, Berlin Heidelberg New York, pp 1–219

Grant D (1977) Chemical structure of humic substances. Nature 270:709–710

Greenland DJ, Oades JM (1975) Saccharides. In: Gieseking JE (ed) Soil components Vol 1. Springer, Berlin Heidelberg New York, pp 213–261

Griffith SM, Schnitzer M (1976) The alkaline curpic oxide oxidation of humic and fulvic acids extracted from tropical volcanic soils. Soil Sci 122:191–201

Haider K, Martin JP, Rietz E (1977) Decomposition in soil of ^{14}C-labelled coumaryl alcohols; Free and linked into dehydropolymer and plant lignins and model humic acids. Soil Sci Soc Am J 41:556–562

Hatcher PG, Maciel GE, Dennis LW (1981) Aliphatic structure of humic acids; A clue to their origin. Org Geochem 3:43–48

Havens JR, Koenig JL, Painter PC (1982) Chemical characterization of solid coals through magic angle ^{13}C nuclear magnetic resonance. Fuel 61:393–396

Hayes MHB (1960) Subsidence and humification in peats. Ph D Diss, Ohio State Univ

Hayes MHB (1980) The role of natural and synthetic polymers in stabilizing soil aggregates. In: Berkeley RJ et al. (eds) Microbial adhesion to surfaces. Horwood, Chichester, pp 263–296

Hayes MHB (1984) Structures of humic substances. In: Int Conf on Organic Matter and Rice, Los Banos 1982 IRRI (in press)

Hayes MHB, Mortensen JL (1963) Role of biological oxidation and organic matter solubilization in the subsidence of Rifle peat. Soil Sci Soc Am Proc 27:666–668

Hayes MHB, Swift RS (1978) The chemistry of soil organic colloids. In: Greenland DJ, Hayes MHB (eds) The Chemistry of Soil Constituents. Wiley and Sons, Chichester, pp 179–320

Hayes MHB, Swift RS (1981) Organic colloids and organo-mineral associations. Int Soc Soil Sci Bull 60:67–74

Hayes MHB, Stacey M, Swift RS (1975) Techniques for fractionating soil polysaccharides. Trans 10th Int Congr Soil Sci Moscow (1974) 12:75–81

Jackson MP, Swift RS, Posner AM, Knox JR (1972) Pholic degradation of humic acid. Soil Sci 14:75–78

Jenkinson DS (1981) The fate of plant and animal residues in soil. In: Greenland DJ, Hayes MHB (eds) The Chemistry of Soil Constituents. Wiley and Sons, New York, pp 505–561

Kononova MM (1966) Soil Organic Matter. Its nature, its role in soil formation and in soil fertility. Pergamon, New York London

Kononova MM (1975) Humus of virgin and cultivated soils. In: Gieseking JE (ed) Soil Components Vol 1. Springer, Berlin Heidelberg New York, pp 475–526

Kovda VA (1974) Biosphere, soils and their utilization. Report of the President of ISSS, 10th Int Congr Soil Sci, Moscow

Maciel EG, Sullivan MJ, Petrakis L, Grandy DW (1982) ^{13}C nuclear magnetic resonance characterization of coal materials by magic-angle spinning. Fuel 61:411–414

Malcolm RL (1981) International standard collection of humic and fulvic acids. Int Soc Soil Bull 60:30–31

Martin JP, Haider K (1976) Decomposition of specifically carbon-14-labelled ferulic acid: Free and linked into model humic-acid type polymers. Soil Sci Soc Am J 40:377–380

Martin JP, Haider K (1977) Decomposition in soil of specifically ^{14}C-labelled DHP and corn stalk lignins, model humic acid-type polymers, and coniferyl alcohols. Proc IAEA-FAO-Agrochem Symp Braunschweig 1976 2:23–32

Martin JP, Haider K, Saiz-Jimenez C (1974) Sodium amalgam reductive degradation of fungal and model phenolic polymers, soil humic acids and simple phenolic compounds. Soil Sci Soc Am Proc 38:760–765

Maximov OB, Shvets TV, Elkin YuN (1977) On permanganate oxidation of humic acids. Geoderma 19:63–78

O'Callaghan MR (1980) Some studies in soil chemistry. Ph D Thesis, Univ Birmingham

Olness A, Clapp CE (1975) Influence of polysaccharide structure on dextran adsorption by montmorillonite. Soil Biol Biochem 7:113–118

Piper TJ, Posner AM (1972) Sodium amalgam reduction of humic acid. I. Evaluation of the method, II. Application of the method. Soil Biochem 4:513–531

Schnitzer M (1974) Alkaline cupric oxide oxidation of a methylated fulvic acid. Soil Biol Biochem 6:1–6

Schnitzer M (1978) Humic substances: chemistry and reactions. In: Schnitzer M, Khan SU (eds) Soil Organic Matter. Elsevier/North-Holland, Amsterdam Oxford New York, pp 1–64

Schnitzer M (1982) Quo vadis soil organic matter research. Trans 12th Int Congr Soil Sci (New Delhi) 4:67–78

Schnitzer M, Khan SU (1972) Humic substances in the environment, Dekker, New York

Schwertmann U, Taylor RM (1981) The significance of oxides for the surface properties of soils and the usefulness of synthetic oxides as models for their study. Int Soc Soil Sci Bull 60:62–66

Stevenson FJ, Marks JD, Varner JW, Martin WP (1952) Electrophoretic and chromatographic investigations of clay adsorbed organic colloids. I. Preliminary investigations. Soil Sci Soc Am Proc 16:69–73

Stotzky G (1956) Carbon and nitrogen transformations during decomposition of muck soil as affected by addition of rye tissue. Ph D Diss, Ohio State Univ

Stotzky G, Mortensen JL (1958) Effect of addition level and maturity of rye tissue on the decomposition of a muck soil. Soil Sci Soc Am Proc 22:521–524

Swift RS (1968) Physico-chemical studies on soil organic matter. Ph D Thesis, Univ Birmingham

Swincer GD, Oades JM, Greenland DJ (1969) Extraction, characterization and significance of soil polysaccharides. Advan Agron 21:195–235

Theng BKG (1979) Formation and properties of clay-polymer complexes, Elsevier/North-Holland, Amsterdam Oxford New York

Turchenek LW, Oades JM (1979) Fractionation of organo-mineral complexes by sedimentation and density techniques. Geoderma 21:311–343

Waldron AC, Mortensen JL (1961) Soil nitrogen complexes. II. Electrophoretic separation of organic components. Soil Sci Soc Am Proc 25:29–32

Wilson MA (1981) Applications of nuclear magnetic resonance spectroscopy to the study of the structure of soil organic matter. J Soil Sci 32:167–186

Wilson MA, Barron PF, Goh KM (1981) Differences in structure of organic matter in two soils as demonstrated by ^{13}C cross polarisation nuclear magnetic resonance spectroscopy with magic angle spinning. Geoderma 26:323–327

Worobey BL, Webster GRB (1981) Indigenous ^{13}C-NMR structural features of soil humic substances. Nature 292:526–529

10. Behavior of Pesticides in the Unsaturated Zone Adsorption and Transport Phenomena

R. CALVET

10.1 Introduction

In cultivated areas, where pesticides are applied, the soil is frequently unsaturated. Thus, it is important to know the behavior of organic agrochemicals in the unsaturated zone because of their potential pollution hazards.

Many observations in the field have shown that the efficiency of a treatment is dependent on the hydration status of the soil, both before and after the pesticide application. Unfortunately, they have rarely been accompanied by measurements on adsorption, degradation, and transport, so that it is difficult to get a complete description of chemical behavior from reported results. Nevertheless, variations of biological effects of pesticides are partially attributed to adsorption and transport. This explains the interest in this field, however, one must emphasize the relatively small number of studies, owing largely to experimental difficulties which are encountered.

The purpose of this chapter is to discuss some aspects of adsorption and transport of organic pesticides. Although published results are referred to, this is not intended to be a review of the entire subject.

10.2 Adsorption

Effects of the water content of an adsorbent medium (soils or soil constituents) have not been extensively studied. Literature shows two kinds of experimental results, firstly, those concerning saturated media with high soil-solution ratios and, secondly, those concerning unsaturated media. The former have been obtained in order to define the most suitable conditions for adsorption measurements and they show that adsorption increases as the soil-solution ratio increases (Grover and Hance 1970, Goring and Hamaker 1972, Calvet and Chassin 1983). However, they are not strictly relevant to the subject and they will not be discussed further. On the contrary, the results of the latter will be analyzed in some detail.

10.2.1 General Considerations

In general, four mechanisms may be used to describe adsorption, taking into account water molecules. They are presented diagrammatically in Table 1, where relevant energy terms are also indicated. Mechanisms may be divided in two categories according to the nature of adsorption sites.

1. Pesticide molecules and water molecules have different adsorption sites. Two situations have to be distinguished:
- *Mechanism 1:* adsorption from the vapor on a dry or a partially hydrated surface, which implies very low water contents, the pesticide having high vapor tension.
- *Mechanism 2:* adsorption from a solution on a hydrated surface. This mechanism occurs during adsorption on hydrophobic sites (on organic matter) as it is the case for molecules, such as $BrCH_3$, Lindane, and DDT. However, it is also possible that other molecules (triazines, ureas, carbamates...) can interact with hydrophobic sites (Calvet et al. 1980).

Table 1. Adsorption mechanisms

Mechanisms	Interaction energies	Possible examples
Vapor (W, W, P) $+P\,(Vapor) \leftrightarrows S-P$	Surface–pesticide	$Br\ CH_3$ Lindane
Solution (W W, W W, W P, W W) $+P\,(Solution) \rightleftarrows S-P$	Surface–pesticide Water–pesticide	$Br\ CH_3$ Triazines Ureas
Vapor or solution (W, W, W, P) $-W+P\,(Vapor)\quad \rightleftarrows S-P+W\,(Vapor)$ $-W+P\,(Solution) \rightleftarrows S-P+W\,(Liquid)$	Adsorption from the vapor Surface–pesticide Surface–water Adsorption from the solution Surface–pesticide Surface–water Water–pesticide Water–water	Parathion Trifluralin Triazines Ureas
Solution (W, W, W W, P, W) $-W-W+P\,(Solution) \rightleftarrows S-W-P+W\,(Liquid)$	Adsorbed water–pesticide Water–pesticide Water–water	Triazines Ureas

S = adsorbent surface; W = water molecule; P = pesticide molecule

2. Pesticide molecules and water molecules have the same adsorption sites. In this case, adsorption from the vapor or the solution takes place by exchange between water molecules and pesticide molecules.
   - *Mechanism 3:* adsorption sites belong to the adsorbent surface.
   - *Mechanism 4:* adsorption sites are adsorbed water molecules.

Amounts of adsorbed pesticide depend on pesticide partial pressure (adsorption from vapor) or on pesticide concentrations (adsorption from solution). They also depend on the various interaction energies (given in Table 1) according to the operating mechanism.

Thus, on the basis of the above description, it appears that the water content of an adsorbent material can influence adsorption in several ways:

 - by modifying the partial pressure of water in the vapor phase or the pesticide concentration in the soil solution,
 - by determining the competition between pesticide molecules and water molecules, either for the approach of the adsorbent surface (mechanisms 1 and 2), or for the interaction with adsorption sites (mechanisms 3 and 4),
 - by modifying the thermodynamic activity of water which can change pesticide-water interactions (the solubility) and pesticide-water exchange (mechanisms 2–3).

This is probably negligible when the soil hydration is close to the saturation and when the water content variations are small. In this case, adsorption through mechanism 2 would not be affected by the water content.

Entropy effects have not been discussed because they are not yet clearly understood for pesticide adsorption.

10.2.2 Some Published Experimental Results

Effects of water content on adsorption are illustrated by several experimental results and the preceding analysis can be used as a general frame of discussion. Three cases will be considered: concentration effects, competition effects, and no effects.

Case 1: Concentration Effect

The effect of concentration was first proposed by Lambert (1966) to explain the influence of water content on adsorption of diuron. Later, this was also studied by Green and Obien (1969) and noted by Hance (1976) and Dao and Lavy (1978).

Using the notation of Green and Obien (1969), the pesticide concentration, C_w, in the soil solution can be expressed as:

$$C_w = \frac{Q}{m(W_e + K)},$$

where: $Q(\mu g)$ = total herbicide applied to the system
  $m(g)$ = total soil weight, oven-dry basis
  K = distribution coefficient
  $W_e(gg^{-1})$ = effective water content, oven-dry basis.

This expression is valid for linear, reversible adsorption. The first condition is frequently met in soil-pesticide systems, but the second is not, since adsorption isotherms are not single-valued owing to hysteresis phenomenon. Another point to discuss is the notion of effective water content, which is defined as the water available for dissolution of the pesticide and is considered as equal to the total water present in the medium. However, this is not necessarily the case, since a water content decrease can have two consequences:

- The ability of water molecules to accommodate pesticide molecules in solution can decrease because of the effect of water-surface interactions on water thermodynamic activity. This would lead to a decrease of solubility and, thus, a decrease of adsorption.
- The modification of pesticide properties: this is possible for molecules able to ionize by protonation. It is well-known that adsorbed water molecules are more ionized than free water molecules (Mortland et al. 1963) and it was shown that molecules, such as triazines can be protonated in soils or clays at low water contents (Cruz et al. 1968). Further, protonation increases the solubility and, thus, increases also the adsorption. This was observed by Calvet and Terce (1975) for atrazine-calcium montmorillonite systems.

As a result of this discussion, it appears that the definition and evaluation of the effective water content is not a simple matter and is probably dependent on the soil composition.

Concerning the concentration effect, one must note that all experimental results explained on this basis were obtained with media which are either saturated or unsaturated with higher water content. On the contrary, the problem remains to be solved at low water contents, because of the difficulties associated with experimental studies.

Case 2: Competition Effects

These are illustrated for mechanism 1, but they are only proposed as possible explanations for mechanisms 2–4.

Competition effects of water molecules are revealed by results concerning adsorption from a vapor phase. Examples are: adsorption of ethylene bromide on a soil (Call 1957) and adsorption of methyl bromide on a sandy soil (Arvieu 1982). As an illustration, Fig. 1 represents the variation of distribution coefficient K_d of $BrCH_3$ and of water vapor pressure as a function of soil water content. Other experiments are related to these situations and they concern adsorption from organic solutions. In order to study the effect of water content on parathion adsorption, Yaron and Saltzmann (1972) used hexane solutions. In the same way, Hance (1977) studied adsorption of atraton and monuron from 2,2,4-trimethyl pentane solutions. Figure 2 gives corresponding results and shows clearly the increase of the distribution coefficient as the water content decreases. Similar observations were also made by Van Bladel and Moreale (1977) for the adsorption of aniline on a soil from benzene solutions. However, the extrapolation of these experiments to field situations is questionable, because interaction energies associated with adsorption are different with organic solvents as compared to water.

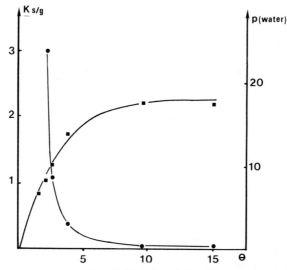

Fig. 1. Adsorption of $BrCH_3$ on a sandy soil. Variation of the solid-gas partition coefficient K s/g (●), the partial water pressure p (water) expressed in mm/Hg (■) against soil water content θ (weight basis) (Arvieu 1982)

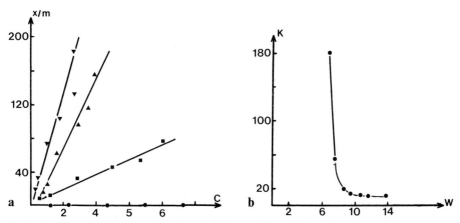

Fig. 2 a, b. Effect of water content on adsorption from solution. **a** Adsorption of parathion on a sandy clay loam from hexane at several relative humidities: 15% (▼), 32% (▲), 50% (■); x/m: $\mu g \cdot g^{-1}$ and C: $\mu g \cdot cm^{-3}$ (Yaron and Saltzman 1972). **b** Adsorption of atraton on Triangle soil. Variation of the distribution coefficient K against the water content θ (% – weight basis) (Hance 1977)

Case 3: No Effects

No effects can be observed in two situations: Firstly, when the soil water content is sufficiently high for both thermodynamic activitiy variations and concentration effects to be neglected. This can explain some results reported in the literature for atrazine by Goring and Hamaker (1972); secondly, when there a high affinity exists between the adsorbent surface and the pesticide molecule as in the case of mechanism 2. An example is given by the adsorption of methylbromide on soil organic matter (Arvieu 1982), which does not depend on the soil water content.

10.3 Transport Phenomena

The analysis of pesticide transport phenomena can be conveniently made by distinguishing:

– transport without flow of the soil solution: molecular diffussion only,
– transport with flow of the soil solution: convection-dispersion.

10.3.1 Transport with Flow of the Soil Solution

Numerous works have been done on hydrodynamic dispersion, essentially in saturated media, less frequently in unsaturated conditions. The description of the transport is not specific to pesticides and the theories and difficulties encountered are the same whatever the solute under investigation.

Variations of soil water content partially determine the size and the shape of flow pathways and the relative proportions of mobile and immobile liquid. This last effect has often been taken into account for explaining experimental results in saturated media. This is more difficult in unsaturated media, since the relationships between soil structure, mobility of water, and solute transport are not well described. Furthermore, the difficulties are increased by the fact that published experimental results do not provide a clear view on transport in unsaturated soils. An example of this situation is the effect of the soil water content on the distribution of chemicals during infiltration. Studying the transport of aldicarb Jamet et al. (1974) found that initial water content plays an important part; the greater it was, the smaller was the leaching. However, other papers report some results which do not show any effects of the soil hydration on transport. This is, for example, the case of 2,4-D in sand columns (Atef Elzeftawy et al. 1976) and for monuron in a sand soil (Upchurch and Pierce 1958). This may or may not be contradicting, since the behavior of unsaturated media is not yet completely understood for solute transport.

The influence of pore size distribution has been investigated in saturated media through its effect on the distribution of water between mobile and immobile phases and is now quite well-understood, particularly in saturated media. Nevertheless, pore dimensions are clearly shown to play a role in transport by observations, such as those of Dekkers and Barbera (1977), who studied the effect of aggregate size on leaching of metribuzin (the leaching being greater with small aggregates). Adsorption-desorption phenomena also have to be taken into account, but the same difficulties are encountered for both saturated or unsaturated media. This may be due to the noninstantaneous and nonlinear adsorption and to hysteresis of the desorption (Hornsby and Davidson 1973, Kay and Elrick 1967, Leistra and Dekkers 1976). Another interaction between the solid phase and solute, which has to be introduced, is anion exclusion. It is possible to deal with in saturated media, but is more difficult to describe in unsaturated conditions (Bresler 1973).

10.3.2 Molecular Diffusion

Molecular diffusion has a noticeable role in soil pesticide dynamics, although it concerns short distances. For pesticides in the vapor phase, molecular diffusion controls to a large degree losses by volatilization. For pesticides in solution, it determines the solute movements between immobile and mobile liquid, playing an important part in the chemical distribution in the soil profile. A detailed theoretical analysis of this process has been given by Letey and Farmer (1974).

Variation of the soil water content can influence the effective molecular diffusion in two ways:

1. by determining the air-filled porosity, thus controling the gas diffusion/liquid diffusion ratio for a given pesticide (given solubility and vapor tension values);
2. by modifying pesticide adsorption. The general trend is a decreasing adsorption as water content increases and this effect is probably greater at low water contents, because of the more pronounced competition between pesticide and water molecules.

The relationship between pesticide diffusion and water content are complex, since they result from combinations of properties of pesticide and soil. According to published results, several cases can be described (Fig. 3).

Case A. Low Vapor Tension. The pesticide is entirely in the soil solution so that diffusion occurs only in the liquid phase. Increasing water content increases the effective diffusion coefficient essentially through its effect on tortuosity (Nye

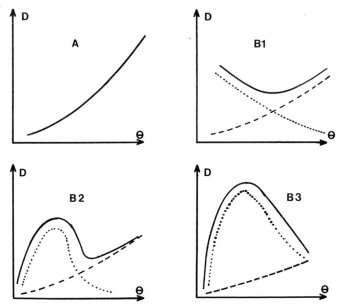

Fig. 3. Possible types of effects of the soil water content θ on the apparent diffusion coefficient of pesticides D. —— Variation of the apparent diffusion coefficient. – – – Variation of the diffusion coefficient in the solution phase. ...... Variation of the diffusion coefficient in the vapor phase

1979). The value of this geometric characteristic is reduced when pores are filled with water, because diffusion pathways become more rectilinear. Due to modifications of the water thermodynamic activity, an influence of adsorption on the diffusion is also possible, although there is no experimental information about this. Examples of case A, are diffusion in soil core of dimethoate (Graham-Bryce 1969), metribuzin (Scott and Paetzold 1978), prometone, atrazine, simazine, prometryne, chlorpropham and diphenamid (Scott and Phillips 1972).

Case B. The vapor tension is high enough to allow some pesticide to diffuse in the vapor phase. Since gas diffusion in soil is closely related to the geometric characteristics of the pore space, bulk density is a factor to be taken into account (Ehlers et al. 1969, Bode et al. 1973 a, b). Concerning adsorption-diffusion relationships, two situations can be described according to the magnitude of adsorption from the vapor.

B1. Low adsorption from the vapor phase. As water content decreases, effective diffusion in solution decreases (increasing tortuosity) and effective diffusion in the vapor phase increases, because the pore volume accessible to gas gets greater. If adsorption is low, this increase can be limited, but not completely ruled out. As a result, the apparent diffusion coefficient of the pesticide goes through a minimum value when water content is changing. An example of such behavior is reported by Graham Bryce (1969) for the diffusion of disulfoton.

B2, B3. High adsorption from the vapor phase. In this case, pesticide adsorption increases as soil water content decreases. Thus, the apparent (effective) diffusion coefficient in the vapor phase increases, to a maximum value, then decreases. Under these conditions, the resulting variation of the apparent diffusion coefficient depends on the relative magnitude of the diffusion coefficient in solution and in the vapor as is shown in Fig. 3. For lindane (Letey and Farmer 1974), the increase of solution phase diffusion corresponds to an increase of the apparent diffusion coefficient (B_2), while for trifluralin (B_3), it does not fully compensate the decrease of the vapor diffusion coefficient (Bode et al. 1973 a, b).

10.4 Conclusion

Diffusion of pesticides in unsaturated media has been described in some detail, although more information is needed on the effect of adsorption. Convection-dispersion of pesticides in unsaturated zones is certainly less understood since experiments are difficult and theoretical treatments are not yet complete. Directions in which progress could be made concern mainly adsorption effects and the relationships between transport, soil structure, and mobile and immobile liquid proportions.

It appears also that more work has to be done on adsorption of pesticides in unsaturated media. Furthermore, one must emphasize that desorption was not treated in this paper. The reason being the impossibility of finding any results on the effect of water content on this phenomenon. This is not at all surprising, because desorption is not well-understood even under water saturated conditions. So, there is probably a need for future researches in this field also.

In relation to the physicochemical processes, which control pesticide dynamics in soil, it is useful to note that a decrease of soil water content can lead to crystallization, because of the frequently low solubility of the chemicals. This was scarcely mentioned, although this phenomenon may sometimes be significant for the unsaturated zone pollution.

Attention has also to be drawn to the effect of decreasing water content on pesticide degradation a subject which has not been examined in this chapter.

References

Arvieu JC (1982) Etude de la dynamique du bromure de méthyle dans le sol. Rapport de fin d'étude. Convention de Recherches „Interactions Pesticides-Sol-Eau". Minist Environ 77:62

Atef Elzeftawy, Mansell RS, Selim HM (1976) Distributions of water and herbicide in lakeland sand during initial stages of infiltration. Soil Sci 122:297–307

Bode LE, Day CL, Gebhardt MR, Goring CE (1973a) Mechanism of trifluralin diffusion in silt loam soil. Weed Sci 21:480–484

Bode LE, Day CL, Gebhardt MR, Goring CE (1973b) Prediction of trifluralin diffusion coefficients. Weed Sci 21:485–489

Bresler E (1973) Anion exclusion and coupling effects in non steady transport through unsaturated soils. I.Theory. Soil Sci Soc Am J 37:663–669

Call F (1957) The mechanism of sorption of ethylene dibromide in moist soils. J Sci Food Agric 8:639–649

Calvet R, Chassin P (1983) Observations sur l'adsorption de l'atrazine par les sols. Ann Fac Ph Montpellier (to be published)

Calvet R, Terce M (1975) Solubilité de l'atrazine dans des gels de montmorillonite calcique. CR Acad Sci (Paris) 281:247–250

Calvet R, Terce M, Arvieu JC (1980) Adsorption des pesticides par les sols et leurs constituants. Ann Agron I. 31:33–62, II. 31:125–162, III. 31:239–257, IV. 31:385–411, V. 31:413–427

Cruz M, White JL, Russell JD (1968) Montmorillonite s-triazines interaction Isr J Chem 6:315–323

Dad TH, Lavy TL (1978) Atrazine adsorption on soil as influenced by temperature, Moisture content and electrolyte concentration. Weed Sci 26:303–308

Dekkers WA, Barbera F (1977) Effect of aggregate size on leaching of herbicide in soil columns. Weed Res 17:315–319

De Smedt F, Wierenga PJ (1979) Mass transfer in porous media with immobile water. J Hydrol 41:59–67

Ehlers W, Letey J, Spencer WF, Farmer WJ (1969) Lindane diffusion in soils. I. Theoretical considerations and mechanisms of movement. Soil Sci Soc Am Proc 35:501–508

Goring CAI, Hamaker JW (1972) Organic chemicals in the soil environment. Vol 1. Dekker, New York

Graham-Bryce IJ (1969) Diffusion of organophosphorus insecticides in soils. J Sci Food Agric 20:489–494

Green RE, Obien SR (1969) Herbicide equlibrium in soils in relation to soil water content. Weed Sci 17:514–519

Grover R, Hance RJ (1970) Effect of ratio of soil to water on adsorption of linuron and atrazine. Soil Sci 109:139–138

Hance RJ (1976) The effect of soil aggregate size and water content on herbicide concentration in soil water. Weed Res 16:317–321

Hance RJ (1977) The adsorption of atraton and momuron by soils at different water contents. Weed Res 17:137–201

Hornsby AG, Davidson JM (1973) Solution and adsorbed fluometuron concentration distribution in a water-saturated soil: experimental and predicted evaluation. Soil Sci Soc Am Proc 37:823–827

Jamet P, Piedallu MA, Hascoet M (1974) Migration et dégradation de l'aldicarbe dans différents types de sols. In: Comparative Studies of food and environmental Contamination. IAEA Vienna, pp 393–415

Kay BD, Elrick DE (1967) Adsorption and movement of lindane in soils. Soil Sci 104:314–322

Lambert SM (1966) The influence of soil-moisture content on herbicidal response. Weeds 14:273–275

Leistra M, Dekkers WA (1976) Computed effects of adsorption kinetics on pesticide movement in soils. J Soil Sci 28:340–350

Letey J, Farmer WJ (1974) Movement of pesticides in soil. In: Pesticides in soil and water. Soil Sci Soc Am, Madison, pp 67–94

Mortland MM, Fripiat JJ, Chaussidon J, Uytterhoeven J (1963) Interaction between ammonia and expanding lattices of montmorillonite and vermiculite. J Phys Chem 67:248–254

Nye PH (1979) Diffusion of ions and uncharged solutes in soils and soil clays. Adv Agron 31:225–272

Scott HD, Paetzold RF (1978) Effect of soil moisture on diffusion coefficients and activation energies of tritiated water, chloride and metribuzin. Soil Sci Soc Am Proc 42:23–27

Scott HD, Phillips RE (1972) Diffusion of selected herbicides in soil. Soil Sci Soc Am Proc 36:714–719

Upchurch RP, Pierce WC (1958) The leaching of monuron from Lakeland Sand Soil. Part II. The effect of soil temperature, organic matter, soil moisture and amount of herbicide. Weeds 6:24–35

Van Bladel R, Moreale A (1977) Aspects physico-chimiques de la pollution des sols: rôle et prédiction du phénomène d'adsorption. Pédologie XXVII 1:44–66

Yaron B, Saltzman S (1972) Influence of water and temperature on adsorption of parathion by soil. Soil Sci Soc Am J 36:536–583

11. Nonbiological Degradation of Pesticides in the Unsaturated Zone

S. SALTZMAN and U. MINGELGRIN

11.1 Introduction

Due to their increasing agricultural use, the fate of the enormous variety of pesticides which reach the unsaturated zone is a matter of general concern regarding environmental pollution hazard. The unsaturated zone may serve as a sink for soil-applied pesticides and a source from which pesticides and their metabolites can move into the water, atmosphere, and food chain.

The most important processes affecting the fate of pesticides in soils are adsorption-desorption, chemical reactions either in homogeneous or heterogeneous phases, and microbial metabolism. With a few exceptions (some chlorinated hydrocarbons, for example), microbial degradation of most pesticides in soils was demonstrated and extensively studied (Audus 1960, Helling et al. 1971, Kaufman 1974). Fewer studies were concerned with nonbiological degradation, which was shown to play an important role for a few groups of pesticides, mainly organophosphates and s-triazines (Konrad et al. 1969, Armstrong and Konrad 1974, Yaron and Saltzman 1978). Both chemical and microbial degradation pathways occur simultaneously and are affected by the same environmental factors, such as soil reaction, cation exchange capacitiy, organic matter content, moisture, temperature, and aeration. Hence, it is generally difficult to establish unequivocally the relative role of each of these degradation mechanisms in the fate of pesticides in the unsaturated zone.

One of the main features of this zone, which could affect degradation processes, is the high ratio of solid-liquid interface to homogeneous liquid phase. This makes surface reactions particularly important. Therefore, the present discussion will focus on some aspects of pesticides degradation in the heterogeneous phase, namely, surface-catalyzed reactions. This discussion is based both on studies carried out by our research group and on examples from the pertinent literature.

11.2 Properties of the Solid-Liquid Interface Affecting Pesticide Degradation

The solid interface of the unsaturated zone is formed by inorganic and organic surfaces, which are often associated in organic matter-clay complexes. The relative importance of available mineral and organic surfaces is variable in all the range between totally mineral and totally organic.

It is generally accepted that in the upper layer of most soils pesticide adsorption is related to a large extent to the organic matter content. On the other hand, the organic matter content is generally decreasing with depth so that its importance in the adsorption of pesticides in the unsaturated zone is more limited. The effect of organic surfaces on pesticide degradation is little understood; reactions involving free radicals as well as nonbiological degradation at such surfaces are supposed to occur (Weed and Weber 1974). However, organic surfaces were shown to be much less active than mineral ones, in inducing chemical degradation of pesticides (Mortland and Raman 1967). Thus, organic coating of mineral particles may actually hinder surface-catalyzed reactions by preventing the access of pesticide molecules to the more active clay mineral surfaces.

The relative importance of mineral surfaces in the unsaturated zone increases with depth. Clay-catalyzed reactions were frequently described and were shown to occur with several chemical groups of pesticides as organophosphates (e.g., Mingelgrin and Saltzman 1979), s-triazines (e. g, Russel et al. 1968 a), and chlorinated hydrocarbons (e. g., Fleck and Haller 1945). Among the main properties of the mineral surfaces relevant to their catalytic ability are: the source and quantity of their surface charge (which is determined predominantly by the type of the present clay minerals), the nature of the exchangeable cation, and the hydration status of the surface. The continuous change in moisture content and the often prevailing "dry" conditions, which are characteristic of the unsaturated zone, may affect both the phase distribution of pesticides and the properties of the solid-liquid interface.

As shown by Theng (1974), the catalytic activity of clays (which are probably dominant among the minerals present in the unsaturated zone in inducing surface reactions) is often correlated to their acid strength, that is, to their ability to act as proton donors (Bronsted acids) or electron acceptors (Lewis acids). The first type of acidity, which derives from the enhanced dissociation of water molecules at the vicinity of the surface (for example, those molecules coordinated with exchangeable cations), is strongly affected by the moisture content and the nature of the compensating cation. At sufficiently high moisture contents, the polarizing effect of the cations and of the surface in general becomes less important in determining the fate of many pesticides in the unsaturated zone. As the moisture content decreases, the polarization effect of the cation increases in significance and the dissociated or polarized water molecules become more important in determining the fate of pesticides. At moisture contents of a few percent, the pH at the solid-liquid interface near the dominant clay minerals present in the unsaturated zone may be several units lower than in the bulk solution (Theng 1974). At even lower moisture contents, the clay surfaces themselves may act as proton-donors (e. g., Russel et al. 1968 b).

Water molecules associated with exchangeable cations were shown to be the main active sites in the adsorption and conversion of many organic molecules (e. g., Bowman et al. 1970, Saltzman et al. 1974). A strong correlation between surface acidity of mineral surfaces and the surface-catalyzed degradation rate of several pesticides has been observed (Theng 1974).

11.3 Importance of Surface-Catalyzed Degradation Processes in the Unsaturated Zone

Chemical vs Microbiological Degradation. The microbiological degradation pathway of pesticides in soils is often considered the major degradation mechanism. Many of the studies which contributed to the formation of this opinion were based on comparisons between the degradation rate of pesticides in natural and sterilized soils. The decrease in the degradation rate following soil sterilization was generally considered a proof of the predominance of microbiological degradation. In many cases, the isolation of specific microorganisms was either not attempted or unsuccessful. On the other hand, several studies have demonstrated that the decrease in the degradation capacity of soils upon sterilization is not necessarily evidence for biological degradation, but it rather reflects the modification of chemical and physical properties of soils upon sterilization.

A relevant example was provided by the work of Kaufman et al. (1968) on soil degradation of 3-amino-1,2,4-triazole (amitrole), a herbicide that is considered stable to chemical degradation. Figure 1a shows the rate of amitrole degradation in a silty clay loam soil and the effect of different sterilization procedures on that rate. It was ascertained that all the sterilization procedures were effective. The degradation rate in the nonsterile soil was indeed the fastest (more than 60% degradation in 7 days). Although the rate decreased in the potassium azide and ethylene oxide treated samples, the degradation was still fast, about 40% and 25%, respectively, over the same period. Amitrole degradation was almost completely inhibited by autoclaving. This effect was irreversible, as reinoculation of autoclaved soil with microorganisms isolated from an amitrole treated soil had no significant effect on the rate of degradation (Fig. 1b). These results, as suggested by the authors (Kaufman et al. 1968), indicate that amitrole degradation in the soils tested was mainly a chemical process. Different sterilization procedures resulted in different changes in the soil properties, which affected differently amitrole degradation. Kaufman et al. (1968) demonstrated that amitrole degradation was greatly enhanced in the presence of several free radical generating systems.

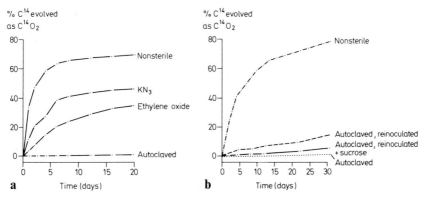

Fig. 1 a, b. The effect of soil sterilization on amitrole degradation. **a** The effect of different sterilization treatments; **b** the effect of reinnoculation of the autoclaved soil (Kaufman et al. 1968)

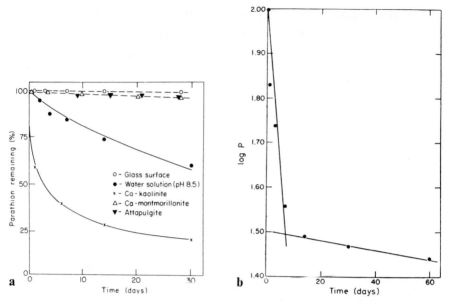

Fig. 2 a, b. The rate of parathion degradation. **a** as affected by mineral surfaces; **b** degradation kinetics on Ca-kaolinite surface. (*P* percent parathion remaining) (Yaron and Saltzman 1978)

The low moisture content often found in the unsaturated zone, the low organic matter content, and the limited gas exchange cause a reduction in microbial activity in the lower unsaturated zone, as compared to the surface soil layer. This in turn increases the importance of nonbiological degradation in the unsaturated zone.

The Rate of Chemical Conversion. The rate of chemical degradation of pesticides under conditions prevailing in the unsaturated zone is an important indication of the potential role of such reactions in the fate of pesticide residues there. Although a very large amount of work has been invested in the investigation of this topic, the reported results are often contradictory. This is due to the sensitivity of such reactions to even minor changes in the environmental conditions. For example, the insecticide 0,0-diethyl 0,p-nitrophenyl phosphorothioate (parathion) is generally considered as nonpersistent in soils (half-life of no more than a few weeks). However, higher persistence, up to 6–8 months and even for several years has been reported (Wolfe et al. 1973). The presence of some mineral surfaces, under "dry" conditions, may increase the degradation rate of parathion appreciably even as compared to parathion degradation in an alkaline aqueous solution. Other surfaces may be considerably less effective in promoting this degradation (Yaron and Saltzman 1978). As shown in Fig. 2 a, both Ca-montmorillonite and the attapulgite clays did not induce parathion degradation, while the degradation rate was significantly increased in parathion-Ca-kaolinite systems. After 5 days, for example, the fraction of parathion, which degraded, was about sixfold larger in the presence of kaolinite than in aqueous solutions of pH 8.5. Kaolinite-induced parathion degradation apparently proceeds in two stages with the first stage being very fast (see for example Fig. 2b, where 76% of the total amount of parathion disappeared after 7 days). Fast degradation rates promoted by clay surfaces were also reported for several other pesticides. The attapulgite clay for example, induced fast degradation of malathion, chlordane, heptachlor, diazinon, and ronnel (Gerstl and Yaron 1981).

These few examples demonstrate the fact that surface-catalyzed degradation of some pesticides could be fast enough to play an important role in the fate of pesticides in the unsaturated zone.

11.4 Factors Affecting Surface-Induced Pesticide Degradation

Most of the studies concerning pesticide-clay interactions suggest that the main factors determining the rate and mechanism of the surface-induced degradation of a given pesticide are the nature of the surface, the saturating cation, and the hydration status of the surface.

The Nature of the Clay Surface. Studies carried out with organophosphate pesticides-clay complexes showed that the degradation of these pesticides in the absence of a liquid phase is a surface reaction requiring specific adsorption sites (Saltzman et al. 1974, Mingelgrin et al. 1975, Mingelgrin et al. 1977). Despite the fact that the reaction sites were shown to be the same in all cases – the ligand water of the exchangeable cations   the catalytic effect of different clays was appreciably different, even if the saturating cations and hydration status were similar.

For example, the half-life of pirimiphos-ethyl (2-diethylamine-6-methyl-pirimidin-4-yl diethyl phosphorothionate), which is susceptible to alkaline hydrolysis, is 9 days in aqueous solutions at pH 10 and 47 °C. Under the same temperature conditions, the degradation rate of pirimiphos-ethyl adsorbed on an air-dried Na-bentonite was faster (half-life 5 days), but somewhat slower with the oven-dried clay (half-life 10 days), as compared to the basic aqueous solution (Fig. 3a). Nakaolinite had a much weaker catalytic effect. With this air-dried clay, the first degradation stage (which fits first-order reaction kinetics) had a half-life of 25 days and was even slower with the oven-dried clay (Fig. 3b). Parathion, on the other hand, decomposed at a fast rate on oven-dried Ca-kaolinite, while bentonite under the same conditions was almost inactive (Yaron and Saltzman 1978).

The nature of the clay surface could affect not only the degradation rate, but also the degradation mechanisms. Mingelgrin and Saltzman (1979) showed that under some conditions, the predominant degradation pathway of parathion was isomerization on bentonite surfaces and hydrolysis on kaolinite surfaces. Parathion isomerization was also observed with attapulgite clays. No isomerization occurred on an organic attapulgite, namely, attapulgite whose exchange complex was saturated with hexadecyltrimethylammonium ion (Gerstl and Yaron 1981a).

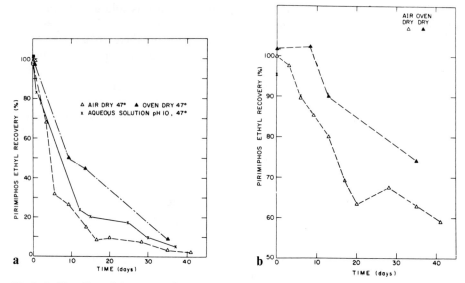

Fig. 3 a, b. The effect of the nature of clay surfaces on pirimiphos-ethyl degradation. a Na-bentonite, b Na-kaolinite (Mingelgrin et al. 1975)

These results may point out the important role of steric factors in the clay-catalyzed conversion of pesticides. Both the surface configuration of the adsorbent and the molecular geometry of the adsorbate are relevant; minor changes could considerably affect the conformation of the organic molecule on the surface and, hence, its chemical reactivity, even when the active site (e.g., the exchangeable cations) is identical (Mingelgrin and Saltzman 1979).

The Exchangeable Cation. Studies carried out with homoionic clays showed that the exchangeable cation could strongly affect the catalytic effect of clays on pesticide decomposition. The rate of dieldrin degradation by homoionic kaolinite, decreased in the order: $Al \cong H > Ca > Na$ (Theng 1974). Homoionic bentonite induced the degradation of ronnel (0,0-dimethyl-0-2,4,5-trichlorophenyl phosphorothioate) at room temperature, the order being $Al^{3+} > Fe^{3+} > H^+ > Mg^{2+} > Ca^{2+}$ (Fig. 4 from Rosenfield and Van Valkenburg 1965). In both cases, the order of the degradation rate could be explained by the differences in the hydration energy of the cations, the catalytic effect being generally stronger with the more polarizing cations.

The rate of parathion degradation on homoionic kaolinite surfaces increased in the order $Ca > Na > Al$ (Saltzman et al. 1974). This suggests that other effects, apart from the extent of polarization of the hydration water, could be of importance in determining the degradation rate. The relatively low catalytic activity of Al-kaolinite in the above example, may be due to the small radius of this cation, which may force the adsorbed molecule into an unfavorable conformation and hinder degradation. It was also suggested that one of the degradation products was strongly adsorbed by Al. Under conditions where desorption could have been

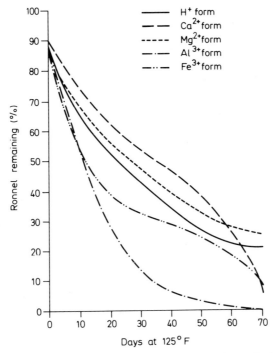

Fig. 4. The effect of the exchangeable cation on the degradation of ronnel adsorbed on bentonite (Rosenfield and Van Valkenburg 1965)

significant, the number of active sites available would thus be reduced (Saltzman et al. 1974).

The Hydration Water. As was already mentioned, one of the main features of the unsaturated zone relevant to pesticide degradation is the fact that the moisture content varies within the range in which changes may bring about significant modifications in the polarization and dissociation of the water and, thus, induce changes in the rates of possible surface reactions. At high moisture content, free water, which is present in the soil or sediments pores, may decrease the importance of surface-catalyzed reactions. Competition of water molecules with adsorbate molecules on the adsorption sites and partition of the adsorbate into the aqueous phase are among the factors which bring about this decrease. As the moisture content decreases, most of the water present is adsorbed water held on the solid surfaces. The structure and properties of the resulting interface depend on the interactions between the adsorbed water and the surface. The exchangeable cation is a major factor affecting these interactions.

An example of the strong moisture dependence of a surface-catalyzed reaction is provided by parathion hydrolysis on kaolinite at different hydration levels (Fig. 5 from Saltzman et al. 1976). The relationship between the degradation rate and the moisture content was qualitavely similar for the various homoionic kaolinites. The curves obtained show several inflection points, which probably

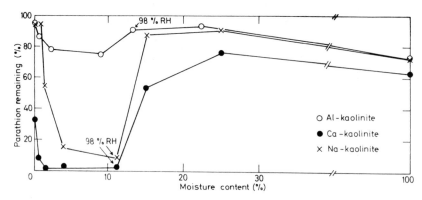

Fig. 5. Parathion hydrolysis on kaolinite, as affected by the clay moisture content (Saltzman et al. 1976)

correspond to qualitative changes in the state of the adsorbed water. With both Na- and Ca-saturated kaolinites, a sharp increase in the degradation rate was obtained by increasing the hydration level up to a moisture content equivalent to the upper limit of bound water (in equilibrium with 98% relative humidity). A sligh increase in moisture content above this value resulted in a steep decrease in the degradation rate. This effect was stronger with Na- and Ca-kaolinites and less pronounced with the Al-saturated clay.

11.5 Conclusion

This chapter has dealt with some aspects of pesticide degradation in the unsaturated zone. It has demonstrated the fact that a precise understanding of the nature of pesticide reactions at the solid-liquid interface is difficult due to the complexity of the solid surfaces and their interaction with water, as well as to the large variety of pesticides reaching this zone. The examples presented indicate, however, that surface-induced degradation could play a dominant role in the fate of pesticides in the unsaturated zone.

Acknowledgements. The authors wish to thank the United States-Israel Binational Agricultural Research and Development Fund (BARD) for its support of part of the research reported in this chapter.

References

Armstrong DE, Konrad JG (1974) Nonbiological degradation of pesticides. In: Guenzi WD (ed) Pesticides in soil and water. Soil Sci Soc Am Madison, Wisc, pp 123–131

Audus LJ (1960) Microbiological breakdown of herbicides in soils. In: Woodford EK, Sager GR (eds) Herbicides and the Soil. Blackwell, Oxford, pp 1–8

Bowman BT, Adams RS Jr, Fenton SW (1970) Effect of water upon malathion adsorption onto five montmorillonite systems. J Agric Food Chem 18:723–727

Fleck EE, Haller HL (1945) Compatibility of DDT with insecticides, fungicides and fertilizers. Ind Eng Chem 37:403–405

Gerstl Z, Yaron B (1981 a) Attapulgite-pesticide interactions. Residue Rev 78:69–99

Gerstl Z, Yaron B (1981 b) Stability of parathion on attapulgite as affected by structural and hydration changes. Clays and Clay Minerals 29:53–59

Helling CS, Kearney PC, Alexander M (1971) Behavior of pesticides in soils. Advan Agron 23:147–240

Kaufman DD (1974) Degradation of pesticides by soil microorganisms. In: Guenzi WD (ed) Pesticides in soil and water. Soil Sci Soc Am Madison, Wisc, pp 133–202

Kaufman DD, Plimmer JR, Kearney PC, Blake J, Guardia FS (1968) Chemical versus microbial decomposition of amitrole in soil. Weed Sci 16:266–272

Konrad JG, Chesters G, Armstrong DE (1969) Soil degradation of malathion, a phosphorodithioate. Soil Sci Soc Am Proc 33:259–262

Mingelgrin U, Saltzman S (1979) Surface reactions of parathion on clays. Clays and Clay Minerals 27:72–78

Mingelgrin U, Gerstl Z, Yaron B (1975) Pirimiphos ethyl-clay surface interactions. Soil Sci Soc Am Proc 39:834–837

Mingelgrin U, Saltzman S, Yaron B (1977) A possible model for the surface-induced hydrolysis of organophosphorus pesticides on kaolinite clays. Soil Sci Soc Am J 41:519–523

Mortland MM, Raman KV (1967) Catalytic hydrolysis of some organic phosphate pesticides by copper (II). J Agric Food Chem 15:163–167

Rosenfield C, Van Valkenburg (1965) Decomposition of (0,0-dimethyl-0-2,4,5-trichlorophenyl) phosphorothioate (ronnel) adsorbed on bentonite and other clays. J Agric Food Chem 13:68–72

Russell JD, Cruz M, White JL (1968 a) Mode of chemical degradation of s-triazines by montmorillonite. Science 160:1340–1342

Russell JD, Cruz M, White JL (1968 b) The adsorption of 3-aminotriazole by montmorillonite. J Agric Food Chem 16:21–24

Saltzman S, Yaron B, Mingelgrin U (1974) The surface catalized hydrolysis of parathion on kaolinite. Soil Sci Soc Am Proc 38:231–234

Saltzman S, Mingelgrin U, Yaron B (1976) The role of water in the hydrolysis of parathion and methylparathion on kaolinite. J Agric Food Chem 24:739–743

Theng BKG (1974) The Chemistry of Clay-Organic Reactions. Hilger, London, pp 136–206, 261–281

Weed SB, Weber JB (1974) Pesticide-organic matter interactions. In: Guenzi WD (ed) Pesticides in soil and water. Soil Sci Soc Am Madison, Wisc, pp 39–66

Wolfe HR, Staiff DC, Armstrong JF, Comer SW (1973) Persistence of parathion in soil. Bull Environ Contam Toxicol 10:1–9

Yaron B, Saltzman S (1978) Soil-parathion surface interactions. Residue Rev 69:1–34

12. The Effect of Solid Organic Components of Sewage on Some Properties of the Unsaturated Zones

B. YARON, A. J. VINTEN, P. FINE, L. METZGER, and U. MINGELGRIN

12.1 Introduction

Application of organic solid wastes to soils occurs during land disposal of waste waters and sludges. While waste waters are characterized by a relatively low level of solids ($< 1\%$), sewage sludges and manures contain higher levels of solids, sometimes up to 90%. Highly toxic organic wastes may also come into contact with the soil when sealed off into waste disposal sites or as a result of accidental spills. These wastes will not be dealt with in any detail in this paper. The effects of solid organic wastes on the properties of the unsaturated zone depend on their persistence, distribution with depth, and on their chemical and physical interaction with the soil solid phase. The effects on soil properties will also depend on the solids composition, rate of solids disposal, and the management of the waste disposal process.

Waste waters represent a water resource, especially in arid areas, and are used extensively for irrigation. Disposal by large scale recharge of aquifers through the soil (Idelovitch, this vol.; Bouwer, this vol.), or in septic tanks is also practiced. Waste waters also contact the soil during storage in waste water ponds. Sewage sludges are applied to land by spreading as liquid or solid material. Often the applied sludges are incorporated into the soil. These sludges represent a valuable nutrient source and can improve soil structural and physical properties. The effect of organic solids in the wastes on the bulk of the soil is likely to be more pronounced for sewage sludges because larger amounts of solids are applied. Yet the effect on the properties of the soil surface (particularly infiltration rate) may be more pronounced in the case of the waste waters because of the frequent application on the soil, which may induce clogging of the soil's surface. The effect of organic wastes on soil infiltration properties are, therefore, discussed primarily with reference to waste waters, while the effect on bulk soil properties are discussed mainly with reference to sewage sludges. Figure 1 presents a schematic distribution of waste-borne solids in the unsaturated zone.

The soil is the upper layer of the unsaturated zone. Effects of sludges on soil physical properties have been recently reviewed by Khaleel et al. (1981). Application was shown to alter bulk density, water-holding capacity, and hydraulic conductivity. The clogging of soils by application of waste-waters has been studied, among others, by Thomas et al. (1966) and Daniel and Bouma (1974). The chemistry of organic wastes in the unsaturated zone (e.g., carbon content, nitrogen and phosphate balance, and heavy metals concentrations) have been reviewed in recent publications (Loehr 1977, L'Hermite and Ott 1981).

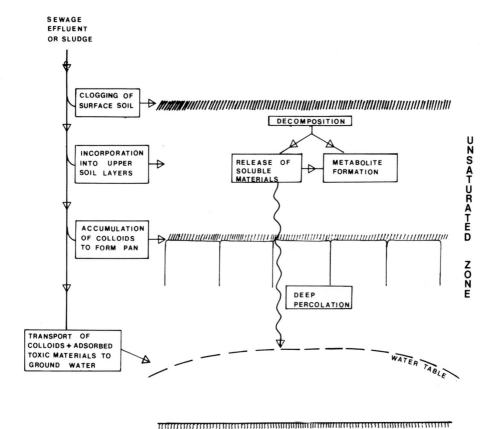

Fig. 1. Distribution of waste-borne solids and decomposition products in the unsaturated zone

There is much data in the literature concerning the persistence of organic solid components of sewage in various media and under various environmental conditions, although little relates specifically to sewage water. Gilmour and Gilmour (1980) developed a computer model simulating sludge decomposition and found fair agreement with experimental data of Terry et al. (1979). The model assumes a rapid decomposition phase followed by a slow one, both described by first-order kinetics. The decomposition rate of the rapidly decomposable fraction is strongly influenced by temperature. The effect of temperature on the slow decomposition stage is difficult to asses as the half-life in the systems used to establish the parameters for the above model varied from 32–56 months so that the effect of the temperature was integrated over several years. Schaumberg et al. (1980) demonstrated the importance of pH and water content in determining the rate of decomposition of sewage sludge in soil by performing IR analysis of mixtures incubated over a period of 100 weeks. IR bands corresponding to polysaccharides, proteinaceous material, sulfonate or sulfate compounds disappeared from the saturation extracts of the soil samples incubated at $^{1}/_{3}$ bar moisture tension after about 10 weeks of incubation. Bands corresponding to carbohydrate and nitrate groups emerged during this period. These changes were much slower under saturated conditions as well as in an acid environment.

Redistribution of organic solids from the site of application, which is the upper part of the unsaturated zone, may occur either by transport with percolating water to the water table or with runoff water. These processes constitute potential mechanisms for pollution with pesticides, pathogenic microorganisms, heavy metals, or other toxic compounds, which may reach the aquifer or surface water bodies adsorbed on the suspended solids.

The aim of this paper is to discuss the effects of solids in organic wastes on the properties of the unsaturated zone. New as well as previously published data will be used to illustrate the effects discussed.

12.2 Composition of Organic Solid Components of Sewage

In order to understand the effect organic solid wastes on the unsaturated zone, information on their composition is required. Table 1 gives a characteristic range of concentrations of organic and inorganic solid components in effluents and

Table 1. Some properties of selected waste effluents and sludges. (After Hunter and Kotalik 1974, McCalla et al. 1977, Kladivko and Nelson 1979, Metzger 1983)

	Waste effluents	Sludge solids
pH	5.3 –11.7	
EC (mmho cm^{-1})	1.0 – 2.2	7.1– 14.0 (SP)[a]
CEC (mEq/100 g)	–	42 –250
Total solids (% w/v)	0.03 – 0.9	<90
Total suspended solids (%)	0.01 – 0.7	
Total volatile solids (%)	0.008– 0.6	<80
Fixed suspended solids (%)	0.001– 0.1	<20

[a] SP: Saturated paste

Table 2. Chemical fractionation of organic solids from waste effluents and sludges (%). (After Manka et al. 1974, Levi-Minzi et al. 1980, Metzger 1983)

	Waste effluent in solids	Sludge solids
Ether extractable	3.7–16.3	10.0 –19.8
Resins	1	2.85– 8.20
Water soluble polysacharides		10.0 –14.4
Hemicellulose		4.0 –10.0
Cellulose		3.2 –12.0
Protein (crude)	21 –23	8 –44
MBAS	12.2–16.6	
Lignin humus	1.0– 2.1	11 –27
Fulvic acid	24 –26.6	
Humic acid	6.1–14.7	
Hymatomelanic acid	4.8– 7.7	

Table 3. Chemical characteristics of selected liquid sludges and waste effluents. (After Neal 1964, Kardos and Sopper 1974, Sommers 1977, Thomas and Law 1977, Libhaber et al. 1979, Dan Region SRP 1982, Fine 1983)

	mg l^{-1}	% dry matter
Total phosphorus	0.5 – 3.83	1.8 – 2.3
Total sulfur	0.44 – 0.66	1.1
Major constituents	mEq l^{-1}	
K^+	0.2 – 2.6	0.27–0.3
Na^+	1.7 – 9.7	0.12–0.27
Ca^{2+}	0.03 – 3.2	3.8 – 5.31
Mg^{2+}	0.04 – 2.8	0.45–0.46
Cl^-	1.3 – 9.3	
HCO_3^-	2.8 –10	
Selected trace elements	mg l^{-1}	mg kg^{-1} dry matter
Fe	0.1 – 830	8,000–11,000
Al	<360	<4,000
Pb	0.02 – 100	652
Zn	0.02 –1,990	1,740 – 1,800
Cu	0.02 – 210	< 850
Ni	0.02 – 100	82 – 190
Cd	0.005– 11	16 – 20
Cr	0.01 – 100	890 – 906
Mn	0.15 – 128	< 200

sludges and some relevant properties of the corresponding wastes. In both the effluents and the sludges, the total volatile organic solids constitute the major component of the material disposed. The large range in pH, EC, and CEC demonstrate the variability of the waste composition. The content of some groups of organic compounds in the solid fraction of selected effluents and sludges is given in Table 2. Table 3 presents data on the range of concentrations of some elements in the solid components of representative effluents and liquid sludges, which suggest both their beneficial potential and their pollution hazard for the unsaturated zone.

12.3 Physical Effects of Organic Solids

The composition of the organic solids in sewage effluents in sludges are similar. However, the methods and rates of organic solid application to the unsaturated zone differ. Clogging of the surface soil layers is, therefore, discussed with reference to application of waste waters (by irrigation, by flooding of recharge ponds, etc.), while effects on the soil bulk properties are discussed with references to organic solids applied as sewage sludge, where much larger amounts are added and are generally well incorporated into the soil.

12.3.1 Waste Waters

Waste waters contain up to 1% w/w solids in suspension. These solids are primarily organic, but contain insoluble inorganic material as well (see Table 1). The major soil physical parameter which is altered by the application of waste waters is the hydraulic conductivity (HC) of the surface soil. Consequently, the infiltrability of the soil declines continuously following the application of effluents even under saturated conditions. It will be demonstrated below that the suspended solids (SS) may be the main component responsible for reduction in HC. Figure 2 shows infiltration data from an experiment by Thomas and co-workers (1966). Effluent (SS 155 mg l^{-1}, BOD 393 mg l^{-1}) was applied at an infiltration rate of 2.05 cm day^{-1} to 11 sand-filled lysimeters. A very pronounced reduction in the infiltration rate was observed up to a limiting value (about 0.1 cm h^{-1}; Fig. 2), which is lower by more than two orders of magnitude than the initial infiltration rate. The gradual reduction in infiltration rate upon application of effluents for extended periods was also observed by others (Jones and Taylor 1965, Chang et al. 1974, Okubo et al. 1979), Piezometric data (Jones and Taylor 1965), visual observation, or sectioning of columns (Jones and Taylor 1965, Chang et al. 1974, Vinten et al. 1983) show that the major head reduction in the columns occurs in the top few centimeters, demonstrating that clogging of the surface zone occurs as a result of effluent application.

Clogging Mechanism. The dominant mechanism of clogging depends on the composition of the waste, the physical and chemical properties of the soil, the method of application of waste water, and the time scale involved. Chemical clog-

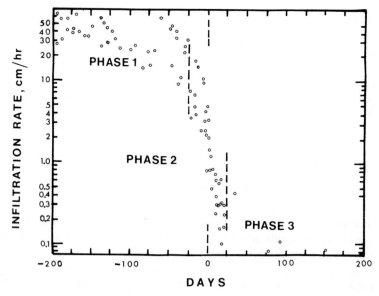

Fig. 2. Reduction in infiltration rate as affected by waste water (SS = 155 mg l^{-1}) application (Thomas et al. 1966)

ging may occur, for example, as a result of precipitation of iron phosphate (Thomas et al. 1966) or ferrous sulfide (Laak 1970), but this is probably of minor importance in a time scale shorter than decades. High sodium contents in the water may also have a deleterious effect on soil physical properties. In the short term, physical clogging of pores by suspended solids (SS) in the upper part of the soil to form a surface mat is a major mechanism of hydraulic conductivity decrease. Accumulation of finer suspended solids in pores below the soil surface as the suspension moves through the soil also occurs. This depth filtration process leads to reduction in soil HC, but at a slower rate than clogging of surface pores. The effect of colloidal suspensions, which are readily leached, needs also to be considered as their accumulation at depth (at the wetting front or at a zone of finer texture) may cause pan formation in the long run. The experiments of Vinten et al. (1983) show the relative deposition of solids in the top 50 mm following leaching with about 100 cm of waste water (98 mg l^{-1} SS), in three different soils (Fig. 3). In one of them (silt loam), the water was filtered before use, giving a SS concentration of 38 mg l^{-1}. There is a strong influence of the soil type on the rate of clogging of the soil, the relative saturated HC of the sandy soil declining much less rapidly than that of silt loam despite the filtering of the effluent before application to the silt loam. This last work demonstrates the importance of surface clogging by the SS of the effluent in the reduction of the hydraulic conductivity of the soils.

Figure 4 shows SEM photographs of the top sections of sandy loam and silt loam columns treated with effluent containing 98 mg l^{-1} suspended solids. The picture of the sandy loam surface after leaching with 1,000 mm of unfiltered effluent shows an accumulation of solids at the surface (Fig. 4 A). A coarse matrix of deposited solids occurs at the surface, filling spaces between the soil aggregates (see for example the three aggregates forming a triangle in the center of the picture). However, the matrix is quite porous and holes and cracks of the order of 10 μm occur regularly. In the silt loam, a uniform surface mat occurs and the holes in it are much smaller than in the sandy loam (Fig. 4 B). This difference in the behavior of the two soils may explain the clogging process. In Fig. 4 C, a longitudinal view of the same top section of the silt loam is shown. A well-defined layer of deposited solids of about 60 μm thickness overlaying the soil matrix is clearly shown. A similar distinct boundary between surface mat and soil is absent in the case of the sandy loam. Swartzendruber and Uebler (1982) developed theoretical "clogging coefficients", which characterize the physical clogging susceptibility of a system, based on physical parameters of the soil and the sewage water, but they did not test the theory satisfactorily for sewage water.

Piezometric data show that as this surface mat develops the hydraulic head gradient across it increases and the degree of unsaturation of the soil beneath increases. This may contribute to increased mat resistance by causing compaction of the mat under the pressure gradient (Rice 1974). The final result is the rapid decrease in infiltration, which is illustrated in Fig. 3 and may also be reflected in the results of Thomas et al. (1966, see Fig. 2).

Where infiltration occurs for extended periods, as in Fig. 2, biological clogging can also be important. The accumulation of a microbial population, which uses deposited suspended solids (SS) and dissolved organics in the water as substrate, has been demonstrated by Avnimelech and Nevo (1964) and Frankenberger et al.

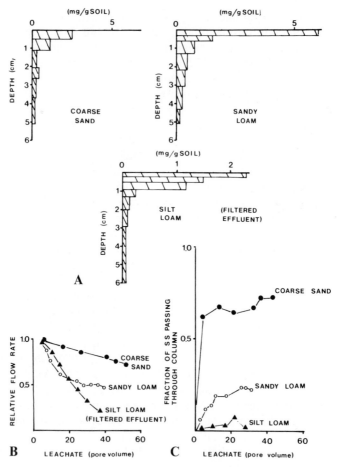

Fig. 3A–C. Deposited solids distribution (**A**) in 3 soils leached with waste water (98 mg l^{-1}): unaltered effluent; 38 mg l^{-1} SS: filtered effluent). The effect on the soil hydraulic conductivity (**B**) and suspended solids transport through the soils (**C**) (Vinten et al. 1983)

(1979). If anaerobic conditions develop, this clogging is greatly accelerated (Kristiansen 1981). This may also be a reason for the rapid sealing effect observed in the case of the continuous flow for long periods (Fig. 2). Clogging rates are greatly reduced by repeated wetting and drying cycles as compared to systems under continuous inundation (Jones and Taylor 1965, De Vries 1972) as the physical mat is disrupted by drying and by maintenance of aerobic conditions. Bacterial polysaccharide production can also cause high clogging rates (Avnimelech and Nevo 1964, Chang et al. 1974), probably by acting as binding material for microbial cells and debris. Some microbial excretions may adsorb on the surfaces of the capillary pores reducing the contact angle and consequently the HC. The rate of growth of bacteria, production of polysaccharides, and hence, biological clogging, depends strongly on the C/N ratio of the SS and organic solutes in the sewage water (Avnimelech and Nevo 1964).

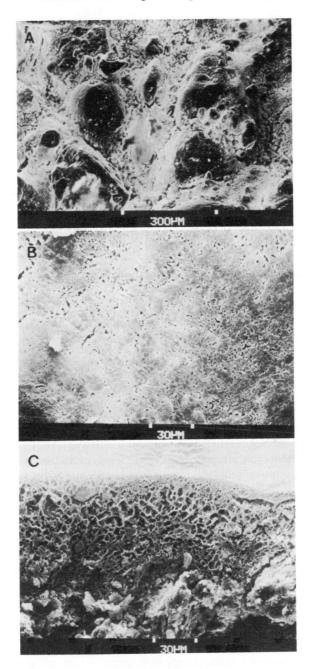

Fig. 4 A–C. Scanning electron micrographs of soils after leaching with effluent. **A** Surface of sandy loam after 1,000 mm of unfiltered effluent; **B** surface of silt loam soil after 700 mm of filtered effluent; **C** vertical section of **B**

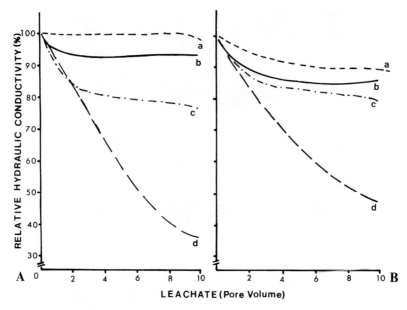

Fig. 5 A, B. The effect of leaching with waste water (50 mg l^{-1} SS) after different degrees of filtration on the relative hydraulic conductivity of sandy loam (**A**) and loamy sand (**B**). *a* model solution, *b* effluent filtered through a 0.45 μm millipore filter, *c* effluent filtered through a Whatman No. 91 filter paper, *d* unfiltered effluent. Initial flow rate. (Metzger et al. 1983)

Reduction in relative saturated hydraulic conductivity (RHC) occurring on irrigation with sewage water also depends on the quality of the water applied. Figure 5 shows the reduction in relative hydraulic conductivity (RHC) during the saturated flow of effluents containing four different levels of SS through a loamy sand and sandy loam. The unfiltered effluent (50 mg l^{-1} SS, pH 8) caused a steady and continuous decline in RHC. Filtering the effluent through a Whatman No. 91 filter paper (effluent c) greatly decreased the extent of the reduction in RHC and a steady value of flow rate is reached more rapidly. Millipore filtered effluent b causes yet a smaller reduction in HC. There is probably little clogging of the surfaces with effluents b and c and the reduction in HC is due mainly to depth filtration in the soil matrix reducing the pore cross section available for water flow.

Clogging of the surface layer of the soil has several important consequences when considering irrigation with sewage water or groundwater recharge. Firstly, the reduced infiltration rate increases the time required for application and the chance of runoff, leading to poor water distribution and possible contamination of surface water if the infiltrability of the soil is low initially. Secondly, the water content in the transmission zone during infiltration is controlled by the conductance of the clogged zone (similarly to crust-topped soils, Hillel and Gardner 1969). Thus, the depth of wetting during infiltration of a constant volume of water increases as the conductance of the clogged zone decreases. This may lead to increased potential for groundwater pollution (though after redistribution differences may be small) and to altered water availability to plants. Development of the surface mat in recharge systems, if the decrease in conductance is sufficient, could lead to unsaturated conditions beneath the clogged zone and unstability of the wetting fronts (Hillel 1980). Such surface clogging may also reduce the maximum rate of discharge per unit area of the discharge field.

During groundwater recharge with sewage effluents, potential hazards exist which are related to solids present in sewage effluent. Column experiments with sewage effluents do not, unfortunately, regularly report SS concentrations in leachates, partly because of the problem of distinguishing effluent SS from those released by the soil. ^{14}C-DDT labeling of organic SS (Vinten et al. 1983) allows the transport of organic SS in soils to be monitored. Figure 3 C shows that 70% of the input effluent SS concentration remains in the leachate at 5 cm depth in the sand and 25% in the sandy loam when columns were leached with effluent (98 mg l^{-1}). Jones and Taylor (1965) found 25% of the input suspension SS concentration in leachate from a 25 cm (10 inc.) column of sand. Baillod et al. (1977) in an evaluation of a sewage water rapid infiltration site, found evidence of suspended solids in well water.

These data suggest the possibility that suspended solids may be transported with drainage water or accumulate in a layer in the soil profile below the surface, which is either of finer texture or is at the leaching depth. Such accumulation may cause pan formation. This could subsequently lead to impeded drainage and eventual failure of recharge systems. It should be noted that the long-term hazard of accumulation is mainly associated with the inorganic solids fraction present in the waste.

A second potential hazard of transport of toxic materials adsorbed on the suspended solids also exists. This process will be discussed in more detail in a later section.

12.3.2 Sewage Sludges

The effect of incorporation of sewage sludge on soil physical properties has been studied in terms of the following physical parameters: porosity and bulk density, water-holding capacity and available water, infiltration rate and hydraulic conductivity, and susceptibility to erosion.

Porosity and Bulk Density. Table 4 shows results of recent research done by Pagliai et al. (1981) on the effect of manure on the total porosity and pore-size distribution. Four months after application of 50 t ha^{-1} organic manure to a

Table 4. The effect of addition of manure on sandy loam soil porosity (%). (After Pagliai et al. 1981)

Time after application	Control	50 t ha^{-1} manure	150 t ha^{-1} manure
4 months	18.4	25.4–30.5	27.1–30.6
12 months	17.5	24.3–31.4	24.4–33.6

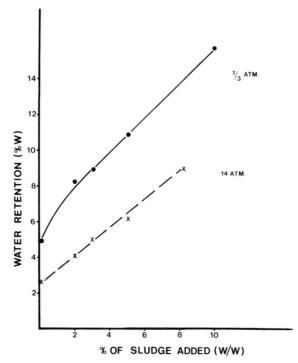

Fig. 6. Effect of sludge incorporation on water-holding capacity of a sandy soil at $^1/_3$ bar and 14 bars
soil moisture tension

sandy loam soil, an increase in total porosity to 25%–31% occured. Further decomposition of organic matter did not cause the total porosity to reduce by much.
One year after application, the increase in porosity as compared to the control
was about the same as after 1 month. The addition of sewage sludge reduced the
proportion of fissures (>0.5 mm according to the Greenland 1977 classification)
and increased the proportion of transmission (0.5 mm 50 µm) and storage pores
(<0.5 µm). Bulk density decreases upon addition of organic waste due to the dilution effect of mixing low density organic matter with the more dense mineral
fraction of the soil and because of the increase in porosity mentioned above.

Water-Holding Capacity and Available Water. Addition of organic solids to
soil has been shown to increase water-holding capacity (WHC) by Gupta et al.
(1977) for a coarse sand, by Klute and Jacob (1949), and by Unger and Stewart
(1974). Figure 6 shows the effect of sludge addition to a sandy soil (Metzger
1983). There is a large increase in water retention at $^1/_3$ bar moisture tension (field
capacity) upon sludge addition. However, most of this increase is in water retained even at 14 bar tension (wilting point). Thus, the available water (AW),
which is the difference between the two lines in Fig. 6, does not increase greatly,
yet some increase in the available water was observed. Gupta et al. (1977) found
no appreciable change in the amount of available water upon addition of up to

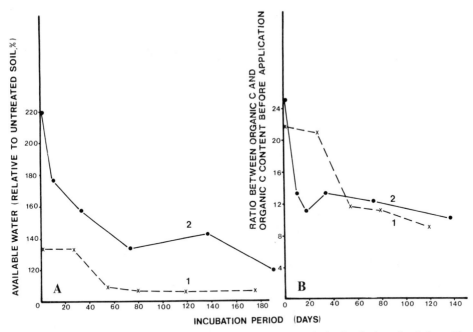

Fig. 7 A, B. Changes in available water (**A**) and organic carbon (**B**) during incubation of a silt loam (*1* Epstein 1975) and sandy soil (*2* Metzger 1983) after application of sludge

450 t ha^{-1} anaerobic sludge. Khaleel et al. (1981) analyzed experimental results in a broad range of soils and various organic wastes by using an exponential multiple regression analysis. They stated that WHC increases at both field capacity and wilting point. For fine-textured soils, the increase in WHC at field capacity is greater than at wilting point; for coarse-textured soils WHC at wilting point increases more than at field capacity. On the other hand, the tabulated data presented by Khaleel et al. (1981) do not support their above-stated conclusion. The effect of addition of sewage sludge on available water is not yet fully understood. The increase in WHC may be primarily due to increased surface area of the soil when organic matter is added. At 14 bars, most of the water will be surface adsorbed and this is in agreement with the fact that most of the increase in WHC is associated with high tension sites. Increase in WHC at $^1/_3$ bar may be caused additionally by the reduction in the size and number of fissures and pores in the soil, which drain above field capacity and thus increase in the amount of smaller pores.

Figure 7 presents the change in available water of a sandy and a silt loam soil during 6 months after application of sludges and the change in organic carbon content during that period. The data of Metzger (1983) show a rapid decrease in available water during the first month and then a continued slow decrease. Data of Epstein (1975) show a less rapid initial decrease and a final value of available water only slightly above that of the control. In both cases, correlation between the total carbon content of the soil and available water is observed.

Hydraulic Conductivity. Little information exists on the effect of addition of solid organics on the hydraulic properties of the bulk soil (as opposed to the effect of soil clogging with SS in effluents described above). Gupta et al. (1977) showed that the saturated HC increased after addition of sludge to a coarse sand and a silty clay loam, but this was not a lasting effect, the HC returning to that of the original soil after some months. Gupta et al. (1977) also showed that the unsaturated HC is lowered by the addition of sludge in a coarse sand. This is a distinct advantage for very sandy soils.

12.4 Chemical Effects of Organic Solids

When organic solid waste is disposed on the upper layer of the unsaturated zone, the following effects may occur:

a) Enrichment of the zone with metabolites resulting from decomposition of the solid waste, which is released and moves as solutes are transported downward adsorbed on suspended solid surfaces.
b) Modification of the pH and EC of the soil solution and unsaturated zone water, which strongly affects precipitation-dissolution and other equilibria.
c) Changes in the composition of the exchange complex of the unsaturated zone solid phase.

Some recent results, which illustrate the above effects are presented below.

12.4.1 Enrichment by Chemicals

Heavy Metals Release. The presence of potential ligands in the organic waste disposed on lands or its decomposition products may lead to the complexation of heavy metals and their release into solutions from the soil-sludge mixture. Table 5 shows, as an example, the water extractable copper and cadmium content of sludges at different stages of processing. The large fluctuation in soluble Cu content suggests continuous changes in Cu speciation in the sludge. At some stages of drying, the content reached levels which may be ecologically harmful.

Table 5. Soluble Cd and Cu in Davis sludge at different stages of processing. (After Mingelgrin and Biggar, pers. comm.)

Processing stage	Concentration ($\mu g \, l^{-1}$)	
	Cd	Cu
Raw sludge	~1	~ 10
Digested sludge	Not determined	100– 200
Saturation extract of sludge dried for 2 months	~1	~ 10
Saturation extract of sludge dried for 12 months	~5	1,000–2,000
Saturation extract of sludge dried for 12 months and leached by heavy rains	Not determined	70– 80

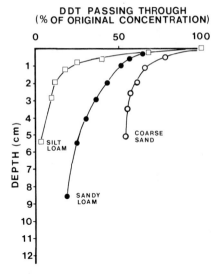

Fig. 8. [^{14}C]-DDT transport during saturated flow of labeled sewage water through 3 soils leached with 40–50 cm of sewage (Vinten et al. 1983)

Chemicals Transported on Suspended Solid Surfaces. If the chemical is strongly adsorbed on colloidal suspended solids, it can move together with its carrier and pollute the deeper layers of the unsaturated zone. Vinten et al. (1983) showed that DDT – an organic chlorinated pesticide, which is very persistent and strongly adsorbed on organic surfaces – moved into the soil together with the suspended organic solids from sewage effluent. Figure 8 shows the downward transport of adsorbed DDT in three soils during saturated flow. These data are drawn from the experimental work shown in Fig. 3. The transport through the silt loam is limited, only 5% penetrating deeper than 5 cm. Yet 20% of the applied DDT is transported below the 5 cm depth in the case of the sandy loam soil, while over 50% is transported below 5 cm in the sand. These data suggest that pesticides or other harmful species may be transported into the soil where they may accumulate in zones where the soil texture changes or may be transported to the groundwater. This leaching is more likely to occur in coarse textured soils. Data in the literature show that considerable quantities of heavy metals, pesticides, and other pollutants are associated with sewage and transported in the adsorbed state and may, therefore, constitute an important source of contamination in the unsaturated zone and groundwater.

Metabolites Release. Biological decomposition is the main process affecting the enrichment of the unsaturated zone by chemicals. This process is extensively described in the literature (e.g., Elliot and Stevenson 1977). The kinetics of the mineralization of organic nitrogen from activated sludge is described in Fig. 9. The experiment was carried out on a sludge-calcareous silty loam soil mixture at a ratio of 5:95 incubated for 60 days (35 °C; 85% of the water content at $^1/_3$ bar tension aerated with moistened air). Ammonification was the governing process

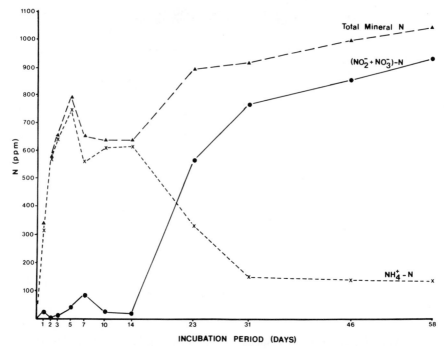

Fig. 9. Net mineral N content in Gilat loessial silt loam-activated sludge mixture (95:5, water content, 14%) vs incubation time

for the first few days, leading to accumulation of NH_4^+-N up to 800 μg g^{-1} soil. This was followed by nitrification leading to N-$(NO_3 + NO_2)$ accumulation. The quantities of the various forms of mineral nitrogen released were high, the actual values depending on the soil-sludge ratio and water content. At 15:85 sludge:soil ratio and 17% moisture content, NH_4^+-N accumulated was 2,000 μg g^{-1} soil. In the field, both nitrates and ammonium may be transported toward the groundwater by irrigation or rainwater as solutes. The pattern of nitrogen accumulation and the amount of nitrates and ammonium leached into the unsaturated zone or the groundwater is affected by the water flow, the physical and chemical properties of the unsaturated zone, and its microbiological activity.

12.4.2 Changes in pH and Total Salt Content

Figure 10 presents the dynamics of pH and electrical conductivity changes in a sludge-soil mixture (5:95 ratio). The changes in pH in the Gilat loessial silty loam fit very well the pattern of change in the NH_4^+ and NO_3^- contents observed in the incubation mixtures (Fig. 9). The pH of the native Gilat is 8.0 (20% $CaCO_3$) and that of the noncalcareous Netanya sandy loam is 7.6. These values decrease upon sludge addition (the pH of the saturation extract of the sludge itself being approximately 6.5). In both soils, the high rates of organic matter decomposition lead to an increase of pH, probably due to ammonification. The observed increase

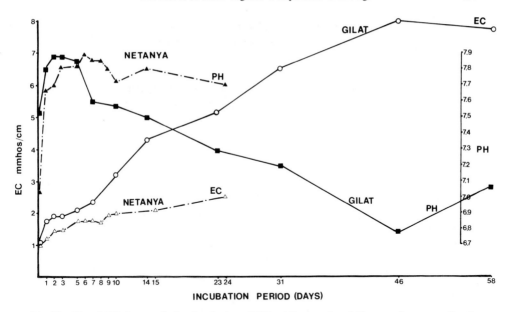

Fig. 10. pH and EC changes during incubation of Gilat (silty loam) and Netanya (loamy sand) soils – activated sludge mixtures (95:5). Measurements were made in 1:1 water extract (for incubation conditions see text)

is the net product of various opposing processes. While ammonification induces an increase in the pH value, volatilization, for example, may cause a decrease (Avnimelech and Laher 1977). The drop observed for Gilat soil (Fig. 9) for both NH_4^+-N and total mineral N values after 5–7 days of incubation, may result from enhanced ammonia volatilization due to the elevation of both NH_4^+-N concent and pH. This volatilization in turn might have been the main cause for the sharp drop in pH in that 5–7 day incubation period. At the onset of nitrification after 14 days of nitrification, the pH is gradually declining. The noncalcareous Netanya sandy soil exhibits a much more pronounced initial pH elevation (Fig. 10). However, owing to the lower nitrification rate as compared to the Gilat soil, the pH reduction which follows is much smaller than expected for a noncalcareous, low CEC soil. This probably reverses on more prolonged incubation.

The electrical conductivity (EC) of the soil extracts increased in both soils during the incubation period following sludge addition (Fig. 10). When sludge was added to Gilat soil at the ratio of 5:95, the salinity of the system, after incubation of 2 months, increased up to 8 mmhos in a 1:1 extract. This is equivalent to a TDS value of approximately 0.5 normality in the soil solution. With higher ratios of sludge addition, the salinization is even more pronounced (Fine et al. 1983). This may strongly affect the infiltration and hydraulic properties of the sludge amended soils. The EC increase is not only because of the nitrate accumulation, but also because soluble salts are released as the pH decreases. Lack of $CaCO_3$, low nitrification capacity, and the lack of the fast decrease in pH resulting from low nitrification rate may explain the relatively low increase of the EC exhibited by the Netanya soil during 3 weeks of incubation (Fig. 10).

12.4.3 Effect on the Exchange Complex and
Dissolution-Precipitation Processes

The ammonium produced from sludge, when applied to soil, may replace some of the exchangeable cations adsorbed on soil constituents. By subtracting water extractable NH_4^+ values from 1 N KCl extractable NH_4^+ values over a range of sludge additions, an estimate of the exchangeable NH_4^+ content of sludge soil mixtures can be obtained. Figure 11 shows the relationship between the estimated exchangeable NH_4^+ and total extractable NH_4^+ for a calcareous silt loam-sludge mixture. Water extraction was done in a 1:1 extract shaken for 60 min at room temp. The 1 N KCl extractable NH_4^+ was determined in a 1:5 extract under the same conditions. In the soil-sludge mixtures examined, about 85% of the total KCl extractable soil NH_4^+ is adsorbed on the soil complex, exchanging the initially adsorbed cations (Fine et al. 1983). The exchangeable NH_4^+ was also determined by a standard procedure (USDA – Handbook 60, 1954) comprising the above results. Ca^{2+} is the main exchangeable cation on the Gilat soil exchange complex. Considerable exchange with NH_4^+ could, therefore, be anticipated for the period of high NH_4^+ production. Table 6 shows estimates of the proportions of the exchange complex occupied by NH_4^+ cation over a range of soil-sludge ratios and incubation conditions. Yet soluble Ca^{2+} concentration does not increase during that period (Fig. 12, 1–7 days of incubation). The reason for this fact may be as follows: when NH_4^+ is produced, much of it becomes exchangeable (Fig. 11). Ammonification induces a rise in pH, which causes the precipitation of the Ca^{2+} removed from the exchange complex as carbonate and phosphate. At

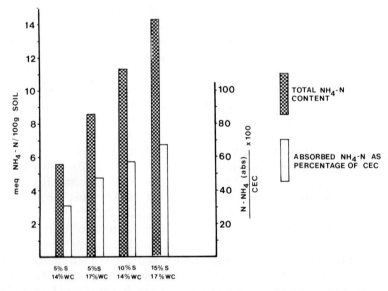

Fig. 11. The effect of sludge-soil mixing ratio and water content on NH_4^+-N, highest contents reached during incubation and on the percentage of CEC occupied by NH_4^+ (5–14 day incubation period, depending upon treatment)

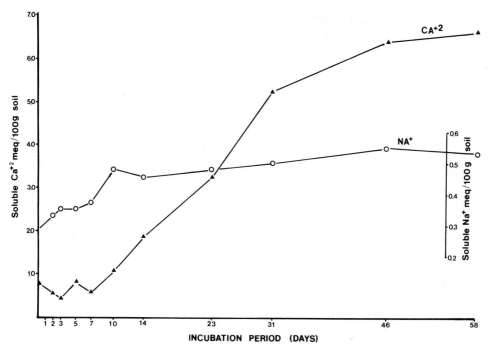

Fig. 12. Soluble Ca^{2+} and Na^+ concentration in Gilat loessial silt loam soil activated sludge mixture (95:5 ratio) vs incubation. Measurements made in 1:1 water extract

Table 6. The effect of sludge-soil mixing ratio and water content on peak NH_4^+-N contents during incubation (peaks attained after 5–14 day incubation)

% of sludge in mixture	Estimated CEC[a]	Water content during incubation (%)	Total NH_4^+ content (mEq/100 g soil)	Calculated % of CEC occupied by NH_4^{+b}
5	15.2	14	5.7	31
5	15.2	17	8.6	48
10	16.6	14	11.3	57
15	18.1	17	14.3	67

[a] Estimation done by adding the CEC of the sludge to the measured CEC of unammended soil according to their ratios in the mixtures. CEC of Gilat soil = 13.8 mEq/100 g. CEC of activated sludge = 42.2 mEq/100 g

[b] Calculation of the adsorbed fraction of the NH_4^+ is made according to the obtained linear fit, i.e., $(NH_4^+)_{ads} = 0.853\,(NH_4^+\text{-N})KCl$ extracted $+9$ (concentration in ppm)

the onset of nitrification with its accompanying acidifying effect, the Ca^{2+} concentration at equilibrium with its slightly soluble salt increases. The increase in Ca^{2+} concentration is proportional to the decrease in pH (Figs. 10 and 12). As a result, Ca^{2+} may reoccupy more of the exchange complex and thus may release Na^+ and NH_4^+ into the solutions.

Due to the occupation of a large portion of the exchange complex by a monovalent cation and the pH elevation during the ammonification stage, deflocculation and soil dispersion may occur. Whether these phenomena will be observed depend on soil factors, such as CEC, composition of exchangeable cations, buffer capacity, and $CaCO_3$ content, on sludge factors, such as mixing ratio, composition of the sludge, amount of NH_4^+ produced or initially present, initial pH of the mixture, and on environmental conditions.

Acknowledgements. This study was supported by the Research Grant WG WT of the Israel National Council for Research and Development and the German Bundesministerium für Forschung und Technologie.

References

Avnimelech Y, Laher M (1977) Ammonia volatilization from soils: Equilibrium considerations. Soil Sci Soc Am J 41:1080–1084

Avnimelech Y, Nevo Z (1964) Studied variation in C:N ratio and OM content effect on biological clogging of sands. Soil Sci 98:222–226

Baillod CR, Waters RG, Iskandar IK, Urga A (1977) Preliminary evaluation of 88 years rapid infiltration of raw municipal sewage effluent at Columet, Michigan. In: Loehr RC (ed) Land as a waste management Alternative. Ann Arbor Sci, pp 489–511

Chang AC, Olmestad WR, Johanson JB, Yamashita G (1974) The sealing mechanism of wastewater ponds. J Water Poll Contr Fed 46:1715–1721

Dan Region Sewage Reclamation Project (1982) Water quality forecast. Tahal Ltd October 1982

Daniel TC, Bouma J (1974) Column studies of soil clogging in a slowly permeable soil as a function of effluent quality. J Environ Qual 3:321–326

De Vries J (1972) Soil filtration of waste water effluent and the mechanism of soil clogging. J Water Pollute Control Fed 44:565–573

Elliott LF, Stevenson FJ (eds) (1977) Soils for Management of Organic Wastes and Waste Waters. SSSA, ASA, CSSA, Madison, Wisc

Epstein E (1975) Effect of sewage sludge on some soil physical properties. J Environ Qual 4:139–142

Fine P (1983) Ph D Thesis, Tel Aviv Univ (Progress report)

Fine P, Yaron B, Mingelgrin U (1983) (In preparation)

Frankenberger JR, Troeh WT, Dumenil FR (1979) Bacterial effects on hydraulic conductivity of soils. Soil Sci Soc Am J 43:333–338

Gilmour JT, Gilmour CM (1980) Assimilation model for sludge decomposition in soil. J Environ Qual 9:194–199

Greenland DJ (1977) Soil damage by cultivation. Phil Trans R Soc B 281:193–208

Gupta SC, Dowdy RH, Larson WE (1977) Hydraulic and thermal properties of a sandy soil as influenced by incorporation of sewage sludge. Soil Sci Soc Am J 41:600–605

Hillel D (1980) Application of Soil Physics. Academic Press, London New York

Hillel D, Gardner WR (1969) Steady infiltration into crust topped profiles. Soil Sci 108:137–142

Hunter JV, Kotalik TA (1974) Chemical and biological quality of sewage effluents. In: Sopper WE, Kardos LT (eds) Conference on recycling treated municipal wastewater through forest and cropland. Penn State Univ Press, University Park PA, pp 6–27

Jones JH, Yalor GS (1965) Septic tank effluent percolation through sands under laboratory conditions. Soil Sci 99:301–309

Kardos LT, Sopper WE (1974) Renovation of municipal wastewater through land disposal by spray irrigation. In: Sopper WE, Kardos LT (eds) Conference on recycling treated municipal wastewater through forest and cropland. Penn State Univ Press, University Park PA, pp 131–145

Khaleel R, Reddy KR, Overcash MR (1981) Changes in soil physical properties due to organic waste applications. J Environ Qual 10:133–141

Kladivko EJ, Nelson DW (1979) Changes in soil properties from application of anaerobic sludge. J Water Pollut Contr Fed 51:325–332

Klute A, Jacob WC (1949) Physical properties of sassafras silt loam as affected by long-term organic matter additions. Soil Sci Soc Am Proc 14:24–28

Kristiansen R (1981) Sand-filter trenches for purification of septic tank effluent. I. The clogging mechanism and soil physical environment. J Environ Qual 10:352–357

Laak R (1970) Influence of domestic wastewater pre-treatment on soil clogging. J Water Pollut Contr Fed 42:1495–1500

Levi-Minzi R, Sartori F, Riffaldi R (1981) Caracterisation de la fraction organique et de la fraction minerale Cristalline de boves d'epuration. In: L'Hermite P, Ott H (eds) Characterization, treatment and use of sewage sludges. Reidel, Dordrecht Netherlands, pp 291–298

L'Hermite P, Ott H (1981) (eds) Characterization, treatment and use of sewage sludge. Reidel, Dordrecht Netherlands

Libhaber M, Kary S, Shevah Y (1979) Land treatment of excess activated sludge, feasibility study. Tahal Consulting Eng, Tel-Aviv, Israel

Loehr RC (ed) Land as a Waste Management Alternative. Ann Arbor Sci Publ Ann Arbor, Mich

Manka J, Rebhum M, Mandelbaum A, Bortinger A (1974) Characterization of organics in secondary effluents. Environ Sci Techn 8:1017–1020

McCalla TM, Peterson JR, Lue-Hing C (1977) Properties of Agricultural and numicipal wastes. In: Elliott LF, Stevenson FJ (eds) Soils for management of organic wastes and waste waters. ASA, CSSA, SSSA, Med Wisc, pp 11–43

Metzger L (1983) Ph D Thesis Hebzew-Univ (progress report)

Metzger L, Yaron B, Mingelgrin U (1983) Soil hydraulic conductivity as affected by physical and chemical properties of effluent. Agronomie 3:771–779

Neal JH (1964) Advandec wast treatment by distillation. AWTR-7 US Publ Health Serv 999-WP-9

Okubo T, Matsumoto J (1973) Effect of infiltration rate on biological clogging and water quality changes during artificial recharge. Water Resour Res 15:1536–1542

Pagliai M, Gindi G, La Marca M, Giachetti M, Lucamante G (1981) Effect of sewage sludges and composts on soil porosity and aggregation. J Environ Qual 10:556–561

Rice RC (1974) Soil clogging during infiltration of secondary effluent. J Water Pollut Contr Fed 46:708–716

Schaumberg GD, Le Vesque-Madore CR, Sposito G, Lund LJ (1980) Infrared spectroscopic study of the water soluble fraction of sewage sludge-soil mixtures during incubation. J Environ Qual 9:297–303

Sommers LE (1977) Chemical composition of sewage sludges and analysis of their potential use as fertilizers. J Environ Qual 6:225–232

Swartzendruber D, Uebler RL (1982) Flow of kaolinite and sewage suspensions in sand and sand-silt. II. Hydraulic conductivity reduction. Soil Sci Soc Am J 46:912–916

Terry RE, Nelson DW, Sommers LE (1979) Decomposition of anaerobically digested sewage sludge as affected by soil environmental conditions. J Environ Qual 8:342–347

Thomas RE, Law JP (1977) Properties of waste waters. In: Elliott LF, Stevenson FJ (eds) Soils for management of organic wastes and waste waters. ASA, CSSA, SSSA, Mad Wisc, pp 47–72

Thomas RE, Schwartz WA, Bendixen TW (1966) Soil chemical changes and infiltration rate reduction under sewage-spreading. Soil Sci Soc Am Proc 30:641–646

Unger PW, Stewart BA (1974) Feedlot waste effects on soil conditions and water evaporation. Soil Sci Soc Am Proc 38:954–957

Vinten AJA, Mingelgren U, Yaron B (1983) The effect of suspended solids in waste water on soil hydraulic conductivity. II. Vertical distribution of suspended solids. Soil Sci Soc Am J 47:408–412

D. Biological Processes

Introductory Comments

J. GOLDSHMID

The unsaturated zone, especially its upper few meters, is a large microbiological reactor. It is infested with microorganisms and many biochemical, as well as chemical, physical, and physiochemical reactions take place in this zone. These reactions are as important to the proper functioning of our globe as one can imagine and it is impossible to envision life in its present form without this huge biochemical reactor.

One can divide the microorganisms, as well as the reactions, into two groups: (1). the allochthonic microorganisms – which enter the unsaturated zone through contamination – such as sewage, solid waste, etc. and (2) the autochthonic soil microorganisms – those native to the soil.

The first group is mainly negative from our point of view and includes the pathogenic microorganisms which form a trail to the groundwater supplies, if not removed before reaching the aquifer. To them, the unsaturated zone is a trap, hopefully efficient enough to protect the groundwater. This group, its survival time, and its interaction with the soil, is discussed by Pekdeger (Chap. 14). Use of the unsaturated zone as a filter for pathogenic microorganisms is discussed in Part F.

The second group, which concentrates in the root zone, influences plant growth directly. Metabolites, such as polysaccharides, can improve the soil texture. Nutrient cycling by enzymic processes influence plant growth as well as groundwater pollution. Pesticides are commonly degraded by mixed communities of microorganisms, each doing its share in the degradation process. All these are reviewed in detail by Lynch (Chap. 13).

The two chapters thus present the two sides of the coin and review the complex microbiological activity in the unsaturated zone, as well as the interaction between the microorganisms, especially bacteria and viruses, and the unsaturated zone building blocks, namely, the soil and the water.

Many view the unsaturated zone as the final stage in sewage treatment, as well as the last barrier which protects the groundwater supplies from contamination. But the unsaturated zone is also the media on which we grow crops and is the firm basis to human activity. Our knowledge of the processes which take place in the unsaturated zone is meager and forces us to design our activities with a great margin of safety; and this margin of safety means an economic burden. Thus, the more we learn about the movement and survival of pathogenic microorganisms in the soil body, the better we can utilize the various reactions and processes, which take place in the unsaturated zone, to our advantage and the smaller will be the necessary safety margin.

13. Biochemical Processes in Unsaturated Zones

J. M. LYNCH

13.1 Introduction

The unsaturated zones of soil will be the focus of this chapter. However, it must be emphasized that field soils cannot always be clearly defined biologically in this respect because the levels of water tables usually fluctuate. In the saturated zones, oxygen becomes depleted from the soil atmosphere and the activities of anaerobes become significant. Fermentative metabolites accumulate under such conditions because oxygen is often required for their catabolism. In the unsaturated zones, anaerobic microsites exist, but the activity of aerobic organisms is relatively greater. The activities of facultative organisms and associations between aerobes and anaerobes are commonly of most significance in the interface between the two zones, that is, in the vicinity of the water table.

13.2 Substrates for Microbial Growth

Roots and plant residues provide the major substrate input to the soil microflora. The native soil organic matter (humus), which is usually the major component of the soil organic matter, has a half-life in excess of 1,000 years (Paul et al 1964, Perrin et al 1964) and can therefore be regarded as recalcitrant. Fresh living roots provide substrates to microorganisms which infect them; the infections include the beneficial mycorrhiza and root nodule bacteria, as well as the pathogens. It is also clear now that less obvious infections, particularly pseudomonads, occur in the root cortex and these can be considered as endorhizosphere organisms. Living roots release carbon in various forms: *Exudates* leak from roots, *secretions* are actively pumped, *lysates* are passively released during autolysis, *mucigel* is composed of *plant mucilage*, produced from epidermal and root cap cells, and *microbial mucilage*, produced from microbial growth around the root. This total flow of carbon can satisfactorily be assessed only by growing plants exclusively on a source of $^{14}CO_2$ with roots in a sealed pot of soil and measuring the ^{14}C released into the soil and the respiratory $^{14}CO_2$ generated from microbial metabolism (Barber and Martin 1976, Sauerbeck and Johnen 1977). This shows that, for cereals, about 20% of the dry matter production of the plant is released into the soil. A vast range of substances has been found in the carbon released under aseptic conditions (Rovira 1965) as would be expected from the release of cell contents. The C/N ratio is about 30:1, based on the amino

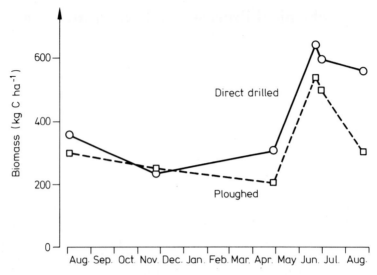

Fig. 1. Seasonal variations in the soil biomass between August 1977 and August 1978 (Lynch and Panting 1980). A winter wheat crop was sown in a clay soil on 10 October 1977 and harvested on 24 August
1978

acid content (Barber and Gunn 1974), which is much greater than that of microbial cells. Hence, microbial growth in the rhizosphere will tend to immobilize soil
or fertilizer N.

As the plant ages, root senescence increases (Deacon and Lewis 1982). The
dead cells also provide substrates for microbial growth and ultimately the whole
root system is available for saprophytic colonization. In Fig. 1, the time when the
soil microbial biomass is maximal is at the time of maximum root production.

After harvest of the crop, the plant residues are returned to provide further
substrates for microbial growth. These residues are matured and usually have a
large content of lignocellulose. Cereal straw is an example and one of the principal
forms of agricultural waste. There appears to be at least two phases in the decomposition of straw lignocellulose (Fig. 2). Primary degradation is of the cellulose
(polymers of glucose), and hemicellulose (polymers of xylose and other sugars),
which is relatively available; however, the polymeric nature means that it is decomposed more slowly than the water soluble components of root derived carbon. The cellulase enzyme complex is responsible for the decomposition.

$$\underset{\text{cellulose}}{(C_6H_{12}O_6)} \sim 3000 \xrightarrow{\text{endo-1,4-}\beta\text{-D-glucanase}} \underset{\text{reactive cellulose}}{(C_6H_{12}O_6)} \ll 3000$$

$$\xrightarrow{\text{1,4-}\beta\text{-D-glucan cellobiohydrolase}} \underset{\text{cellobiose}}{(C_6H_{12}O_6)_2} \xrightarrow{\text{cellobiase}} \underset{\text{glucose}}{C_6H_{12}O_6}.$$

The second phase appears to result from the decomposition of the cellulose,
which is more intimately bound to the lignin. Even during this phase, however,
there is little degradation of lignin, the decomposition rate constant of the latter

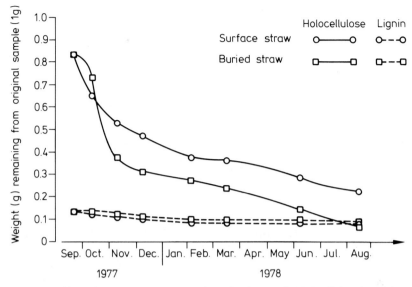

Fig. 2. Decomposition of straw oat components in a clay loam soil. Holocellulose is cellulose plus hemicellulose, which comprised 72% w/w of the straw. Lignin was 17% w/w of the straw (Harper and Lynch 1981)

being less than 10% of that of cellulose (Harper and Lynch 1981). Lignin, which is polyphenolic, may never become a true growth substrate for the soil microflora. More probably, it will act as a transformation substrate such that end groups of the polymer are modified and the breakdown products and remaining skeleton are incorporated into the soil humic fraction.

13.3 Microbial Metabolites in Relation to Plant Growth

The vast array of microbial species present in soil and the complexity of the substrates available to them means that a great range of metabolites are produced in soil. Of these metabolites, many could potentially inhibit plant growth, while others could be stimulatory. Commonly, some substances, both natural and xenobiotic (foreign to nature), are stimulatory in one range of concentrations, especially with exposure to a limited part of the root system (Gussin and Lynch 1982) and are inhibitory in greater concentrations. Thus, it is critical that the concentration of the substance be measured in soil and also that mild extractants are used, preferably just water, to minimize the risk of chemical modification of the materials during the extraction. Typically, for example, alkalis can degrade polymeric substances from soil. Having determined the concentrations present, the significance must be determined by measuring the effect of the pure compound on plant growth over a range of concentrations.

Very few reports have satisfied the necessary conditions to establish that the microbial metabolite has significance to the plant growing in soil. Notable exceptions are the production of the volatile fatty acids, particularly acetic, and hydrogen sulphide (Drew and Lynch 1980). However, both of these products are produced in fermentative metabolism by anaerobes, and whereas they may be produced in anaerobic microsites of unsaturated zones, their production is likely to be more characteristic of saturated zones, provided the necessary energy substrates are available. Besides their production being favoured anaerobically, restricted aeration prevents their subsequent metabolism. This applies to many metabolites, although under severely reducing conditions, which can result from a lack of aeration, some metabolites are further metabolized. For example, acetic acid is reduced to methane at redox potentials below zero. Thus, accumulations of metabolites in the unsaturated zone is a balance between their production and subsequent metabolism.

13.4 Microbial Effects on Soil Structure

Whereas it has long been recognized that microorganisms are responsible for improving the aggregate stability of soil, the mechanisms involved are still unclear. Bacteria, like clay particles, carry a net electronegative charge on their surfaces, but by polarization of the charge and metal ion bridging, adhesion takes place between the cells and particles. However, they also produce extracellular polysaccharide, which provides an alternative source of adhesive force. Table 1 demonstrates that for a yeast, the extracted polysaccharide is almost as effective as the total cellular material in improving the aggregate stability of soil. Fungi cannot always enter the smaller pore spaces of soil and they tend to bind around particles, providing an alternative to the cementing action of bacteria. These processes of aggregate stabilization are important in the unsaturated zone especially when the zone is subject to periodic saturation from surface irrigation or rainfall. Falling water droplets will disperse unstable aggregates on the soil surface to produce "capping", which hinders crop establishment. Destruction of aggregates in the bulk soil results in the various forms of water erosion. It is important to maintain a substrate input to soils, either as roots or plant residues, in order to "fuel"

Table 1. Comparison of the effect of *Lipomyces starkeyi* and its polysaccharide on the aggregation of a silt loam (Hamble series) (Lynch 1981)

Form of polysaccharide added	Suspended soil (% w/w)
Cells	3.6
Purified extract	2.9
None (control)	4.7
Least significant difference (P=0.005)	1.0

The same weight of polysaccharide was added in each treatment (1 mg dry wt g^{-1} dry soil) on the assumption that 50% of the cell dry weight is polysaccharide. Lack of water stability is indicated by larger amounts of soil in suspension

the microbial biomass to produce aggregating agents. In this respect, it therefore seems unwise to burn important substrates, such as straw residues, in areas where the soil is subject to water erosion.

13.5 Nutrient Cycles

Due to its economic importance in crop productivity, the biochemistry of the nitrogen cycle has been studied most intensively. Nitrogenase, the enzyme responsible for the fixation of dinitrogen, has probably received the greatest attention. It is present in root nodule bacteria (e.g., *Rhizobium*), free-living aerobic (e.g., *Azotobacter*) and anaerobic (e.g., *Clostridium*) bacteria and blue-green algae. The photosynthetic blue-green algae are only of significance when the soil surface is wet. In all organisms, the nitrogenase enzyme is oxygen-sensitive and needs respiratory protection. The microbial protein-N introduced into the soil and the other sources of organic matter are eventually ammonified to produce a source of N for plants. This is brought about by the following enzymic sequences:

$$\text{proteins} \xrightarrow{\text{protease}} \text{peptides} + \text{amino acids}$$

$$H_2O + \text{amino acids} \xrightarrow{\text{amino acid hydrolase}} \text{aspartate} + NH_3.$$

(e.g., asparagine)   (e.g., *asparaginase*)

Urea, originating from fertilizer or animal waste is also hydrolyzed.

$$NH_2CONH_2 \xrightarrow[H_2O]{\text{urease}} NH_2COOH + NH_3$$

$$\rightarrow CO_2 + 2NH_3 \xrightarrow{H_2O} H_2CO_3 + 2NH_3.$$

The ammonium ion also provides a substrate for the oxidases from autotrophic bacteria, which produce nitrite (*Nitrosomonas* spp.) and nitrate (*Nitrobacter* spp.). Thus, these enzymic processes contribute to the pools of nitrate present in soil, which is also added to by fertilizers. Many facultative heterotrophic anaerobes have nitrate and nitrite reductases for conversion of the N back to ammonia, while nitrite can also be metabolized by denitrifiers through nitrous oxide and to dinitrogen. The net balance of all these enzymic reactions is still relatively poorly understood and, therefore, biological contributions to any possible nitrate pollution is uncertain.

Transformations involving phosphorus are usually less important to carbon decomposition than those involving N. A large proportion of the soil P can be present as stable inorganic or organic forms. Bacteria can be important in solubilizing rock phosphate. When free phosphate is in the rhizosphere, bacteria can promote uptake by roots over short (30-min) periods, but decrease it over long (24-h) periods (Barber et al. 1976). Mycorrhizal fungi can transport P directly into the plant. Simulation models of P cycling indicate that the annual flow through the microbial biomass was greater than that through above and beneath ground plant P (Cole et al. 1977). The major controlling factors in these models were inorganic P solubility, soil water content and rates of diffusion.

Microorganisms can also be involved in the many other nutrient cycles in the soil. The sulphur cycle has been studied extensively in sediments, but relatively little in soils. The biochemical steps are uncertain, but seem to involve adenine-5'-phosphosulphate (APS) and parts of the cytochrome system (cyctochrome c and c_3) with NAD, NADP or ferrodoxin (Campbell 1977). Reduction also involves 3'-phosphoadenosine-5-phosphosulphate (PAPS):

Assimilatory reduction (microorganisms, higher plants):

$$ATP + SO_4^{2-} \rightleftharpoons APS + PO_4^{2-}$$
$$ATP + APS \rightleftharpoons PAPS + ADP$$
$$PAPS + NAD(P)H \rightleftharpoons PAP + SO_3^{2-} + NAD(P)$$
$$NADH + SO_3^{2-} \rightleftharpoons SH^- + NAD.$$

Dissimilatory reduction (bacteria):

$$ATP + SO_4^{2-} \rightleftharpoons APS + PO_4^{2-}$$
$$APS + 2e^- \rightleftharpoons SO_3^{2-} + AMP.$$

Oxidation (bacteria):

$$2AMP + 2SO_3^{2-} \rightleftharpoons 2APS + 4e^-$$
$$2APS + PO_4^{2-} \rightleftharpoons 2ADP + 2SO_4^{2-}.$$

Organic S can be released as either sulphate or hydrogen sulphide, depending on the aeration conditions. The hydrogen sulphide from sulphate reduction at a redox potential of about -180 mV or anaerobic decay may precipitate metal sulphides. Sulphides may be oxidized by chemosynthetic bacteria at positive redox potentials or photosynthetic sulphur bacteria at negative potentials. In oxidizing the various forms of S, *Thiobacillus* spp. produce acid and the increased pH can increase the availability of other elements. The pH changes can decrease the flocculation of clay particles and hence worsen soil structure. Where there is extensive S oxidation, lime must be added to correct the pH drift.

The cycles of iron and manganese have received some attention. Microbial production of organic acids can release iron and manganese into soluble forms (Lynch and Gunn 1978). Some bacteria, such as *Pedomicrobium* spp., accumulate Fe and Mn from organic matter and deposit it as the hydroxide in their slime capsules. The truly chemosynthetic iron bacterium, *Thiobacillus ferrooxidans*, has a very low pH optimum (1.7 to 3.5), but its cellular pH is about 5. The oxidation of Fe^{2+} to Fe^{3+} yields about 11.5 Kcal, which is used to reduce (fix) CO_2. There is some evidence for iron reduction in waterlogged gley soils, which causes the characteristic blue-green colour.

13.6 Pollutant/Microorganism Interactions

An accumulation of certain inorganic ions, such as nitrate, in the unsaturated zones is sometimes viewed with concern as these could subsequently leach into water courses, thereby becoming a pollutant. The source of the nitrate could be

Table 2. (A) Nitrogen uptake into harvested parts of a grass crop for year in which nitrogen-15 labelled fertilizer was applied to lysimeters. (B) Loss of nitrogen by leaching in winter after fertilizer addition. (Data of R.J. Dowdell, Letcombe)

	Clay	Silt loam
	kg N ha^{-1}	
Labelled nitrogen applied as Ca (NO$_3$)$_2$	400	400
(A) Nitrogen uptake by crop		
Labelled N	183	184
Unlabelled N	141	145
Total	324	329
(B) Loss of nitrogen by leaching		
Labelled N	25	22
Unlabelled N	19	16
Total	44	38

in the form of fertilizers or from nitrification following the application of organic wastes to the land. In any event, on the rare occasions where chemisorption has not taken place, the activity of denitrifiers is important in preventing the potential pollution. Animal wastes reach the soil from feedlot wastes or farm livestock wastes (slurries). The soil is generally regarded as a site for disposal, although low density applications are made for their fertilizer value. Few changes appear to occur in the soil microbial population balance (fungi, aerobic and anaerobic bacteria, *Escherichia coli* – types, nitrifiers and denitrifiers) following the application of cattle feedlot wastes to soil (Davis et al. 1980). Little NO$_3$-N is leached down the soil profile (Burford et al. 1976). Although in winter, drainage water contained about 60 mg N l^{-1}, only about 6% of this originated from the animal waste (Burford 1976). Some of the N in the drainage water originates directly from fertilizer N. However, about half of the applied N enters the soil organic N pool (Table 2) and it is the subsequent nitrification of this N which reaches drainage water (Dowdell 1982).

An important question on the use of inorganic nitrogen fertilizers is not only their direct potential action in the pollution of water courses, but whether they are detrimental to other biochemical processes in soil. Overall, because of their high C/N ratios compared with microbial cells, the decomposition of fresh plant tissue or plant residues has been considered to be limited by the availability of N. Thus, application of inorganic N would tend to hasten these processes and, therefore, increase the loss of soil C. However, little experimental evidence has been provided to support this view and indeed recent evidence (Knapp et al. 1983) shows that respiration during straw decomposition is increased more from the addition of a suitable soluble carbon source, glucose, than by the addition of soluble N. Thus, on that basis, it would seem unlikely that increased use of N fertilizer causes a decrease in soil organic matter, including that which is responsible to soil stability. However, that argument is in itself tortuous. Increased fertilizer use can increase the N content of the straw at harvest and recent studies (L. F. Elliott and

J. M. Lynch, unpubl.) indicate that the smaller the N content of straw, the more microbial polysaccharide is produced during its decomposition, resulting in better soil aggregation. The significance of these observations in the field has yet to be evaluated.

A clear negative effect of fertilizer N on biochemical processes in soil is that it represses the action of the enzyme responsible for N_2-fixation, nitrogenase (Postgate and Hill 1979). This is a particular problem with legumes and in flooded soils where blue-green algae can contribute significantly to the N economy of soils. In much of the western world, where the energy inputs to agriculture are large, this effect tends to be ignored.

Microbial interactions with organic pollutants has received greater attention than those with inorganics. The microbiologist is concerned with two distinct aspects. Firstly, do microorganisms have the power to degrade the chemical, thereby avoiding biomagnification or bioaccumulation in the soil? Then, does the chemical affect the activities of specific microbial groups and, thus, modify the population balance in soil? Most commonly pesticides, including herbicides, have been studied in these respects, but the concepts for study are equally relevant to any potential pollutant, such as factory effluents.

The herbicide 2,4-D (dichlorophenoxy acetic acid) is in widespread use and is degraded by several genera of soil bacteria. However, the introduction of an extra chlorine atom into the benzene ring to form 2,4,5-T (trichlorophenoxy acetic acid) makes the molecule resistant to degradation (recalcitrant). Although this is an undesirable characteristic in a herbicide, it was removed from most markets because of dioxin contamination. Besides chemical recalcitrance, there are many other factors which can cause recalcitrance, such as (a) absence of water or oxygen; (b) presence of toxins, such as acetic acid; (c) low temperature and sometimes, high hydrostatic pressure; (d) chemical combination, such that substrates are immobilized; (e) lack of access spatially; and (f) substrate concentration is too small.

Some pesticides are degraded directly, usually by induced enzymes, whereas others are broken down by the common alternative route of co-oxidation or cometabolism. The latter can be defined (Dalton and Stirling 1982) as the transformation of a nongrowth substrate in the obligate presence of a growth substrate or another transformable compound. A nongrowth substrate is a compound unable to support cell replication as opposed to an increase in biomass. Xenobiotic compounds are often degraded by pure cultures in the laboratory, but a more efficient degradation often occurs with mixed cultures or communities (Bull 1980, Slater and Bull 1982); the latter is of course most likely in soil. The members of the community can cooperate in producing adaptive enzymes to deal with otherwise recalcitrant molecules. A common activity of such communities degrading pesticides is in the provision of dehalogenases as many xenobiotics are halogenated analogues of natural compounds. There is a range of dehalogenases, showing specificity for the various substrates. Gratuitous biodegradation is the metabolism of xenobiotics by enzymes evolved for natural compounds, whereas fortuitous biodegradation is the combination of gratuitous activities found in two organisms to generate a complete pathway. The genes of the catabolic pathways of many compounds are coded in plasmids (loops of DNA separate from the chromosome). Plasmid exchange between organisms can generate single organisms with capacity for the complete metabolism of a compound, whereas the capacity does not occur in either parent. Degradation can also be enhanced by selecting mutants with elevated enzyme activities.

Xenobiotics can affect soil enzymes and the evidence for agrochemicals doing this has been reviewed (Cervelli et al. 1978). It should be noted with caution that sometimes degradation products can be more toxic to animals, plants or microorganisms, than the original chemicals. It should also be recognized that pesticides can sometimes have nontarget effects, such as increasing disease. Some soil-applied herbicides can increase the leakage rate of metabolites from roots and provide substrates for pathogens (Altman and Campbell 1977), while others can promote the colonization of roots by subclinical pathogens, such as pseudomonads (M. P. Greaves, pers. comm.).

13.7 Conclusion

Microbial ecology and soil biochemistry have developed recently such that we can make better quantitative assessments of the biochemical activities of microorganisms in stabilizing soil structure and affecting the growth and nutrition of plants. In any agricultural system or ecosystem, it is important that these activities should be monitored, alongside physical, chemical and botanical measurements. However the absolute value of such measurements is often of less interest that the comparative value between soils and between soil conditions. Particularly, it is of interest, for example, to know how flooding influences the processes. Also, it is of paramount importance to measure how any potential pollutant affects the biochemistry of the soil, both in terms of crop productivity and leaching of undesirable chemicals through saturated zones into water courses.

References

Altman J, Campbell CL (1977) Effect of herbicides on plant disease. Ann Rev Phytopathol 15:361–385
Barber DA, Gunn KB (1974) The effect of mechanical forces on the exudation of organic substances by the roots of cereal plants grown under sterile conditions. New Phytol 73:39–45
Barber DA, Martin JK (1976) The release of organic substances by cereal roots in soil. New Phytol 76:69–80
Barber DA, Bowen GD, Rovira AD (1976) Effects of micro-organisms on the absorption and distribution of phosphate in barley. Aust J Plant Physiol 3:801–808
Bull AT (1980) Biodegradation: some attitudes and strategies of microorganisms and microbiologists. In: Ellwood DC, Hedger JN, Latham MJ, Lynch JM, Slater JH (eds) Contemporary microbial ecology. Academic Press, London New York pp 107–136
Burford JR (1976) Effect of the application of cow slurry to grassland on the composition of the soil atmosphere. J Sci Food Agric 27:115–136
Burford JR, Greenland DJ, Pain BF (1976) Effect of heavy dressings of slurry and inorganic fertilizers applied to grassland on the composition of drainage waters and the soil atmosphere. In: Agriculture and Water Quality, MAFF Technical Bull 32:432–433
Campbell R (1977) Microbial ecology. Blackwell, Oxford

Cervelli S, Nannipieri P, Sequi P (1978) Interactions between agrochemicals and soil enzymes. In: Burns RG (ed) Soil enzymes. Academic Press, London, pp 251–293

Cole CV, Innis CJ, Stewart JWB (1977) Simulation of phosphorus cycling in a semiarid grassland. Ecology 58:1–15

Dalton H, Stirling DI (1982) Co-metabolism. Phil Trans R Soc Lond B 297:481–496

Davis RJ, Mathers AC, Stewart BA (1980) Microbial populations in Pullman clay receiving large applications of cattle feedlot waste. Soil Biol Biochem 12:119–124

Deacon JW, Lewis SJ (1982) Natural senescence of the root cortex of spring wheat in relation to susceptibility to common root rot (*Cochliobolus sativus*) and growth of a "free-living" nitrogen-fixing bacterium. Plant Soil 66:13–20

Dowdell RJ (1982) Fate of nitrogen applied to agricultural crops with particular reference to denitrification. Phil Trans R Soc Lond B 296:363–373

Drew MC, Lynch JM (1980) Soil anaerobiosis micro-organisms and root function. Ann Rev Phytophatol 18:37–66

Gussin EJ, Lynch JM (1982) Effect of local concentrations of acetic acid around barley roots on seedling growth. New Phytol 92:345–348

Harper SHT, Lynch JM (1981) The kinetics of straw decomposition in relation to its potential to produce the phytotoxin acetic acid. J Soil Sci 32:627–637

Knapp EB, Elliott LF, Campbell GS (1983) Microbial respiration and growth during the decomposition of wheat straw. Soil Biol Biochem 15:319–323

Lynch JM (1981) Promotion and inhibition of soil aggregate stabilization by micro-organisms. J Gen Microbiol 126:371–375

Lynch JM, Gunn KB (1978) The use of the chemostat to study the decomposition of wheat straw in soil slurries. J Soil Sci 29:551–556

Lynch JM, Panting LM (1980) Cultivation and the soil biomass. Soil Biol Biochem 12:29–33

Paul EA, Campbell CA, Rennie DA, McCallum KJ (1964) Investigations of the dynamics of soil utilizing carbon dating techniques. Trans 8th Int Congr Soil Sci 3:201–208

Perrin RMS, Willis EH, Hodge CAH (1964) Dating of humus podzols by residual radiocarbon activity. Nature 202:165–166

Postgate JR, Hill S (1979) Nitrogen fixation. In: Lynch JM, Poole NJ (eds) Microbial ecology. A Conceptual Approach. Blackwell, Oxford, pp 191–213

Rovira AD (1965) Plant root exudates and their influence upon soil microorganisms. In: Baker KF, Snyder WC (eds) Ecology of soil-borne plant pathogens – prelude to biological control. Murray, London, pp 170–186

Sauerbeck DR, Johnen BG (1977) Root formation and decomposition during plant growth. In: Soil organic matter studies Vol 1. IAEA, Vienna, pp 141–148

Slater JH, Bull AT (1982) Environmental microbiology: biodegradation. Phil Trans R Soc Lond B 297:575–597

14. Pathogenic Bacteria and Viruses in the Unsaturated Zone

A. PEKDEGER

14.1 Introduction

The contamination of subsurface water by pathogenic bacteria and viruses have caused large outbreaks of waterborne diseases. The evaluation of case histories shows that outbreaks happened only in a situation where the contaminated infiltrating water could pass through the unsaturated zone. Thus, it can be concluded that the unsaturated zone is a very effective barrier against the infiltration of bacteria and viruses into the groundwater.

The subsurface transport of bacteria and viruses is controlled mainly by the following factors:

1. the persistence of bacteria and viruses under the biological and chemical conditions in the unsaturated zone and groundwater;
2. the physical and physicochemical processes which control the transport of bacteria and viruses.

The most important pathogenic bacteria and viruses which might possibly be transported to the subsurface water are: *Salmonella* sp., *Shigella* sp., *Yersinia enterocolitica, Y.pseudotuberculosis, Leptospira* sp., *Francisella tularensis, Dyspepsia coli,* enterotoxigenic *E.coli* (ETEC), *Pseudomonades, Vibrio* sp., *Legionella* sp., and the viruses, infectious *hepatitis* virus, *polio* virus, *Coxsackie* viruses, *adenovirus, rotavirus,* and *Norwalk-like* virus (Gerba and Keswick 1981).

Using realistic data on the concentration of pathogenic bacteria and viruses in sewage, surface water and drinking water (EPA 1978) it can be concluded that the concentration of these microorganisms in potable water has to be reduced by 7 log units from the concentration typical to sewage (Table 1). This means that the 99.99% elimination of pathogenic microorganisms in the groundwater, broadly used as measure for the efficiency of the elimination processes, is not sufficient (Matthess and Pekdeger 1981, Althaus et al. 1982).

Table 1. *E. coli* and virus concentrations in water (EPA 1978)

	$E.\,coli/100$ ml	Virus units/100 ml
Sewage effluent	10^7–10^8	10^1 –10^2
Treated sewage effluent	–	10^0 –10^1
Contaminated surface water	10^4–10^5	0 –10^0
Drinking water after treatment	<1	10^{-8}–10^{-9}

The main contamination sources of the unsaturated zones are septic tanks, leaky sewer lines, sanitary landfills, waste oxidation ponds, and land application of waste water.

14.2 Persistence of Bacteria and Viruses

Two groups of microorganisms must be differentiated when the subsurface survival of bacteria and viruses is considered:

1. allochthonic pathogenic microorganisms (parasitic bacteria and enterotoxine producing bacteria), which enter the unsaturated zone and the groundwater due to contamination;
2. autochthonic soil and groundwater microorganisms.

The autochthonic microbial soil and groundwater microorganisms fluorish under favorable ecological conditions, developing high population densities ($\gg 100 \, \mathrm{ml}^{-1}$) (Hirsch and Rades Rohkohl 1983). The allochthonic bacteria are usually eliminated in the subsurface environment, but under oligotrophic conditions they may survive without a substantial decrease, or even slightly increasing in number during the first few weeks. After this period, the elimination of bacteria and viruses may be approximateley described by an exponential decay function [Eq. (1); Berg 1967, Merkli 1975].

$$C_{(t)} = C_0 e^{-\lambda(t-t_0)} \tag{1}$$

$t \geq t_0$ and $t_0 \leq 20 \, \mathrm{days}$

$C_0, C_{(t)}$ = initial concentration and concentration at time t

λ = elimination constant = $\dfrac{\ln 2}{\tau_{1/2}}$

t = time

$\tau_{1/2}$ = half-life time.

The bacteria and viruses may survive many weeks in sewage, septic tanks, and waste deposits, e.g., *Salmonella* sp. persists 3–10 weeks.

The persistence of bacteria and viruses in the soil depends on the biological activity of the top soil. *Cholerae* sp. can survive only a few days in a biologically active top soil, but in sterile soils it may survive up to 6 months. Survival times for *Salmonella* sp. range up to 3 months, for *Streptococcus faecalis* up to 2 months, and for coliform bacteria up to 7 months (Althaus et al. 1982).

Bacteria and viruses can survive even longer (>6 months) in the deeper parts of the unsaturated zone and in the groundwater, when oligotrophic conditions are present. This means that under certain ecological conditions, bacteria and viruses may survive a long period in the unsaturated zone until they are transported into the groundwater.

Elimination is a combined effect of the physical (temperature), biological, and chemical conditions. The biological factors are the most important factors for the survival of pathogenic bacteria and viruses. Elimination is faster at high temper-

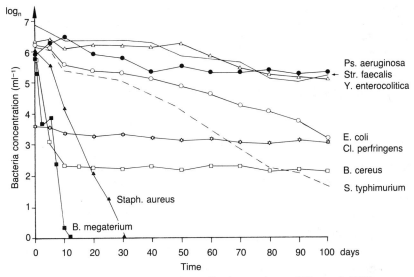

Fig. 1. Persistence of bacteria in sterilized groundwater (Filip et al. 1983)

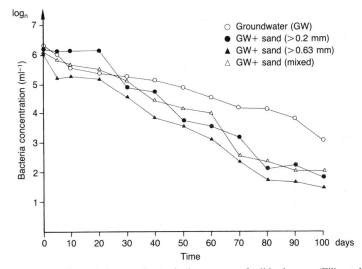

Fig. 2. Persistence of bacteria in groundwater in the presence of solid substance (Filip et al. 1983)

ature (37 °C), at pH-values of about 7, at low oxygen concentration, and at high content of dissolved organic substances. This type of water is encountered in contaminated soils and groundwater. In this case, the autochthonic bacteria are active and act antagonistically to the pathogenic microorganisms. The elimination constant depends on the physical, chemical, and biological parameters mentioned earlier and is specific for the different microbial species (Figs. 1–4). It is not possible to predict with the necessary accuracy the elimination constants on the basis of controlling factors, therefore, they must be measured for each specific species

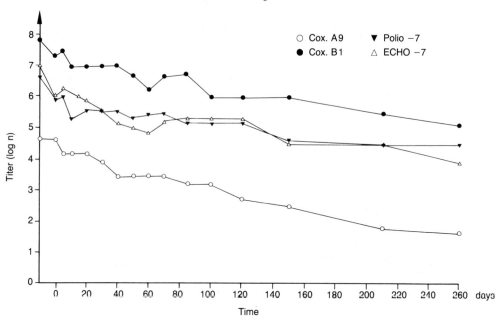

Fig. 3. Persistence of viruses in sterilized groundwater (Dizer et al. 1983)

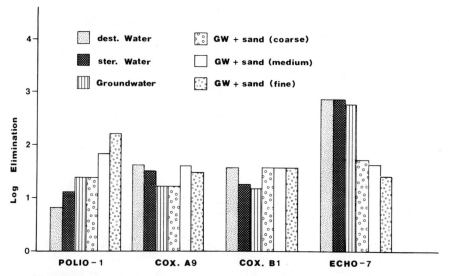

Fig. 4. Persistence of viruses in water in the presence of solid substance (Dizer et al. 1983)

and environment. The published values are contradictory and vary over a broad range (Matthess and Pekdeger 1981, Althaus et al. 1982).

The published data on the removal of bacteria and viruses in the subsurface are not consistent with the long survival times. Therefore, the survival time can not be the only criterion for the purifying effect of subsurface passage of water and the physicochemical transport processes are to be considered.

14.3 Transport Processes

A contaminated point source disperses as it travels through the unsaturated zone and enters the groundwater, which causes a distribution of the contaminants in time and space, so that the concentration in the contaminated plume decreases with time and with the transport distance. This process can be described by the general transport equation, Bear 1972 [Eq. (2)].

$$\frac{\delta C}{\delta t} = \text{div}\left(\frac{D}{R_d} \cdot \text{grad } C - \frac{V_w}{R_d} \cdot \text{grad } C\right) - \lambda C \tag{2}$$

$$
\begin{aligned}
D &= \text{coefficient of hydrodynamic dispersion} = D' + D_d + D_e \\
D' &= \text{coefficient of hydromechanic dispersion} \\
D_d &= \text{diffusion coefficient} \\
D_e &= \text{coefficient of the active mobility of bacteria} \\
\text{grad } C &= \text{concentration gradient} \\
V_w &= \text{average groundwater velocity} \\
R_d &= \text{retardation factor} \\
\lambda &= \text{elimination constant.}
\end{aligned}
$$

The dispersion equation in this form includes a term for the active mobility of bacteria, which can be considered as a random tumbling. The active mobility decreases with decreasing temperature. Maeda et al. (1976) give for *Escherichia coli* velocities of $1-0.1$ m d^{-1} at $20-10$ °C.

The lateral dispersion of bacteria is greater than that of a conservative tracer. Seiler and Alexander (1983) reported a broader lateral distribution of *E. coli* compared with the conservative tracers, Bromine-82 and fluoresceine. On the other hand, the longitudinal distribution pattern of *E. coli* is affected by filtration processes. The bacteria are transported mainly in the large pores, which causes after long travelling distances the formation of a narrow breakthrough curve (Havemeister et al. 1983, Seiler and Alexander 1983). The breakthrough curves of conservative tracers used simultaneously with bacteria are broader, as the tracer travels also in smaller pores with low velocities (Fig. 5).

The velocity of seepage water in the unsaturated zone is generally very slow (Matthess et al. 1979). Even in humid climates, the seepage velocity is rarely faster than 1 m y^{-1} (Moser and Rauert 1980), in arid climates the velocity is slower due to lower annual groundwater recharge. Only after heavy rainfalls or due to sewage reclamation by land treatment, the seepage velocity in coarse soils may be much higher (up to few m/day). Artificial or natural cracks may also allow very fast infiltration. A frost crack in the soil led in 1927, in Southwest Germany to a wide spread typhoid epidemy (Althaus et al. 1982).

When the transport time in the unsaturated zone is low enough (> 200 days), a considerable elimination of the bacteria and viruses can take place.

The groundwater velocities are generally much higher than seepage velocity (Table 2). If the level of the water table changes, the bacteria may also be immobilized in the capillary fringe. These bacteria can be remobilized even after a few months, when the water table rises after heavy rainfall (Fig. 6).

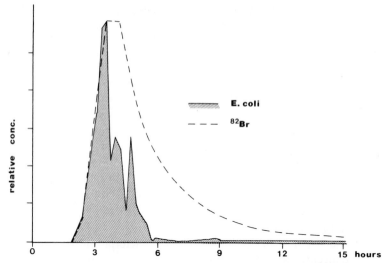

Fig. 5. Breakthrough curves of *E. coli* and ^{82}Br in coarse gravel in a flow distance of 20 m (Seiler and Alexander 1983)

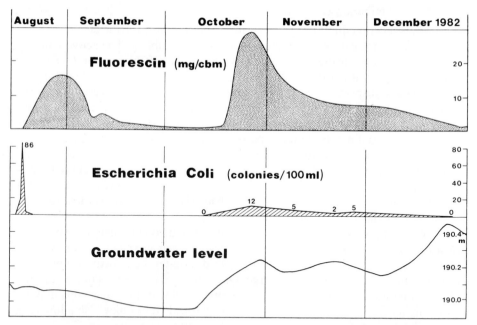

Fig. 6. Effect of groundwater level fluctuations on bacteria transport (Kaess et al. 1983)

A high dilution is expected when the bacteria and viruses pass slowly through the unsaturated zone in a vertical direction and enter the generally faster flowing groundwater in approximately a lateral direction (Anonymous 1978).

The transport velocity of the microorganisms may differ from the seepage and groundwater flow velocity. Microorganisms are subjected to adsorption on soil

Table 2. Groundwater velocities in different aquifers types
(Matthess and Pekdeger 1981)

Sand	$1 - 10$ m d^{-1}
Gravel	$1 - 100$ m d^{-1}
Hard rock	$0.3–8,000$ m d^{-1}
Karstic rocks	$-2,600$ m d^{-1}

particles (Fig. 7). The equilibrium between the concentrations of the suspended (C_s) and adsorbed (C_a) microorganisms (Eq. 3) can be described by the Freundlich isotherm (Berg 1967, Merkli 1975):

$$C_a = k \cdot C_s^n .\tag{3}$$

The empiric constants k and n are assumed to be specific for the investigated soil, rock, and microorganisms. The adsorption of bacteria and viruses takes place quite rapidly (24 h and 2 h, respectively; Althaus et al. 1982). Desorption velocity is less known and should be measured in future investigations.

The continuous adsorption/desorption causes a retardation of the microorganisms with respect to water transport. The retardation is described by the retardation factor R_d, the quotient of mean water velocity v_w to mean transport velocity of microorganisms v_m [Eq. (4)].

$$R_d = \frac{v_w}{v_m}.\tag{4}$$

The retardation factors can be obtained by laboratory and field tests (Matthess and Pekdeger 1981). In laboratory and field experiments, retardation factors between 1 and 2 were found for the indicator bacteria *E. coli* and the tracer bacteria *Serratia marcescens* (Havemeister and Riemer 1983, Havemeister et al. 1983, Jung and Schröter 1983, Schröter 1983), if the scale of the experiments were large

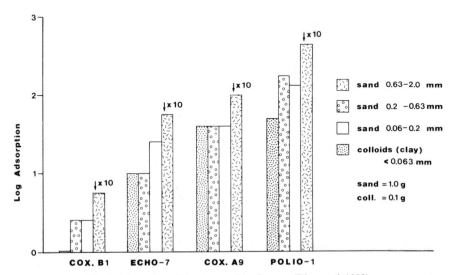

Fig. 7. Adsorption of viruses on solid substance (Dizer et al. 1983)

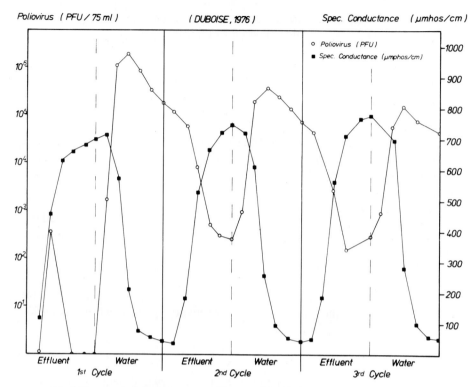

Fig. 8. Transport of viruses in sewage effluent and destilled water (Duboise et al. 1976)

enough (> few meters). In small scale experiments, the used formalism yields an apparently higher retardation factor because a certain portion of the bacteria is held back by the filtration mechanisms, which are reversible over short distances. Most of the pathogenic viruses, especially the polio viruses, have high retardation factors which range up to 500.

Model calculations using the data of known adsorption coefficients and elimination constants, show that the underground passage can provide a very effective protection against virus contamination. However, according to other observations, the viruses can be desorbed again when the chemistry of water changes (Fig. 8; Duboise et al. 1976). For example, the viruses can be desorbed when cation concentrations decreases, e.g., due to a very heavy rainfall, thus, enabling the viruses to travel further. This may also explain why generally in arid climates, the land treatment of sewage is reported to be more effective than in humid regions.

14.4 Filtration Processes

The removal of microorganisms from subsurface water is a complex process. It includes the time depending mechanism described earlier and the filtration mechanism, which depends on the length of the transport path. The filtration efficiency, which is defined as the ratio of the final concentration C_x to the initial concentration C_o can be described by a factor λ_f (Iwasaki 1937).

$$C_x = C_o \cdot e^{-\lambda_f \cdot x}. \tag{5}$$

The microorganism transport may be limited by the pore size of the subsurface material as compared to the size of the microorganisms. However, mechanical filtration is not very effective in sandy and gravelly soil due to the small diameters of bacteria (0.2–5 µm) and viruses (0.25–0.02 µm) compared with the diameters of the pores, which (e.g., in uniform sands) are generally larger than 40 µm. In natural sediments with heterogenic grain size distribution, only a small fraction of the pore diameters can interfere with bacteria transport ($> 10\%$).

The particle accumulation on a solid substance surface is mainly due to sedimentation, flow processes, diffusion, and interception (Fig. 9; Yao et al. 1971): Yao et al. (1971) give the following equations for filtration efficiency [Eqs. (6)–(10)]:

$$\ln \frac{C}{C_0} = -\frac{3}{2} \cdot (1-n) \cdot \alpha \cdot \eta \left(\frac{L}{d}\right), \tag{6}$$

$$\eta = \eta_D + \eta_I + \eta_G, \tag{7}$$

$$\eta_D = 4.04 \, \text{Pe}^{-2/3} = 0.9 \left(\frac{kT}{\mu d_p d v_f}\right)^{2/3}, \tag{8}$$

$$\eta_I = \frac{3}{2} \left(\frac{d_p}{d}\right)^2, \tag{9}$$

$$\eta_G = \frac{(\varrho_p - \varrho)g d_p^2}{18 \mu v_f} \tag{10}$$

with C_0, C = influent and effluent concentration
n = porosity
α = stabilization coefficient
η = single collector efficiency
d = grain size of filter material (m)
d_p = diameter of suspended particles (m)
L = bed depth (m)
η_D, η_I, η_G = single collector efficiencies for diffusion, interception, sedimentation
P_e = Peclet number $= V_a/D_d$
k = Boltzmann constant $(1.38054 \cdot 10^{-23} \, \text{J/K})$
T = absolute temperature (K)
μ = water viscosity $(N \cdot s \, m^{-2})$
v_f = filter velocity $(m \, s^{-1})$
ϱ_p, ϱ = density of suspended particles and water $(kg \, m^{-3})$
g = gravity constant $(0.80665) \, m \, s^{-2}$.

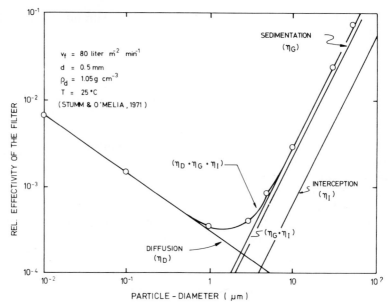

Fig. 9. Filtration parameters in porous media (Yao et al. 1971)

The comparison of filtration efficiency of clay minerals ($< 2\ \mu m$ fraction, $\varrho_p = 2.7\ g\ cm^{-3}$) with that of bacteria ($\varrho_p = 1\ g\ cm^{-3}$) show that bacteria are eliminated more effectively than clay minerals. The filtration efficiency for clay minerals predicted from Eqs. (6)–(10) agrees reasonably well with the experimental results. The reason for the deviation of bacteria removal efficiency from theory is presumably due to the active mobility of bacteria ($\simeq 0.1\ m\ d^{-1}$). The effective diameter of a bacterium is also greater than 1 μm due to its irregular shape and the presence of filaments on their surface.

The intensity of bacteria and virus attachment to the underground solid materials depends on the adsorption mechanism. The generally negatively charged bacteria and viruses are strongly adsorbed by anionic adsorbents and only slightly by cationic adsorbents. The solid substances of the underground are generally negatively charged. Exceptions are the iron and manganese hydroxides and humic substances at low pH values. The negatively charged microorganisms stay in suspension at high pH values (Fig. 10), as the repulsive electrostatical forces are stronger than the Van der Waals forces.

The dissolved cations in water decrease the repulsive forces of the grain surfaces. Monovalent cations are adsorbed by the solid substance and decrease their charge deficiency. Bivalent cations can also cause a positive charge deficiency so that the electrostatic forces can be more efficient in bacteria and virus adsorption. Under these conditions, the mass forces are more effective and an accumulation of particles can take place. This can be demonstrated by the dependence of virus adsorption on the solute concentration in the water mentioned earlier (Fig. 8).

Very important for the filtration efficiency is the duration of the contaminating process and the initial concentration of contaminants. During a continuous

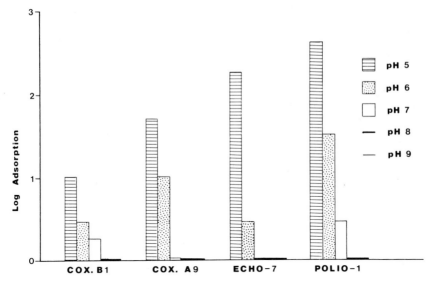

Fig. 10. Adsorption of viruses at different pH values (Dizer et al. 1983)

contamination process by organic substances and microorganisms, the contaminated plume becomes smaller with time because elimination and filtration mechanisms are favored. At very high initial concentrations, flocculation and aggregation can occur at the source of contamination so that only a limited transport into the aquifer can take place.

References

Althaus H, Jung KD, Matthess G, Pekdeger A (1982) Lebensdauer von Bakterien und Viren in Grundwasserleitern. Umweltbundesamt Materialien 1/82: 190, Schmidt Verlag

Anonymous (1978) Jahresbericht des Institutes für Radiohydrometrie der gsf. Neuherberg, München

Bear J (1972) Dynamics of fluids in porous media. Am Elsevier, Environ Sci New York, pp 764

Berg G (ed) (1967) Transmission of viruses by the water route. Wiley-Interscience, New York

Dizer H, Seidel K, Lopez JM, Fiup Z, Milde G (1983) Verhalten enterotroper Viren im Grundwasser unter Modellbedingungen, Forum „Mikroorganismen und Viren in Grundwasserleitern". Proc DVGW Wasserfachl Aussprachetag München '83, March 1–4 (in press)

Duboise SM, Moore BE, Sagik BP (1976) Poliovirus survival and movement in a sandy forest soil. Appl Environ Microbiol 31:536–543

Environmental Protection Agency (1978) Human viruses in the aquatic environment. Environ Res Center, EPA-Rep 570/g-78/006, pp 37

Filip Z, Kaddu-Mulindwa D, Milde G (1983) Überlebensdauer einiger pathogener und potentiell pathogener Mikroorganismen im Grundwasser, Forum „Mikroorganismen und Viren in Grundwasserleitern". Proc DVGW Wasserfachl Aussprachetag München '83, March 1–4 (in press)

Gerba CP, Keswick BH (1981) Survival and transport of enteric bacteria and viruses in groundwater. Stud Environ Sci 17:511–515

Havemeister G, Riemer R (1983) Transport von Bakterien und Viren in verschiedenen porösen Medien, Forum „Mikroorganismen und Viren in Grundwasserleitern". Proc DVGW Wasserfachl Aussprachetag München '83, March 1–4 (in press)

Havemeister G, Riemer R, Schröter J (1983) Lebensdauer und Transport in typischen Grundwasser-
    leitern: Glazifluviatile Sande im Segeberger Forst, Forum „Mikroorganismen und Viren in Grund-
    wasserleitern". Proc DVGW Wasserfachl Aussprachetag München '83, March 1–4 (in press)
Hirsch P, Rades-Rohkohl E (1983) Die Zusammensetzung der natürlichen Grundwasser-Mikroflora
    und Untersuchungen über ihre Wechselbeziehungen mit Fäkalbakterien, Forum „Mikroorganis-
    men und Viren in Grundwasserleitern". Proc DVGW Wasserfachl Aussprachetag München '83,
    March 1–4 (in press)
Iwasaki T (1937) Some notes on sand filtration. J AWWA 29:1591–1602
Jung KD, Schröter J (1983) Lebensdauer und Transport in typischen Grundwasserleitern: Halterner
    Sande, Forum „Mikroorganismen und Viren in Grundwasserleitern". Proc DVGW Wasserfachl
    Aussprachetag München '83, March 1–4 (in press)
Kaess W, Ritter R, Sacré C (1983) Lebensdauer und Transport von Bakterien in typischen Grundwas-
    serleitern: Oberrheinische Schotterebene, Forum „Mikroorganismen und Viren in Grundwasser-
    leitern". Proc DVGW Wasserfachl Aussprachetag München '83, March 1–4 (in press)
Maeda J, Mimae YF, Osawa P (1976) Effect of temperature on mobility and chemotoxis of Escherichia
    coli. Bacteriol 127:1039–1046
Matthess G, Pekdeger A (1981) Concepts of a survival and transport model of pathogenic bacteria and
    viruses in groundwater. Sci Total Environ 21:149–159
Matthess G, Pekdeger A, Rast H, Rauert W, Schulz HD (1979) Tritium tracing in hydrogeochemical
    studies using model lysimeters. Isot Hydrol 1978 2:769–785 IAEA Vienna
Merkli B (1975) Untersuchungen über Mechanismen und Kinetik der Elimination von Bakterien und
    Viren im Grundwasser. Naturwiss Diss ETH, Zürich
Moser H, Rauert W (1980) Isotopenmethoden in der Hydrogeologie. Borntraeger, Berlin
Schröter B (1983) Der Einfluß von Textur- und Struktureigenschaften poröser Medien auf die Disper-
    sivität. Ph D Thesis, Univ Kiel
Seiler KP, Alexander I (1983) Lebensdauer und Transport von Bakterien in typischen Grundwasser-
    leitern – Münchener Schotterebene, Forum Mikroorganismen und Viren in Grundwasserleitern.
    Proc DVGW Wasserfachl Aussprachetag München '83, March 1–4 (in press)
Yao KM, Habibian MT, O'Melia CR (1971) Water and Wastewater filtration: concepts and applica-
    tions vol 5. Dep Environ Sci Techn Univ N Carolina, Chapel Hill, pp 1905–1112

E. Unsaturated Zone and Groundwater: Pollution Sources

Introductory Comments

J. GOLDSHMID

The three chapters which follow all discuss various pollution sources which endanger the groundwater aquifers. They all concentrate on widespread sources, such as agriculture, reclaimed sewage, and animal and household waste. Unfortunately, all three describe semi-arid conditions (Israel and Southern California) and neglect industrial waste.

The unsaturated zone is the buffer between human activity and the groundwater sources. As such, it serves two functions: as reactor and as storage reservoir. Unlike from a storeroom, it is almost impossible to retrieve a pollutant from the unsaturated zone. A pollutant which enters the top soil is transferred by the water movement through the big reactor, and if it does not decompose, or become consumed by vegetation, or attached to the soil material, it will finally reach the aquifer and contaminate our groundwater supplies. Thus one can think of the unsaturated zone as a pollutant-filled time bomb, which ticks slowly, but which will eventually explode.

Letey and Pratt (Chap. 15) concentrate on fertilizers and pesticides as the main agricultural pollutants. In the fertilizer group, they differentiate between the nitrogen and the phosphorus and potassium compounds. Nitrogen may occur in the soil in many organic and mineral forms. Nitrate, which is very mobile, is the final stable product of nitrogen compounds oxidation and endangers the underground water. Phosphorus and potassium compounds are not degraded, but are also not mobile in most soil systems and as such do not form a threat to the underground water.

Almost all pesticides, which are presently used commercially degrade rapidly and/or adsorb strongly to the soil and do not contribute to groundwater pollution. Pesticides that endanger the water supplies have, in almost all cases, been banned from the market.

Ronen, Kanfi, and Magaritz (Chap. 16) try through a material balance to identify the source of the nitrates in Israel's coastal plain aquifer. They conclude that the reclamation of the coastal swamps and the cultivation of virign land during the first half of this century are the major sources of nitrogen. The nitrogen stored in the unsaturated zone of the coastal plain aquifer of Israel is estimated by them to be 100 times greater than the quantity presently dissolved in the groundwater and 500 times greater than the average quantity of nitrate that had reached the groundwater yearly in the last decade.

Waldman and Shevah (Chap. 17) consider the disposal of domestic and agricultural waste and the re-use of treated sewage as the major sources of groundwater pollution. They suggest the use of a fraction of the nitrogen that will reach the

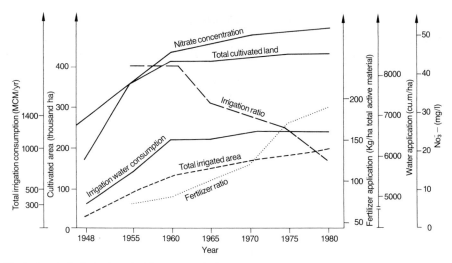

Fig. 1. Trends in irrigation, fertilizer use and nitrate concentration in Israel's coastal plain aquifer (1948–1980)

groundwater, which they name the "potential pollution load", to compare different pollution sources. They advocate the use of engineering methods in the disposal of waste in order to protect the groundwater quality. Two such examples are the preparation of a national masterplan for disposing of domestic waste through landfill operation and the use of animal waste for energy production through anaerobic digestion.

All three chapters discuss nitrate pollution of groundwater. This is natural since many aquifers in various countries became polluted by nitrate in the last century. Letey and Pratt (Chap. 15) and Waldman and Shevah (Chap. 17) both take the well-known stand that fertilizers, waste, and treated sewage are the major sources of nitrate contamination of groundwater. Ronen, Kanfi, and Magaritz (Chap. 16) on the other hand, reach the conclusion that reclamation of swamp land and cultivation of virgin land are the major sources of the nitrate pollution of Israel's coastal plain aquifer. The difference between these two opinions is the same as the difference between being able to control nitrate pollution by such methods as suggested by Waldman and Shevah and not being able to control the addition of nitrogen compounds to the unsaturated zone. Ronen, Kanfi, and Magaritz (Chap. 16) emphasize the major role played, in their opinion, by the unsaturated-saturated interface in the nitrification and denitrification of nitrogen compounds.

It is of interest to plot a Ronen, Kanfi, and Magaritz curve presenting the change of nitrate concentration with time on Fig. 2 of Waldman and Shevah (Chap. 17) (see attached Fig. 1). The similarity in the shapes of the nitrate concentration and the irrigation water consumption curves and the coincidence of the decrease in the irrigation ratio with the sharp decrease in the rate of nitrates build-up is striking. It is, thus, possible that the decrease in irrigation water consumption coupled with the change to drip irrigation are the major factors in the decreased rate of nitrate contamination of Israel's coastal aquifer. The fact that water, in excess of the evaporation and evapotranspiration, is required to move the nitrogen compounds down the unsaturated zone, is well-known and is discussed here by Letey and Pratt.

The protection of the groundwater aquifers is of prime importance to all nations, but even more so in a semi-arid climate. That is why it is so important to identify the sources and mechanisms of aquifer pollution. The chapters presented in Part E are a good example of the research work still needed in order to be able to protect our groundwater resources.

15. Agricultural Pollutants and Groundwater Quality

J. Letey and P. F. Pratt

15.1 Introduction

Agricultural systems of a global basis are highly variable depending upon climate, soil, water availability, crop, mechanization, economic conditions, etc. Contribution of agriculture to groundwater degradation is therefore expected to be site-specific. This report will attempt to deal with the more common agricultural practices and draw general conclusions. Deviations from the general conclusions are to be expected under specific conditions.

Factors affecting groundwater degradation by potential pollutants include application rate, degradation, and movement of the pollutant from the land surface to groundwater. Movement from the land surface to groundwater is dependent upon water flow, which serves as a carrier, and adsorption of the pollutant by soil material, which retards pollutant movement. Except for application rates, these mechanisms have been quantitatively analyzed in previous chapters. The pollutants considered in this workshop can be classified as plant nutrients, heavy metals, synthetic organic chemicals, inorganic salts (salinity), and disease causing organisms (bacteria and viruses). This report will concentrate on the relationship between common agricultural practices and the factors involved in potential groundwater degradation by these various types of water pollutants.

15.2 Application Rates

Plant nutrients (fertilizers) are commonly applied to increase crop production. Nitrogen, phosphorus, and potassium are the major plant nutrients which are applied in by far the greatest amount. Increasing amounts of these major plant nutrients have been applied on a worldwide basis yearly. During 1980, the worldwide application of N, P_2O_5, and K_2O was 57.2×10^6, 30.0×10^6, and 23.9×10^6 Mg, respectively.[1]

Fixation by rhizobium-legume combinations or by free-living organisms add N directly to the crop or the soil in organic forms. Animal manure, sewage sludge, crop residues, and roots also add organic materials which then mineralize at various rates to given NH_4 and NO_3.

---

[1] White, W.C., The Fertilizer Institute, 1015 18th Street, N.W., Washington, D.C. 20036, USA (pers. comm.)

Agricultural practices, therefore, clearly apply relatively large amounts of N, P, and K and these must be considered as potential groundwater pollutants.

Heavy metals are not commonly applied and disease-causing organisms do not commonly result from agricultural practices, so they are not considered as contributors to groundwater pollution from agriculture systems.

Synthetic organic chemicals, usually in the form of pesticides, are frequently applied in agricultural operations. Production and sales of synthetic organic pesticides in the U.S.A. have increased every year between 1957 and 1977 except for 4 years when there were slight declines. The highest year was 1975 when 0.72×10^6 Mg were marketed (Fowler and Mahan 1978). Although there has been a levelling off of farm pesticide use in the United States, consumption in the developing countries is expected to increase by 4% to 5% yearly (Eichers and Serletis 1982). From an application point of view, synthetic organic chemicals must be considered as potential groundwater pollutants resulting from agriculture.

Salts are concentrated in the soil solution through evapotranspiration. Thus, water moving below the root zone toward the groundwater may carry a substantial load of dissolved salts. This is particularly true under semi-arid irrigated agriculture where the irrigation water may be relatively high in electrolyte concentration. Although degradation of groundwater by salts may be severe and occur on a fairly large global land area basis, the topic will not be discussed further in this report, because of space limitation. Furthermore, salinity does not appear to be the main thrust of the workshop.

In summary, plant nutrients and synthetic organic chemicals have been identified as being extensively applied in agricultural operations and must be considered further for their contribution to groundwater pollution. On the other hand, heavy metals and disease-causing organisms have been eliminated from general consideration because of the low contribution of agricultural practices to these pollutants.

15.3 Movement to Groundwater

15.3.1 Water Movement

Water reaching land surface either through precipitation or irrigation in excess of evapotranspiration moves beyond the root zone toward the water table. Rain-fed agricultural systems, where the precipitation approximates or is less than evapotranspiration, have very little water movement toward the water table. Agriculture under these conditions, therefore, contributes little to groundwater pollution. On the other hand, where precipitation exceeds evapotranspiration, water moves to the water table and can serve as a carrier for applied pollutants.

Irrigation has become a prevalent agricultural practice where precipitation is less than evapotranspiration. Water movement to the groundwater is expected under irrigated agriculture. First, some leaching is required to prevent salt build-up in the root system to detrimental levels. Secondly, water addition in excess of the leaching requirement is required for optimum production of most irrigated

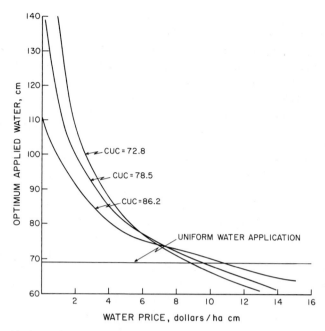

Fig. 1. Relationship between optimum water application to corn and water price for various uniformities of water infiltration

crops except for conditions of uniform water intake rate over an entire irrigated field (a condition not commonly encountered). The curves presented in Fig. 1 illustrate the effect of nonuniformity of water infiltration on optimum water application to a field of corn (Letey et al. 1984). Optimum water application is defined as that amount which maximizes the farmer's profit. Note that for common agricultural water prices, the optimum water application under nonuniform infiltration conditions exceeds that which would be applied under perfectly uniform infiltration. Water which is applied in excess of that indicated for the uniform case moves to the groundwater and serves as a pollutant carrier.

In summary, water flows beyond the root zone toward the water table under common agricultural practices. Exception to this condition is rain-fed agriculture when precipitation does not exceed evapotranspiration.

15.3.2 Adsorption and Degradation

Groundwater may be protected from applied pollutants through degradation, which removes the pollutant from the system or adsorption, which retards movement. These processes, particulary adsorption, are discussed in detail in previous reports. Nevertheless, to make our presentation complete, we will briefly consider these processes as related to the potential pollutants from agricultural practices.

Nitrogen may occur in the soil in various organic and mineral forms. NO_3 is the final stable product of the overall mineralization process in well aerated soils and is of greatest concern from a groundwater quality point of view. Denitrification does transform NO_3 to nitrogenous gas under anoxic conditions with the presence of an energy source for the denitrifying organisms. Complete NO_3 removal through denitrification is not expected, however, under most agricultural conditions. Thus, variable amounts of NO_3 are available for leaching to the groundwater. NO_3 is very mobile, generally moving as rapid as the water. Only in soils containing a significant positive charge can one expect some retardation of NO_3 movement.

Phosphorus and potassium compounds are not degraded and lost from the system. They are, however, not mobile in most soil systems. Phosphate would be expected to move only in sands or organic soils free of carbonate or iron and aluminum oxides. Potassium has very low mobility through the soil and the rate of transport would be inversely proportional to the cation exchange capacity and the amount of micaceous materials.

Summarizing the plant nutrients, only NO_3 would be expected to move to the groundwater except under unique soil conditions. Because of the relatively high amounts of N added to agricultural soils and the high NO_3 mobility, agriculture might be expected to contribute to groundwater pollution with NO_3.

Numerous organic compounds are applied to agricultural soils as pesticides. These chemicals vary greatly in terms of degradation and adsorption. Those pesticides which are either rapidly degraded or highly adsorbed by soil would not be expected to contribute significantly to groundwater degradation. On the other hand, pesticides which are not readily degraded or adsorbed may be transported to the groundwater. Croll (1974) provides a summary on the toxicity, degradation, solubility, and relative mobility of various types of pesticides. The organochlorine insecticides are generally classed as having a very low degradation rate, low water solubility, and relatively high adsorption. Adsorption is not only a property of the chemical in question, but also the soil. Soils high in clay and/or organic content generally have higher adsorptive capacity for pesticides as compared to a low organic matter sandy soil. Organophosphorus insecticides exhibit a wide range of degradation, water solubility, and adsorption and each compound must be judged separately as a potential pollutant. In general, however, organophosphorus compounds are less persistent in soil than the organochlorine insecticides. For example, in the study on the persistence of nine organophosphorus insecticides in surface waters, they were found to be significantly degraded within 1 week and were not detectable after 8 weeks (Eichelberger and Lichtenberg 1971). This relatively high degradation rate would be expected to decrease the organophosphorus insecticides potential for groundwater pollution.

Because of the relatively rapid degradation and/or high adsorption by many applied pesticides, they would not expect to contribute significantly to groundwater pollution. On the other hand, there are unique pesticides which are relatively mobile and not rapidly degraded which can cause groundwater degradation and these will be reported later. It is further recognized that the adsorptive behavior of pesticides is highly dependent upon soil properties and the potential for groundwater degradation is considerably higher under sandy soils than finer textured or organic soils.

15.4 Results

Before reviewing results of measurements of various pollutants in ground-water, it is important to establish criteria to judge the results. In other words, what concentration of a given chemical would be considered as significantly high to classify water a polluted? A polluted water is one in which its properties have been altered to negatively affect its beneficial use. Thus the criterion for water pollution also depends upon its intended use. For purposes of this report, ground-water use will be considered to be primarily for domestic and agricultural uses. Of these, domestic use generally has the more stringent requirement for purity.

The primary drinking water standard in the U.S.A. for nitrate is 10 mg l^{-1} of NO_3-N. A national technical advisory committee proposed the standards presented in Table 1 for various pesticides in public or farmstead water supplies. Results will be interpreted relative to these recommendations.

Detection of pollutants in water is also dependent upon the sensitivity of the measurement. The detection or lack of detection by various investigators must be related to the sensitivity of analysis. Except for NO_3 concentrations which will be reported in mg l^{-1} numercial values for other compounds will be expressed in units of $\mu g l^{-1}$. Detection limits will also be reported for each investigation.

15.4.1 Nitrates

Miller and Smith (1976) calculated a N balance for the southern San Joaquin Valley of California in an area of 1,772,000 ha containing 1,039,000 ha under irrigation. The N harvested in 1971 was 54, 73, 43, 19, and 37% of the fertilizer N inputs, respectively, for field, hay, vegetable, tree, and vine crops, leaving substantial amounts of N to be lost by volatilization, denitrification, or leaching. The average NO_3 leached was estimated as 45 kg ha^{-1} yr^{-1} for the total area or 76 kg ha^{-1} yr^{-1}, if it is assumed to come from irrigated lands. If the N leached was contained in 25 cm of water per year, the concentration would be 30 mg l^{-1}. Measurements by Rible et al. (1979) at 18 irrigated sites in the San Joaquin Valley gave a NO_3-N concentration average of 39 mg l^{-1} in the water beneath the root zone. The range of concentrations was 9 to 163 mg l^{-1} for individual sites.

Lund (1982) reported mean NO_3-N concentrations of 40 to 136 mg l^{-1} in the water of the unsaturated zone beneath several fields in the Santa Maria Valley of coastal California. Data from Rible et al. (1979) for 83 sites show NO_3-N concentrations in the range of 20 to 49 mg l^{-1}.

Clearly, considerable NO_3 is moving below the root zone of agricultural crops in California and most probably other agricultural areas with similar cultural practices, so that there is a significant potential for groundwater pollution.

Avnimelech and Raveh (1976) found that the NO_3 leached below the root zone was less than the difference between N applied and N removed by crop. They provided evidence that considerable denitrification occurred in the root zone, particularly for clay soils and soils receiving high N application. They reported that denitrification below the root zone was not great.

Navone et al. (1963) sampled about 100 groundwater sources at 781 points (in 8 counties) four times a year in southern California. Considering only the highest reading for a given point, 11.4% had greater than 10 mg l^{-1}, which were fairly equally divided among the counties. The source of NO_3 contributing to these high concentrations could not be identified specifically. The authors report that large amounts of N compounds were used for crop fertilization in the vicinity of wells. In other instances, sewage and industrial wastes were being disposed of by percolation into lands overlying some of the groundwater basins studied and some wells studied are in areas where individual household sewage is disposed.

Animal wastes from feed lots also contribute to groundwater pollution from NO_3. Stewart et al. (1968) made measurements in the profiles from feedlots, irrigated fields, and dry land crop fields in the South Platt River Valley in Colorado. They found highly variable results, but the highest concentrations and quantities were detected under feedlots. Estimates on irrigated crop land (excluding alfalfa) were that 28 to 34 kg ha^{-1} yr^{-1} moved beyond the root zone. The concentrations of NO_3-N found in dry land agriculture were similar to those under irrigated land, but there was much lower water flux.

In summary, common application of fertilizer N and high mobility of NO_3 suggests that the probalitity for groundwater degradation by NO_3 is relatively high under agricultural systems. Measurements of NO_3 in water below the root zone under intensively managed croplands and feedlots have demonstrated the high potential for groundwater degradation by NO_3.

15.4.2 Pesticides

Croll (1969) examined 74 samples of underground wells and springs in Kent, England and was unable to detect organochlorine insecticides in any of the samples. Limits of detection ranged from 0.0005 to 0.01 μg l^{-1} for various insecticides. Several of these sites were selected because they were most likely to show contamination because of the regular and intense insecticide application for many years.

Maddy et al. (1982) analyzed water samples collected from 54 wells in five counties in California where pesticides had been used for over 15 years. No samples had detectable residues of 1,3-dichloropropene at a minimum detectable level of 0.1 μg l^{-1}. No residues of 22 chlorinated hydrocarbons or 27 organophosphates were detected at a minimum detectable level of 5 μg l^{-1}. Eight wells were monitored for the presence of chloroallyl alcohol with a detectable limit of 0.6 μg l^{-1}. No chloroallyl alcohol was detected in any of the samples. Maddy et al. (1981 b) sampled 15 domestic supply wells for the presence of atrazine. The sites were selected in three countries of California to represent the "worst possible case" situation. For example, the area selected in San Joaquin County had a history of high use of atrazine and very permeable soils. With a minimum detectable limit of 2 μg l^{-1}, all 15 samples were negative. An investigation on carbofuran, which is a carbamate pesticide, was also conducted by Maddy et al. (1981 a). Six shallow wells were sampled where carbofuran had previously been used. Each water sample was analyzed for carbofuran and two of its metabolites with negative results for all three chemicals each with a minimum detectable limit of 1 μg l^{-1}.

Analysis of 19 pesticides, each with a minimum detectable limit of 0.01 μg l^{-1}, was conducted on the drainage effluent from 25 systems in the Coachella and Imperial Valleys of California. Of the 475 measurements (19 chemicals times 25 sites), 30 were above minimum detectable limit and the highest measured concentration was only 0.03 μg l^{-1} (Eichers 1979).

Samples were collected from 262 wells in California and analyzed for 1,2-dibromo-3-chloropropane (DBCP) by Peoples et al. (1980). A positive measurement was recorded on 90 of these samples with the minimum detection limit of 0.1 μg l^{-1}. Of the 90 positive samples, 2 had concentrations >20, 8 had concentrations between 10.0 and 19.9, 50 had concentrations between 1.0 and 9.9, and 34 had concentrations between 0.1 and 0.9 μg l^{-1}. The California Department of Health Services has set a temporary action level of 1 μg l^{-1} and recommended that water above this level should not be consumed by humans. The U.S. Environmental Protection Agency has under consideration a recommendation for a permanent ceiling level of 0.05 μg l^{-1}. DBCP, therefore, is one pesticide used in agriculture, which has significantly contributed to groundwater pollution. Continued use of DBCP is banned in many cases.

Richard et al. (1975) analyzed various waters collected in Iowa for atrazine, DDE, and dieldrin. Their detection limit was 0.0005 μg l^{-1}. They found atrazine in shallow wells in the alluvial planes of contaminated rivers (0.004 to 0.483 μg l^{-1}), but little or no atrazine in well systems outside of the alluvial plain. A concentration of 0.028 μg l^{-1} of DDE was found in one well in the alluvial plain, but little or no DDE or dieldrin in any of the other wells in Iowa.

Nicholson and Thoman (1965) reported concentration of parathion (an organophosphorus insecticide) at a level of 1 μg l^{-1} in well waters. According to Croll (1974) in his review, it is the only recorded instance of detected organophosphorus insecticide in groundwater. Note that this concentration is well below the permissible level presented in Table 1.

Table 1. Proposed standards for various pesticides in public or farmstead water supplies[a]

Pesticide	Permissible level (μg l^{-1})
Aldrin	17
Chlordane	3
DDT	42
Dieldrin	17
Endrin	1
Heptachlor	18
Lindane	56
Methoxychlor	35
Organic phosphates plus carbamates	100
Toxaphene	5
Herbicides, 2,4-D + 2,4,5-T, +	100

[a] U.S. Federal Water Pollution Control Adm. (1968). Water quality criteria – report of the National Technical Advisory Committee to the Secretary of the Interior

During 1979 and 1980 about 16% of the 8,000 domestic wells within 2,500 feet of potato farms in Long Island, New York were found to contain aldicarb residues and half of those exceeded the 7 $\mu g \, l^{-1}$ standard for total aldicarb residues set by the New York State Department of Health (Trautman 1981). Rothschild et al. (1982) investigated the presence of aldicarb in groundwater in the Central Sand Plain of Wisconsin. They made measurements at various depths under the water table and found the highest concentrations immediately below the water table. Fifty-one percent of the samples from the shallow depth had detectable (1 $\mu g \, l^{-1}$ detection limit) levels of aldicarb. The average concentration in those samples with detectable levels was 26.8 $\mu g \, l^{-1}$ and the range was 3.0 to 85.5 $\mu g \, l^{-1}$. Aldicarb, therefore, represents a significant risk to groundwater quality and already has been taken off the market in New York.

15.5 Case Study

Agricultural tile drainage waters in the San Joaquin Valley of California have been monitored for plant nutrients, pesticides, metals, and other miscellaneous constituents for several years. The California Department of Water Resources, the U.S. Bureau of Reclamation, and the State Water Resources Control Board have all had active involvement in this monitoring program. Inasmuch as the San Joaquin Valley represents diverse, intensive agriculture management, including ample fertilizer and pesticide application, the results from the monitoring program have been chosen to be presented as a case study in this report. Although concentration of various elements in tile drainage water may not precisely reflect these concentrations as they reach groundwater, they probably do represent a "worst case" situation. The information reported herein has been extracted from annual reports of the San Joaquin Valley Drainage Monitoring Program prepared by the San Joaquin District of the State of California, Department of Water Resources.

Samples were collected from the northern, central, and southern parts of the San Joaquin Valley. Average results from the three different areas have been reported separately. Only results from the central and southern sections will be presented in this report. Data from the northern are similar to results from the southern section.

The plant nutrients monitored were NO_3-N and PO_4-P. Results for NO_3-N are illustrated in Fig. 2. The average of the maximum measured concentration as well as the average NO_3-N concentrations are reported for various years. The average of the minimum NO_3-N concentrations ranged from 1 to 3 mg l^{-1} for both the central and southern areas. In general, the NO_3-N concentrations in the central area are higher than for the southern area. Measured concentrations are commonly above the U.S. primary drinking water standard. The quantity of NO_3-N in the tile effluent is also plotted for the two areas in Fig. 2. Average discharge in the central area ranged from 60 to 100 kg ha^{-1} and thus represents a significant quantity of N leached from the root zone. In the southern area, the average is approximately 10 kg ha^{-1}. There appears to be a trend toward decreasing NO_3-N

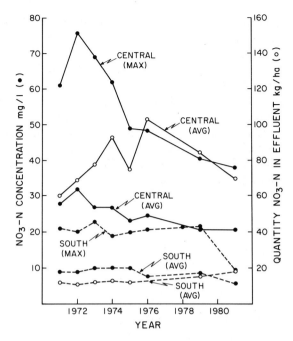

Fig. 2. NO_3-N concentrations and quantities in tile drainage effluents in the central and southern part of the San Joaquin Valley of California during various years

concentration in the central area with time. The decrease in concentration, however, is not accompanied by a decrease in total quantity drained. Apparently, the drainage volume has tended to increase with time resulting in lower concentration, but fairly stable total discharge.

The data illustrate that both NO_3 concentrations and mass emissions can vary over a wide range of values from very low to exceedingly high. Results, therefore, are as could be expected, highly site-specific. Nevertheless, clearly agricultural practices can and do at times contribute to groundwater pollution from NO_3-N. It should be recognized, however, that the NO_3-N collected in tile effluent may not always have been supplied from agricultural sources. Letey et al. (1977) reported data on NO_3-N in effluent from agricultural tile drains in California, which suggested that some alluvial soils on the west side of the San Joaquin Valley contain significant native N, which contributes both to high NO_3-N concentrations and total mass emissions in tile effluent in that area. Irrigation does, however, provide water to transport the N to the water table.

In contrast to NO_3-N, very low concentrations and mass emission of PO_4-P were observed in the subsurface tile effluents. For example, in 1981, the average PO_4-P concentrations were 0.04 and 0.85 mg l^{-1} in the central and southern areas, respectively. The average mass emissions were 0.10 and 2.09 kg ha^{-1} yr^{-1} in the central and southern areas, respectively. These data support the hypotheses that groundwater pollution from PO_4 would not be generally expected.

Table 2. Summation of identified chlorinated hydrocarbon compounds detected in subsurface drain effluents (1965–1981)

Year	Times sampled	Times detected	Percentage of times detected	Concentration (μg 1^{-1})			
				Maximum	Minimum[a]	Average[b]	Average[c]
1965	50	48	96	1.100	0.035	0.189	0.197
1966	105	78	74	1.175	0.003	0.112	0.146
1967	121	48	40	0.320	0.002	0.016	0.044
1968	79	30	38	0.543	0.003	0.040	0.105
1969	51	37	73	1.325	0.003	0.130	0.161
1970	60	46	77	2.850	0.003	0.275	0.375
1971	55	32	58	2.000	0.005	0.142	0.243
1972	44	37	84	0.480	0.003	0.075	0.089
1973	41	26	63	1.600	0.030	0.136	0.243
1974	43	35	81	5.000	0.004	0.361	0.443
1975	114	58	51	13.000	0.005	0.390	0.767
1976	80	46	57	0.540	0.005	0.055	0.096
1977	37	10	27	3.000	0.010	0.150	0.508
1978	37	26	70	0.610	0.010	0.038	0.049
1979	31	8	26	0.770	0.020	0.063	0.243
1980	27	11	41	1.300	0.010	0.130	0.303
1981	11	5	45	0.180	0.030	0.039	0.086

[a] Minimum of detected concentrations
[b] Average value includes "0" values when chlorinated hydrocarbons were not detected, including unknowns as DDT after 1968
[c] Average value includes only the detected chlorinated hydrocarbons, including unknowns after 1968

Table 3. Summary of minor elements detected in subsurface drain (1981)

Element	Standard[a] mg/l	Central area Concentration (mg 1^{-1})				Southern area Concentration (mg 1^{-1})			
		% Times detected[b]	Maximum	Minimum[c]	Average[d]	% Times detected[e]	Maximum	Minimum[c]	Average[d]
Arsenic	0.050	12	0.01	0.01	0.01	100	0.14	0.01	0.05
Barium	1.000	0				0			
Cadmium	0.010	2	0.01	0.01	0.01	0			
Chromium	0.050	98	0.17	0.01	0.04	0			
Lead	0.050	17	0.06	0.01	0.03	11	0.01	0.01	0.01
Mercury	0.002	32	0.004	0.001	0.002	0			
Silver	0.050	2	0.01	0.01	0.01	0			

[a] U.S. primary drink water standard
[b] Out of 41 samples
[c] Minimum of detected concentrations
[d] Average of detected concentrations
[e] Out of 9 samples

Results of measured chlorinated hydrocarbon compounds in the subsurface drain affluence are summarized in Table 2. Note that the average concentration was generally much closer to the minimum values than to the maximum reported value. The maximum values, therefore, apparently occurred far less frequently than the minimum values. Indeed, there were certain percentages of samples each year in which no chlorinated hydrocarbon compounds were detected. In comparing the results from Table 2 to the proposed standards in Table 1, it appears that even the maximum concentrations were well within the range which might be acceptable for public or farmstead water supplies.

Analysis for organic phosphorus compounds in 1981 resulted in detectable concentrations being measured in 2 out of 11 samples. The concentrations were 0.090 and 0.010 $\mu g l^{-1}$. The 1981 data on organic phosphorus compounds is typical of results achieved in previous years which are not presented in detail.

The drainage waters were analyzed for several trace elements. Results of those elements for which there is a U.S. primary drinking water standard established, are reported in Table 3. Except for chromium in the central area and arsenic in the southern area, the trace elements were detected in a relatively small percentage of the samples. Although there were individual samples, which exceeded the drinking water standard, generally, the elements were either not detected or detected in concentrations, which are not considered to be harmful. The prevalance of chromium in the central area is probably a result of the soil and not of agricultural operations.

References

Avnimelech Y, Raveh J (1976) Nitrate leakage from soils differing in texture and nitrogen load. J Environ Qual 5:79–82

Croll BT (1969) Organo-chlorine insecticides in water. Part I. Water Treat Exam 18:255–274

Croll BT (1974) The impact of organic pesticides and herbicides upon ground water pollution. Groundwater Pollut Eur Proc Conf 350–64. Source: JA Cole, Water Information Center, Inc. Port Washington, N.Y.

Eichelberger JW, Lichtenberg JJ (1971) Persistence of pesticides in river water. Environ Sci Technol 5(6)641–644

Eichers TR, Serletis WS (1982) Farm pesticide supply-demand trends, 1982. Agric Econ Rep 485:23

Fowler DL, Mahan JN (1978) The pesticide review 1977. US Dep Agric, Washington DC

Letey J, Blair JW, Devitt D, Lund LJ, Nash P (1977) Nitrate-nitrogen in effluent from agricultural tile drains in California. Hilgardia 45(9):289–319

Letey J, Vaux HJ Jr, Feinerman E (1984) Optimum crop water application as affected by uniformity of water infiltration. Agron J (submitted)

Lund LJ (1982) Variations in nitrate and chloride concentrations below selected agricultural fields. Soil Sci Soc Am J 46:1062–1066

Maddy KT, Richmond D, Siani N (1981 a). A study of the possible presence of carbofuran and its metabolites in groundwater. Report HS-871, June 5, 1981. Div of Pest Management, Environmental Protection and Worker Safety, Calif Dept of Food and Agric Sacramento

Maddy KT, Schneider F, Fong HR, Scott A, Fredrickson (1981 b) A study and analysis of the migration potential of atrazine into selected aquifers in selected counties of California in 1981. Report HS-890, August 20, 1981. Div of Pest Management, Environmental Protection and Worker Safety, Calif Dept of Food and Agric, Sacramento

Maddy KT, Fong HR, Lowe JA, Conrad DW, Fredrickson AS (1982) A study of well water in selected California Communities for residues of 1,3-Dichloropropene, chloroallyl alcohol and 49 organophosphate or chlorinated hydrocarbon pesticides. Bull Environ Contam Toxicol 29:354–359

Miller RJ, Smith RB (1976) Nitrogen balance in the southern San Joaquin Valley. J Environ Qual 5:274–278

Navone, Remo, Harmon JA, Voyles CF (1963) Nitrogen content in ground water in Southern California. Am Water Works Assoc J 55(5):615–618

Nicholson HP, Thoman JR (1965) Pesticide persistence in public water their detection and removal. In: Chichester Co (ed) Research in pesticides. Academic Press, London New York, pp 181–190

Peoples SA, Maddy KT, Cusick W, Jackson T, Cooper C, Fredrickson AS (1980) A study of samples of well water collected from selected areas in California to determine the presence of DBCP and certain other pesticide residues. Bull Environ Contam Toxicol 24:611–618

Rible JM, Pratty PF, Lund LJ, Holtzlcaw KM (1979) Nitrates in the unsaturated zone of freely drained fields. In: Pratt PF (ed) Nitrates in Effluents from Irrigated Lands. Report to NSF, May 1979. Available from Nat Technic Informat Serv, US Dept of Commerice, Springfield, VA pp 297–320

Richard JJ, Junk GA, Avery MJ, Nehring NL, Fritz JS, Soec HJ (1975) Analysis of various Iowa waters for selected pesticides: atrazine, DDE, and dieldrin-1974. Pestic Monit J 9(3):117–123

Rothschild ER, Manser RJ, Anderson MP (1982) Investigation of aldicarb in ground water in selected areas of the Central Sand Plain of Wisconsin. Groundwater 20(4):437–445

Stewart BA, Viets FG Jr, Hutchinson GL (1968) Agriculture's effect on nitrate pollution of groundwater. J Soil Water Conserv 23:13–15

Trautman N (ed) (1981) Progress report of the Long Island Pesticide Steering Committee, December 1981. Prepared jointly through the Res Office, Coll Agric Life Sci, Center of Environ Res, and Water Resour Program of Cornell Univ

16. Nitrogen Presence in Groundwater as Affected by the Unsaturated Zone

D. Ronen, Y. Kanfi, and M. Magaritz

16.1 Introduction

Existing data and new findings on the nitrogen balance in the Coastal Plain aquifer of Israel emphasize the importance of the unsaturated zone as a buffering link between the top soil and the water table. The unsaturated zone is a huge reservoir of nitrogen (Ayers and Branson 1973, Boyce et al. 1976). It also operates as a physicochemical reactor in which portions of both the natural and the anthropogenic nitrogen is mobilized (Reinhorn and Avnimelech 1973, Sollins and McCorison 1981), detained (Gasser 1969, Klein and Brodford 1979), or lost (Lind 1979, Oberman 1982). Thus, the processes occurring in this zone determine the quantity and the chemical forms of nitrogen reaching the water table (Steward et al. 1968).

The retarding property of this zone even for fast-moving anions, such as NO_3^-, is also recognized (Ayers and Branson 1973, Mercado 1980). Due to the lag time between the surface release of nitrates and groundwater contamination, it is difficult to establish a cause and effect relation (Ronen 1972).

The nitrogen-related processes, such as nitrification, denitrification, ammonification, and adsorption in the top soil, have been studied extensively and will not be reviewed in this paper.

This chapter deals with the mass balance of nitrogen in the Coastal Plain aquifer of Israel during this century. Two different nitrogen sources are discussed: exhaustive sources related to the organic matter stored in the top layers of the unsaturated zone and continuous nitrogen sources that are a consequence of ongoing human activity. The amount of nitrogen stored in the unsaturated zone has been calculated and processes modifying this nitrogen pool are presented. The nitrate history of groundwater has been related to inputs and the processes that modify these inputs.

16.2 Geohydrological and Historical Background of the Coastal Plain Aquifer

The Coastal Plain aquifer stretches along 120 km from Mount Carmel to the Gaza Strip; its width varies between 7 and 20 km. It has a wedge-like cross section with a maximum thickness of 180 m near the sea that tapers toward the east until it disappears near the foothills of the Samarian and Judean mountains.

The aquifer is composed of clastic sediments of Pleistocene age overlying impervious clays of Pliocene age. The Pleistocene sequence consists of interfingered continental and marine units composed of sandstone, calcareous sandstone, siltstone, and red loamy soils (Issar 1968). In the eastern half, the aquifer is uniform and phreatic. Toward the west, it is partially divided by intervening clay layers into subaquifers of which the lower ones are confined.

During the last glacial age, 18,000 years ago, when the sea was about 150 m below its present level (Milliman and Emery 1968) and the seashore tens of kms westward (based on the bathimetric map of Ginzburg et al. 1975), the Pleistocene sequence had to have been dry. With the sea level rise between 14,000 to 6,000 years ago (Tooley 1978) and the increase in the amount of precipitation (Gat and Magaritz 1980), the present aquifer began to fill. At the end of this period, the seawater-freshwater interface had moved eastward to its present position and coastal swamps developed (Fig. 1).

Mercado's work (1975) provides much useful data on the aquifer. The thickness of the present day unsaturated zone ranges from 4 m to 98 m, averaging 34 m. The amount of water stored in it is 11×10^9 m^3. The average transit time to the water table is 26.8 years, ranging from 1 to 50 years in most of the area. For some restricted regions higher values up to 800 years were calculated.

The aquifer is replenished by rain, 500 mm yr^{-1} on the average. Under natural conditions, groundwater drains to the sea. However, in recent years (from the 1940's), this flow decreased progressively to about 55×10^6 m^3 yr^{-1}. In some regions, the phreatic water table was lowered by as much as 15 m by over pumpage and in some places groundwater depressions actually developed.

At present, 400×10^6 m^3 yr^{-1} are pumped from groundwater through 1,600 wells. Since 1963, the aquifer has also been artificially replenished by surface waters during the winter months.

The settlement of the Coastal Plain area by the present population began in the late nineteenth century. At that time, most of the area was uninhabited. One of the first activities was swamp reclamation. The most intensive agricultural development, however, took place between 1927 and 1933. At the end of this period, two-thirds of orchards existing today had already been planted (Mercado et al. 1975). Presently, more than 50% of Israel's total population is concentrated on top of the Coastal Plain aquifer.

16.3 Nitrogen in the Coastal Plain Aquifer

Aspects of the nitrogen mass balance in the Coastal Plain aquifer of Israel are schematically presented in Fig. 2. The processes operative in this system will be discussed systematically from the top soil down to the groundwater.

16.4 Nitrogen Sources

Two different nitrogen sources are recognized; exhaustive and continuous. Exhaustive sources are related to the organic matter that accumulated in the top

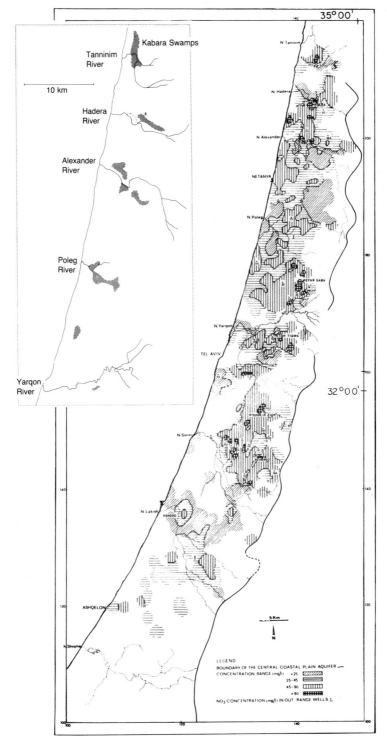

Fig. 1. Nitrate content in groundwater (nitrate as NO_3^-) in the Coastal Plain aquifer of Israel in summer 1981. (After Kanfi et al. 1983). The enclosed map shows the areas covered by swamps in the late nineteenth century. (After Palestine Exploration Fund maps 1878 and 1881)

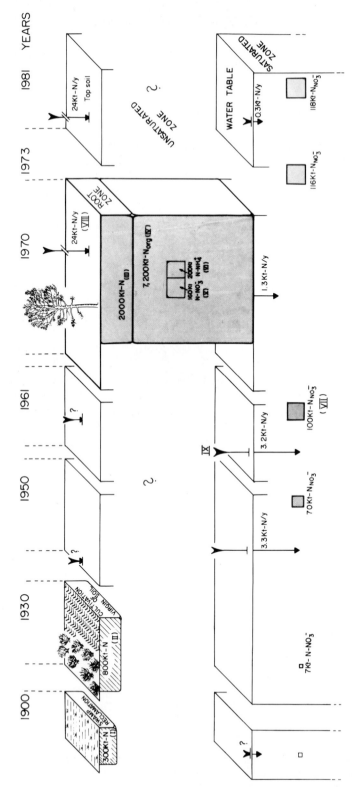

Fig. 2. Nitrogen balance in the Coastal Plain aquifer of Israel. Each time segment from the 1900's to 1981, represents the aquifer system. All quantities expressed as tons N × 10³; fluxes (*black arrows*) as tons N × 10³ yr⁻¹. *Shaded areas* proportionally represent the nitrogen amounts. (*I*) Nitrogen released due to the reclamation of swamps. (*II*) Nitrogen released due to the cultivation of virgin soils (Kanfi et al. 1983). (*III*) Nitrogen stored in the root zone (95% as N$_{Org}$). (Based on the raw data of Raveh et al. 1972). (*IV, V, VI*) N$_{Org}$, N-NO$_3^-$, and N-NH$_4^+$, respectively, stored in the unsaturated zone (root zone excluded). (Based on the raw data of Raveh et al. 1972). (*VII*) N-NO$_3^-$ content in groundwater [these amounts are calculated on the basis of an effective groundwater volume of 10 × 10⁹ m³ (Mercado et al. 1975)]. (*VIII*) Yearly net nitrogen input to the top soil layer (after deduction of nitrogen amounts utilized by plants) from fertilizers, sewage, solid wate, animal excreta rain water, and irrigation water (Saliternik et al. 1972). (*IX*) Average yearly nitrogen input to groundwater (Ronen et al. 1983a)

soil layer of the unsaturated zone during its geological evolution. Continuous nitrogen sources are mainly a consequence of ongoing human activities.

16.4.1 Exhaustive Sources

Reclamation of Swamps. Reclamation of the coastal swamps started at the beginning of the last century with the concomitant decomposition of soils rich in organic matter. The geographical distribution of swamps in 1878 is presented in the Fig. 1 insert. The area that was covered by perennial swamps is estimated to have been 14 km^2 (Palestine Exploration Fund maps, 1878, 1881). In the winter seasons, the flooded regions had extended over an area of 75 km^2 (Kartin, in press). Only in nature reserves do small relicts of these swamps remain. The hydromorphic and grumosol soils of the once swampy areas are presently intensively cultivated (Reifenberg 1947, Dan et al. 1976).

The only available data on the mineralization of nitrogen in the coastal swamps has been obtained from the recently drained Kabara swamps located at the northern edge of the aquifer. We have assumed that these swamps are representative of the whole area.

Marish (1975) reports that in the area formerly covered, by Kabara swamps, peat is no longer found at the soil surface. However, thin peat layers (10–30 cm) are still found in the subsurface. These organic layers are reported to be also in a state of progressive decomposition. Data on the present organic matter and total nitrogen content in soil samples of the reclaimed Kabara swamps are presented in Table 1. This information and data reported by others were used to calculate the total amount of nitrogen released from the decomposition of peat throughout the Coastal Plain.

Table 1. Organic matter and total nitrogen content in soil samples of the reclaimed Kabara swamp

Depth (cm)	C_{Org} (%) Dry basis	N_{Org} (g/g) Dry basis	N/C
0– 30[a]	2.6		
30– 65	8.8		
0– 25[a]	3.7		
25– 45	4.4		
45– 62	14.2		
0– 30[a]	4.6		
30– 75	6.0		
75– 90	4.7		
0– 30[b]	3.28	3,752	0.11
30– 60	2.30	2,938	0.13
60– 90	2.10	2,652	0.13
90–120	2.20	2,192	0.10

[a] Marish (1975). [b] Reinhorn and Avnimelech (1973)

The average content of organic matter in peat is about 25% (Raveh and Avnimelech 1973; Davis 1978, Levanon et al. 1979) and its apparent density is 0.5
g/cm^3 (Raveh and Avnimelech 1973). The present soil organic matter content in
the reclaimed Kabara soils is about 5% and the N/C ratio range from 0.10–0.13.
Assuming that the total thickness of peat in the perennial swamps was 1 m, we
estimate that the 20% decrease in the soil organic matter released 168,000 tons
of nitrogen. An additional nitrogen release of 134,000 tons was calculated for the
winter flooded regions (assuming that the thickness of peat was only 30 cm).

The 300×10^3 tons of nitrogen and $2,500 \times 10^3$ tons of organic carbon were
released at the beginning of the twentieth century. At that time, large amounts
of groundwater still drained to the sea; however, we consider that swamp reclamation contributed to nitrate buildup in groundwater due to transit time in the
unsaturated zone and the low velocity of groundwater flow. This is an analogous
process to the contamination of the Sea of Galilee by the high supply of nitrates
from the upstream reclaimed Hula Valley swamps (Raveh and Avnimelech 1973;
Avnimelech et al. 1978).

Oxidation of Virgin Soils. The cultivation of virgin soils induces the oxidation
of soil organic matter and, consequently, nitrogen mineralization. This has been
reported for soils of the Coastal Plain aquifer (Reinhorn and Avnimelech 1973,
1974) and elsewhere (Kreitler 1975, Young et al. 1976).

According to Reinhorn and Avnimelech (1973, 1974), the release of carbon
and nitrogen followed linear relationships:

$$\Delta C = -1,182 + (2,936 \pm 219)\, C_0 \qquad R = 0.88$$

$$\Delta N = -197 + (0.323 \pm 0.021)\, N_0 \qquad R = 0.90$$

Δ C and Δ N are the amounts of organic carbon and organic nitrogen, respectively, released in $kg/1,000\ m^2$ for a 30 cm layer; C_0 and N_0 are the concentrations
of organic carbon and nitrogen in the virgin soil (%C and $\mu g\ N/g$, respectively).
The above relationship was based on data from 20 pairs of boreholes drilled to
a maximum depth of 90 cm. Application of this formula leads to an estimate that
this source contributed on the average $400\ kg\text{-}N\ 1,000\ m^{-2}$. Part of the nitrogen
was released from the deep soil layers that are not mechanically disturbed
(60–90 cm depth).

The newly cultivated soils approached equilibrium between 3–5 years. No difference was detected in the C/N ratio between cultivated and virgin soils. On the

Table 2. Amount of organic carbon and nitrogen released from cultivated soils in the Costal Plain area.
(After Reinhorn and Avnimelech 1973)

Soil type	Number of boreholes	C_{Org} released kg $1,000\ m^{-2}$		N_{Org} released kg $1,000\ m^{-2}$	
		Range	Average	Range	Average
Sandy loam	5	−2,016 to 3,574	1,373	− 41 to 67	18
Clay loam	10	2,890 to 11,100	6,359	−160 to 1,602	658
Clay	3	− 350 to 23,741	14,212	70 to 3,026	1,753

other hand, it was shown that soils with a nitrogen content lower than 600 ppm and with an organic matter content lower than 3,500 ppm will be enriched in these components by plant growth. The amounts of nitrogen and organic carbon released are clearly related to the different soil types (Table 2). On the basis of the areal distribution of the soil in the Coastal Plain, we calculate that the decomposition of organic matter stored in virgin soils had released about 800×10^3 tons of nitrogen and about 10×10^6 tons of organic carbon.

16.4.2 Continuous Sources

Continuous nitrogen sources, such as fertilization with manure and chemicals, irrigation with nitrogen containing waters and sewage, and solid waste disposal, contributed a net amount of 24×10^3 tons of nitrogen per year in the 1970's (Ronen 1972, Saliternik et al. 1972). (The net nitrogen input was defined as the total amount of nitrogen added to the soil minus the nitrogen amount removed by plants.)

Ronen et al. (1983a) estimated that this amount did not change drastically in the last 10 years. From the history of human activity in the coastal area, it is obvious that earlier nitrogen input from these sources had to have been much smaller.

16.5 Nitrogen Stored in the Unsaturated Zone

Nitrogen compounds have been detected in large quantities in the unsaturated zone below native soils and cultivated areas (Stewart et al. 1968, Young et al. 1976, Klein and Bradford 1979, 1980, Lind 1979, Lucas and Reeves 1980).

Concentrations of organic carbon, organic nitrogen, exchangeable ammonium ion, and nitrate were reported by Raveh et al. (1972) in the unsaturated zone of the Coastal Plain aquifer. In the course of their study, 35 test holes were drilled to a maximum depth of 12 m. Four wells were drilled down to the water table (16–33 m deep). The geological sections studied by Raveh et al. (1972) represent most of the lithologic types of the aquifer. The selected drill sites represent most present land uses in the Coastal Plain: virgin soils, nonirrigated crops, irrigated crops, and citrus groves, sewage irrigated areas, barn yards, sewage and solid waste disposal sites.

Their raw data (summarized in Table 3) and the physical dimensions and water volume content of the unsaturated zone (aquifer surface area $-$ 2,000 km^2; average thickness of the unsaturated zone $-$ 34 m; apparent density of sediments $-$ $1.5\,\mathrm{g\,cm^{-3}}$, and total volume of water stored in the unsaturated zone $-$ 11 $\times 10^9\mathrm{m^3}$) were used to calculate the amount of organic carbon and nitrogen compounds stored in the unsaturated zone.

For the upper 2 m (defined in this paper as root zone), only the organic nitrogen was estimated. Exchangeable NH_4^+ and NO_3^- in the water phase were not calculated as the water content of this zone is difficult to establish. Below the root zone, the total amount of nitrates was calculated considering its concentration in

Table 3. Nitrogen components and organic carbon in the root zone and major lithological units of the Coastal Plain aquifer (in ppm). (Calculated from data of Raveh et al. 1972)

	n	N-NH$_4^+$		N-NO$_3^-$		N$_{Org}$		C$_{Org}$		N/C		C$_{Org}$[a]
		$\bar{C}$	σ	$\bar{C}$	σ	$\bar{C}$	σ	$\bar{C}$	σ	$\bar{C}$	σ	
Root zone	148	4.8	11.8	7.5	9.3	325	485	3,540	6,200	0.10	0.03	
Sand and sandstone	192	2.3	3.4	2.6	3.1	26	23	410	287	0.09	0.10	2,000
Sandy and silty loam and loam	103	2.5	1.8	2.7	2.0	70	50	635	503	0.14	0.11	
Clay loam and clay	126	3.2	2.8	3.3	4.0	148	65	1,700	1,380	0.11	0.04	12,300
Average (root zone excluded)	421	2.6	2.9	2.8	3.2	73	70	850	990	0.11	0.09	9,000

$\bar{C}$ – Average, σ – standard deviation, n – number of samples
[a] Recent Mediterranean sediments near the Israeli shore (Nir, pers. commun., 1982)

the liquid phase. Raveh et al. (1972) do not differentiate between adsorbed and soluble ammonia, therefore, we calculated its total amount in relation to both the liquid and solid phases. The true value for exchangeable NH$_4^+$ stored in the unsaturated zone must be between the two calculated values. Dissolved organic matter and organic nitrogen in solution were not measured, therefore, they are included in the organic estimates. Results of these calculations are presented in Table 4.

As noted by Avnimelech and Raveh (1976), the subsurface layers contain less than 20% of the organic matter und organic nitrogen found in the root zone above (Fig. 3). The various subsurface lithological layers differ considerably in the content of these two components. In contrast, NO$_3^-$ and NH$_4^+$ content of these layers was found to be almost uniform among the layers and measured 40% and 55%, respectively, of that found in the root zone.

Table 4. Amounts of nitrogen components and organic carbon in the unsaturated zone of the Coastal Plain aquifer of Israel (tons × 10³)

Root zone	Sediment	Water
(0 to 1.9 m deep)		
N$_{Org}$	1,950	
C$_{Org}$	21,000	

Unsaturated zone (2 m down to the water table)	Sediment	Water
N-NH$_4^+$	260	140
N-NO$_3^-$		160
N$_{Org}$	7,200	
C$_{Org}$	84,000	

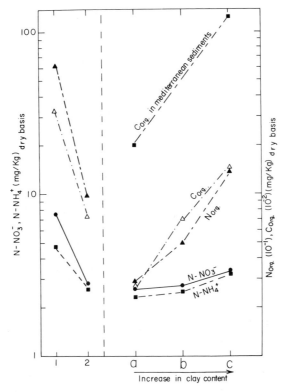

Fig. 3. C_{Org} and nitrogen components in the root zone and sediments of the Coastal Plain aquifer of Israel: *1* Root zone; *2* average in the unsaturated zone (root zone excluded); *a* sand and sandstone (below the root zone); *b* sandy and silty loam and loam (below the root zone); *c* clay loam and clay (below the root zone). The C_{Org} content of recent mediterranean sediments near the Israeli shore (Nir, pers. commun., 1982) is also presented. Note the parallel increase of C_{Org} with increase of clay content in the Mediterranean sediments and in the unsaturated zone (below the root zone)

16.6 Processes Modifying Nitrogen Storage in the Unsaturated Zone

The biological and physiochemical processes affecting the nitrogen fate in the unsaturated zone may be inferred from studies conducted mainly on the root zone. However, this extrapolation is questionable due to differences in environmental variables, such as bacterial populations (Power et al. 1974; Klein and Bradford 1980), oxygen supply (Stewart et al. 1968; Power et al. 1974), and biodegradability of organic matter (Young et al. 1976; Champ et al. 1979). What may appear as slow, unimportant reactions (Gasser 1969, Sollins and McCorrison 1981) for the time scale relevant to agronomic studies of soil processes may have a great importance when geological times are considered.

Raveh et al. (1972) compared the net nitrogen input to the top soil with the amount of NO_3^- leached down from the root zone. They reported that large

Table 5. Percentage of N-NO$_3^-$ washed down from the root zone related to the net nitrogen input to the top soil. (After Raveh et al. 1972)

Area	Number of boreholes	N-NO$_3^-$ washed down / N net input	
		Average (%)	Range (%)
Percolating pits, refuse dump and feedlots	5	< 1	
Irrigated clay soils	3	16	2– 50
Sewage irrigated land	4	18	10– 31
Citrus groves planted on sandy soil	10	35	5–>100

amounts of nitrogen were lost (Table 5). Avnimelech and Raveh (1976) suggest that nitrogen loss in the root zone may be the result of ammonia volatilization mainly in the topmost layer of this zone. However, they postulate that the main mechanism for nitrate removal was denitrification. Denitrification processes in the root zone are reported by others (Gasser 1969, Raveh and Avnimelech 1973, Verstraete 1974, Todd and McNulty 1976). Large numbers of denitrifying bacteria in the top soil layers were found by Klein and Bradford (1980). However, the root zone must not be considered only as a nitrogen reducing interface, but also as a large nitrogen pool from which nitrates and organic nitrogen tend to be mobilized. Such processes have been reported for other areas (Sollins and McCorison 1981).

The organic matter below the root zone is a potential source of NH$_4^+$ and NO$_3^-$. On the other hand, its anaerobic decomposition may lead to the reduction of the existing nitrates.

Klein and Bradford (1979) measured δ^{15}N enrichment in cores obtained from depths of at least 10 m and related this enrichment to denitrification. They also found small numbers of nitrate reducing bacteria down to 30 m depth (Klein and Bradford 1980). Nitrate reduction was found by Lind (1979) at 14 m depth. Denitrification was postulated also by Ayers and Branson (1973) to explain discrepancy between the calculated and measured nitrate content of pore water.

Data on nitrogen mineralization below the root zone is scarce. Sollins and McCorison (1981) found that in situ organic matter was mobilized as nitrates and dissolved organic nitrogen in a Douglas-fir watershed.

The amount of N-NO$_3^-$ stored in the unsaturated zone (Table 4) of the Coastal Plain aquifer is more than 500 times higher than the average amount of N-NO$_3^-$ that had reached groundwater yearly in the last decade (Ronen et al. 1983 a). Nitrates may have been washed down from the root zone as postulated by Raveh et al. (1972) or may be the result of organic matter or exchangeable NH$_4^+$ oxidation at depth. It is difficult to forecast the amount of nitrate that will finally reach groundwater from the nitrogen pool in the unsaturated zone.

Exchangeable ammonium ion is seldom considered as a potential pollutant of groundwater mainly due to the cation exchange capacity of the unsaturated zone. However, evidence exists about NH$_4^+$ mobility, especially in calcareous sandy sediments and in regions under high organic loads, such as feedlots (Stewart et al. 1968, Murphy and Gosh 1970) and sanitary land fills (Raveh et al. 1972, Ronen 1972). The amounts of exchangeable ammonium ion and nitrates stored in the unsaturated zone of the Coastal Plain aquifer are similar (Table 4). The inferred mobility of ammonium ion, the alkaline properties of the sediments in the unsaturated zone, and the zero concentration of ammonia found at the water table (Ronen 1972) and in groundwater (Kanfi and Ronen 1976) suggest that nitrification processes take place at or above the water table.

16.7 The Unsaturated-Saturated Interface

The highly reactive interface between the unsaturated zone and the water table is a dynamic interface whose thickness and position is defined by the fluctuations of the water table and the capillary rise of water. This interface is characterized by drastic periodic changes in moisture content and high concentrations of infiltrated dissolved organic carbon and nitrogen components. Processes occurring at this interface determine the nitrogen flux to groundwater. These processes are the result of bacterial activity operative in a media in which the oxygen supply varies greatly due to the wetting and drying cycles. The importance of these moisture cycles in denitrification and nitrification processes was demonstrated in soil studies (Patrick and Wyatt 1964, Raveh and Avnimelech 1973). The report of Schwille (1976) on degradation of dissolved oil products in groundwater further substantiates the existence of subsurface microbiological processes. Stewart et al. (1968) found large numbers of bacteria near the water table and high concentrations of organic carbon at the water table of a shallow aquifer. Bacteria were also detected in pumping wells of the Coastal Plain aquifer (Fleisher 1983). Raveh et al. (1972) reported that large fluctuations of NO_3^- and NH_4^+ were found above the water table in all four wells that reached groundwater. No significant variations were measured in the concentration of the conservative Cl^-. They could not discern if the operative phenomena were nitrification or denitrification. Groundwater temperatures of 21°–27 °C were found in waters of the Coastal Plain aquifer by Kanfi and Ronen (1976). These values are relatively higher than those reported by other investigators (Power et al. 1974). This factor will surely increase the rate of all microbial processes.

16.8 Nitrogen in Groundwater

The nitrogen flux into groundwater, through this last interface, decreased gradually from an average input of 3,300 tons N yr^{-1} during the 1930's to 300 tons N/yr in the 1970's. The N-NO_3^- amounts stored in groundwater (calculated on the basis of an affective groundwater volume of $10 \times 10^9 m^3$) increased from 7×10^3 tons in the 1930's to 118×10^3 tons in 1981 (Fig. 2). Nitrate buildup in the saturated zone must not reflect strictly the nitrogen fluxes from the unsaturated

234   D. Ronen et al.

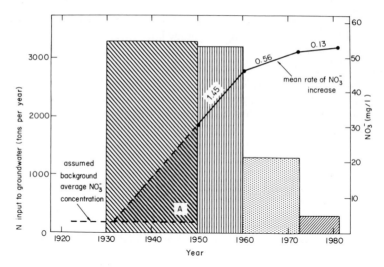

Fig. 4. Reconstruction of average NO_3^- concentration (nitrate as NO_3^-) in the Coastal Plain aquifer of Israel. (After Kanfi et al. 1983) and the yearly average nitrogen input (as nitrogen) to groundwater. (After Ronen et al. 1983a). *Shaded area (A)* represents the possible rates of nitrate increase in the period 1930 to 1950

zone. Denitrification may also take place in groundwater (Schwille 1976, Champ et al. 1979, Oberman 1982, Ronen et al. 1983b).

The average nitrate concentration in groundwater in 1981 was 52 mg l^{-1} (nitrate as NO_3^-) (Kanfi and Ronen 1982). The geographical distribution of nitrate in groundwater is presented in Fig. 1. Kanfi et al. (1983) reconstructed the average nitrate trend in the aquifer since the 1930's (Fig. 4). They concluded that the rate of increase of nitrate in groundwater is decreasing from 1.45 mg l^{-1} yr^{-1} during the 1940's to 0.13 mg l^{-1} yr^{-1} during the 1970's. They postulate that the nitrogen balance in the aquifer is approaching steady state. The veracity of this assumption is critical to management of water resources in Israel.

Acknowledgements. We wish to thank Mr. Y. Raveh and Mr. A. Kartin for providing us their data. We wish to acknowledge Dr. W. Garner of the Israel Water Commission for his considerable help in writing this manuscript and to Prof. Y Avnimelech and Prof. J. R. Gat for their helpful comments.

References

Avnimelech Y, Raveh Y (1976) Nitrate leakage from soils differing in texture and nitrogen load. J Environ Qual 5:79–82

Avnimelech Y, Dasberg S, Harpaz A, Levin I (1978) Prevention of nitrate leakage from the Hula Basin, Israel: a case study in watershed management. Soil Sci 125:233–239

Ayers RS, Branson RL (eds) (1973) Nitrates in the Upper Santa Ana River Basin in relation to groundwater pollution. Univ California, Div Agric Sci Bull 861:59

Boyce JS, Muir J, Edwards AP, Seim EC, Olson RA (1976) Geologic nitrogen in Pleistocene loess of Nebraska. J Environ Qual 5:93–96

Champ DR, Gulens J, Jackson RE (1979) Oxidation-reduction sequences in groundwater flow systems. Can J Earth Sci 16:12–23

Dan Y, Yaalon DH, Koyumdjisky H, Raz Z (1976) The soils of Israel. The Volcani Center, Bet Dagan, Israel, Pamphlet No 159

Davis RA Jr (ed) (1978) Coastal sedimentary environments. Springer, Berlin Heidelberg New York

Fleisher M (1983) Annual report. Central District, Ministry of Health (In Hebrew) (in press)

Gasser JKR (1969) Some processes affecting nitrogen in the soil. Proc Conf Soil Sci Nat Agric Advis Serv 22–23 October, Tech Bull Minist Agric Fish Food 15:15–29

Gat JR, Magaritz M (1980) Climatic variations in the Eastern Mediterranean Sea area. Naturwissenschaften 67:80–87

Ginzburg A, Cohen SS, Hay-Roe H, Rosenzweig A (1975) Geology of Mediterranean shelf of Israel. Am Assoc Pet Geol Bull 59:2142–2160

Issar A (1968) Geology of the central Coastal Plain of Israel. Isr J Earth-Sci 17:16–29

Kanfi Y, Ronen D (1976) The chemical quality of groundwater in the Coastal Plain aquifers of Israel. Water Commiss Off Rep 76/1 (in Hebrew)

Kanfi Y, Ronen D (1982) Nitrate trends in the Coastal Plain aquifer – prediction and reality. Water Pollut Contr Unit Isr Water Commiss Rep 82/1:6 (in Hebrew)

Kanfi Y, Ronen D, Magaritz M (1983) Nitrate trends in the Coastal Plain aquifer of Israel. J Hydrol (in press)

Kartin A (1983) Swamp reclamation in Israel – governmental and Jewish activities during the Mandate Period (1922–1937). Dept Geogr Tel Aviv Univ (in Hebrew) (in press)

Klein JM, Bradford WL (1979) Distribution of nitrate and related nitrogen species in the unsaturated zone, Redlands and vicinity, San Bernardino County, California. US Geologi Survey Water-Resour Invest 79–60:81

Klein JM, Bradford WL (1980) Distribution of nitrate in the unsaturated zone, Highland-east, Highlands area, San Bernardino County, California. US Geologi Survey Water Resour Invest 80–48:70

Kreitler CW (1975) Determining the source of nitrate in groundwater by nitrogen isotope studies. Bur Econom Geol Univ Texas at Austin, Rep Invest 33:57

Levanon D, Henis Y, Okon Y, Dovrat A (1979) Alfalfa roots and nitrogen transformations in peat. Soil Biol Biochem 11:343–347

Lind AM (1979) Nitrogen in soil water. Nord Hydrol 10:65–78

Lucas JL, Reeves GM (1980) An investigation into high nitrate in groundwater and land irrigation of sewage. Prog Water Technol 13:81–88

Marish S (1975) Kabara swamp – soil survey. Survey Conserva Drainage Dep Isr Minist Agric (in Hebrew)

Mercado A (1980) The Coastal Aquifer of Israel: some quality aspects of groundwater management. In: Shuval HI (ed) Water quality management under conditions of scarcity. Academic Press, London New York pp 93–145

Mercado A, Avron M, Kahanovich Y (1975) Groundwater salinity of the Coastal Plain – chloride inventory and assessment of future salinity trends. Tahal Rep 01/75/88:68 (in Hebrew)

Milliman JD, Emery KO (1968) Sea levels during the past 35,000 years. Science 162:1121–1123

Murphy LS, Gos JW (1970) Nitrate accumulation in Kansas groundwater. Project Completion Report, Kans Water Resour Res Inst

Oberman P (1982) Contamination of groundwater in the Lower Rhine Region (FRG) due to agricultural activities. Presented at the Int Symp IAH, Prague, Caechoslavakie, Impact of Agricultural Activities on Groundwater pp 13

Palestine Exploration Fund maps (Surveyed 1878) reprint May 1917. Sheets X (Arsüf) and XIII (Jaffa). Scale 1 inch to 1 mile

Palestine Exploration Fund maps (Surveyed 1881) additions and corrections 1915. Engraved at Standford's Geographical Establishment. Scale 3/8 inch to 1 mile

Patrick WH, Wyatt T (1964) Soil nitrogen loss as a result of alternate submergence and drying. Soil Sci Soc Am Proc 28:647–653

Power JF, Bond JJ, Sandoval FM, Willis WO (1974) Nitrification in Paleocene Shale. Science 183:1077–1079

Raveh A, Avnimelech Y (1973) Minimizing nitrate seepage from the Hula Valley into Lake Kinneret (Sea of Galilee) I. Enhancement of nitrate reduction of sprinkling and flooding. J Environ Qual 2:455–458

Raveh Y, Avnimelech Y, Saliternik C (1972) Nitrates concentration in the soil layer above the Coastal Plain aquifer. Tahal, Rep HR/72/073:149 (in Hebrew)
Reifenberg A (1947) The soils of Palestine. Murby, London pp 18–159
Reinhorn T, Avnimelech Y (1973) Nitrogen release due to the decrease in soil organic matter in newly cultivated soils. Soils Fertilizer Lab, Technion, Haifa Pub 177:1–11
Reinhorn T, Avnimelech Y (1974) Nitrogen release associated with the decrease in soil organic matter in newly cultivated soils. J Environ Qual 3:118–121
Ronen D (1972) Nitrate contamination of groundwater in the Rishon Le Zion-Rehovot district. Tahal Rep HR/72/046:100 (in Hebrew)
Ronen D, Kanfi Y, Magaritz M (1983a) Sources of nitrates in groundwater of the Coastal Plain of Israel – Evolution of Ideas. Water Res (in press)
Ronen D, Kanfi Y, Magaritz M (1983b) Oxidation-reduction sequences in the Yarkon-Tanninin carbonate aquifer of Israel. In: Shuval HI (ed) Developments in ecology and environmental quality vol II, Balaban, Rehovot Philadelphia pp 301–310
Saliternik C, Kahanovich Y, Shevach Y (1972) Groundwater nitrate pollution sources. Tahal Rep HR/72/042:68 (in Hebrew)
Schwille F (1976) Anthropogenically reduced groundwaters. Hydrol Sci Bull XXI:629–645
Sollins P, McCorison FM (1981) Nitrogen and carbon solution chemistry of an old growth coniferous forest watershed before and after cutting. Water Resour Res 17:1409–1418
Stewart BA, Viets FG jr, Hutchinson GL (1968) Agriculture's effect on nitrate pollution of groundwater. J Soil Water Conserv 23/1:13–15
Todd DK, McNulty DEO (1976) Polluted groundwater. A review of the significant literature. Water Informat Center, New York
Tooley JM (1978) Interpretation of Holocene sea level changes. Geol Foren Stockh Forh 100:203–212
Verstraete IW (1974) Biotransformation of nitrogen in soils in relation to environmental health – a review. Proc Seminar on Groundwater Qual Contr Management, Brussels pp 265–309
Young CP, Oakes DB, Wilkinson WB (1976) Prediction of future nitrate concentrations in groundwater. Ground Water 14:426–438

17. Prevention of Groundwater Pollution on a National Scale: Israel as a Case Study

M. WALDMAN and Y. SHEVAH

17.1 Introduction

Applying municipal and agricultural wastes to land has been practiced for many centuries. The beneficial aspects of this traditional means of waste management and its possible detrimental effects on the soil profile and groundwater are still the subject of much discussion (Bouwer and Chaney 1974, Bole et al. 1981, Khaleel et al. 1981).

Specific studies have been made regarding the major sources of pollution, with the emphasis on such sources of environmental concern as mineral fertilizer application and the disposal of animal manure (Stewart et al. 1968, Bolton et al. 1970, Hillel et al. 1980, and Hergert et al. 1981). This concern has been intensified by the trends toward larger feedlots, higher stocking densities, and higher rates of mineral fertilizer application (Viets 1971). Similarly, the increased use of landfills for disposal of solid wastes and the necessity for treatment of leachate has been discussed by Winant et al. (1981).

Recent experiments indicate that land disposal can be an effective mean of waste treatment and can meet environmental protection standards if properly designed. Potential pollutants can also become nutrients for plant growth, as shown by Feigin (1975) and Day et al. (1981) in investigating the use of wastewater as a source of irrigation water and plant nutrients for commercial cotton production. Similarly, Kipnis et al. (1979) and Waisman et al. (1981, 1982) have demonstrated its use in the irrigation of Rhodes grass, a perennial forage crop that not only absorbs a large quantity of potential pollutants, but also produces a high yield of dry matter for animal feeds.

17.2 Wastes and Potential Pollutants

17.2.1 Physical Background

Israel is characterized by a Mediterranean climate varying from semiarid in the north to arid in the south. The annual rainfall, coming mostly during five winter months, ranges from about 1,000 mm in the north to about 200 mm in the south and less than 100 mm in the extreme south.

The total area of 22,000 km^2 can be conveniently divided into five main regions, namely: Kinneret Watershed, Central and Western Galilee, Coastal Plain, mountains and foothills, and Negev (Fig. 1).

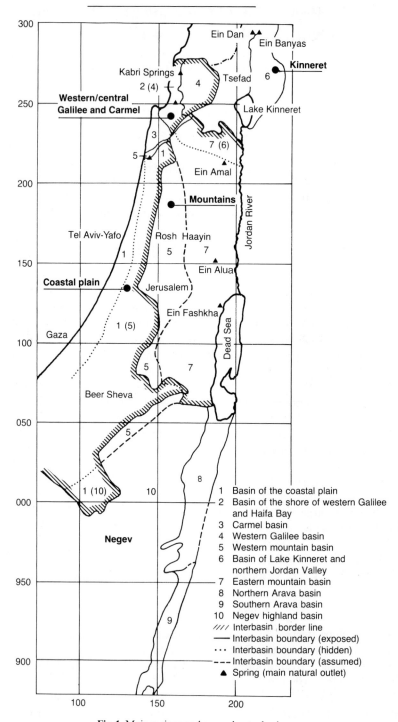

Fig. 1. Main regions and groundwater basins

The whole state is underlain by aquifers at varying depths, ranging from about 20 m in the Coastal Plain to about 100 m in the mountains and several hundred meters in the Negev region.

Natural precipitation results in an annual replenishment of about 1,750 MCM, of which some 950 MCM is groundwater and the rest is surface water. The surface water originates mostly in the Kinneret Watershed region drained by the Jordan River.

17.2.2 The Potential Pollutant Load

Israel is characterized by very intensive human activity. The amount of activity has increased rapidly over the years as shown in Table 1, which focuses on the changes in population and agricultural activity over the last 30 years. These trends are expected to continue in the future, resulting in a substantial growth in domestic and agricultural wastes. In Israel, these are the main sources of pollution, since industries producing toxic wastes are either confined to areas where the impact is localized or are too small for their effluent to be differentiated from domestic waste flows.

The distribution of population and agricultural activity within the five regions is uneven. There is a disproportionate concentration of population and agricultural activity in the Coastal Plain, which comprises only 10% of the State's area, but 52% of the population (910 persons km^{-2}) and about 30% of the agricultural activity (Table 2). The mountain region is less populous, but has a substantial number of domestic animals. At the other extreme, the Negev region comprises 60% of the State, accounts for only 8% of the total population (23 persons km^{-2}), and supports agricultural activity mostly of the dry farming type.

The intensity of human activity can be expressed in terms of specific potential pollutants. In this case, the fraction of nitrogen that will reach the groundwater from the various wastes has been calculated and used to derive the potential pollutant load in the various regions as given in Table 3. The total quantity of nitrogen currently leached through the soil profile is estimated at about 56,000 tons of which about 40% derives from human wastes and about 60% from agricultural wastes. Solid refuse and wastewater contribute, respectively, 15% and 85% of the nitrogen originating from human waste; and mineral fertilizer and domestic animals contribute, respectively, 40% and 60% of the nitrogen originating from agricultural wastes. Regional data indicate that 40% is generated in the Coastal Plain, with each of the other regions accounting for between 12% and 18%.

Table 1. Population and agricultural activity

	1950	1981
Population (cap)	1,370,000	3,977,900
Cultivated land (ha)	248,000	427,000
Cattle (head)	47,835	292,000
Sheep (head)	52,000	383,000
Poultry (million birds)	2.9	30.0

Table 2. Population and livestock by region, 1981. (Compiled from *Statistical Abstract of Israel*, 1982, Central Bureau of Statistics)

Region	Area (km^2)	Population (1,000)	Population density (Pers. km^{-2})	Cultivated Area (1,000 ha)	Cattle (10^3)	Sheep (10^3)	Poultry (10^6)	Water exploitation (MCM yr^{-1})
Kinneret	2,700	200	75	37	52	11	8	580 [a]
Western Galilee and Carmel	2,600	800	310	73	64	98	4	220
Coastal Plain	2,300	2,100	913	138	88	27	10	375
Mountains	1,600	600	375	58	60	11	5	420
Negev	12,800	300	23	129	26	243	3	30
Total	22,000	4,000	180	425	290	390	30	1,625

[a] Surface water

Table 3. Potential pollutant content of human and agricultural wastes in terms of nitrogen (1,000 ton yr^{-1})

Region	Refuse [a]	Wastewater [b]	Total	Fertilizer [c]	Livestock [d]	Total	Total
Kinneret	95	975	1,070	1,540	4,009	5,549	6,619
Western Galilee and Carmel	692	3,900	4,592	1,640	3,634	5,274	9,866
Coastal Plain	1,834	10,238	12,072	5,160	5,552	10,712	22,784
Mountains	599	2,925	3,524	1,910	3,045	4,955	8,479
Negev	200	1,462	1,662	2,620	3,804	6,424	8,086
Total	3,420	19,500	22,920	12,870	20,044	32,914	55,834

(Human wastes: Refuse, Wastewater, Total; Agricultural wastes: Fertilizer, Livestock, Total; Total)

[a] 3.5 kg N/ton of solid waste as given in Solid Waste Disposal Master Plan, 1976
[b] 75 mg N/l; 65 m^3/capita/yr
[c] 30 kg/ha, weighted average for rainfed area (190,000 ha), irrigated crops (160,000 ha), and plantations (75,000 ha)
[d] 50 gm, 25 gm and 1 gm N per head of cattle, sheep, and poultry, respectively

The regional differences are reflected in the quality of groundwater extracted from the various aquifers. An average of 52 mg NO_3 l^{-1} has been reported in water extracted from the coastal sandstone aquifer, which underlies the most active region (the Coastal Plain), as compared with an average of less than 10 mg NO_3 l^{-1} in water extracted from the limestone aquifer (Ronen et al. 1983).

17.3 Plans and Trends in Land Disposal

17.3.1 Agricultural Wastes

Past trends in agricultural activity (Fig. 2) indicate a radical increase over the last 30 years, expressed mainly by the increase in the total irrigated area.

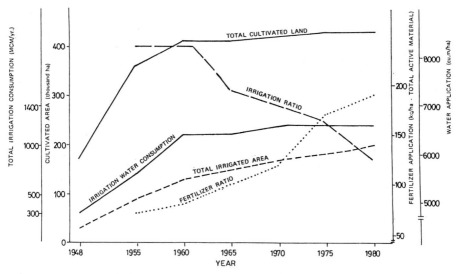

Fig. 2. Trends in irrigation and fertilizer use (1948–1980)

This increase has been achieved by tremendous improvements in irrigation techniques, and particularly by the recent transition from sprinkler irrigation to drip irrigation. This has permitted the irrigation ratio to decline from about 11,000 m³ ha⁻¹ in 1960 to about 6,000 m³ ha⁻¹ in 1981, representing a 40% increase in efficiency in the 20-year period (Table 4). Marketing sources estimate that drip irrigation is already in use in about 30% of the irrigated area.

The introduction of drip irrigation has had a marked effect on the amount of excess percolating below the root zone as return flow, as well as on the amount of leachate moved down through the soil profile. This is because the low dosages and application rate obtainable with drip irrigation allow more time for uptake, absorption, and transformation processes to occur in the upper soil, thus reducing the amount of leachate reaching the unsaturated zone.

It is to be noted, however, that the impact of the reduction in leaching and return flow due to drip irrigation has to some extent been counterbalanced by the increase in the use of fertilizers and animal manure. The application rate for N. P.K. fertilizers has increased from about 83 kg ha⁻¹ in 1960 to about 190 kg ha⁻¹ in 1981, with a consequent increase in nutrient and other pollutant residues in the soil.

Table 4. Irrigated area, irrigation water consumption, and fertilizer application

	1950	1981
Total cultivated land (ha)	248,000	427,000
Total irrigated land (ha)	37,500	203,000
Irrigation water consumption (MCM yr⁻¹)	413	1,200
Irrigation ratio (m³ ha⁻¹)	11,013	5,910

The trends in water and fertilizer application have contradictory implications for groundwater quality. This can perhaps partially explain the conflicting observations and projections regarding the nitrate content of groundwater from the shallow coastal sandstone aquifer presented by Mercado (1983) and Ronen et al. (1983).

The pollution load from animal wastes may be changed drastically in future if current research and development in the recycling of animal wastes prove commercially applicable. Extensive research and field experiments currently underway in many locations in Israel have shown that anaerobic digestion processes can be used to convert animal wastes to energy (biogas) and to a slurry having a variety of practical applications (Marchaim and Criden 1981). The liquid fraction of the slurry is used as a high-value fertilizer, while the solid fraction may be used as feed for fish or livestock or to replace peat moss as a substrate in nurseries.

17.3.2 Wastewater Treatment

All urban and rural settlements in Israel have pipe-borne water. Central sewerage systems are also highly developed, with over 90% of the urban areas currently served by such systems. However, the development of proper treatment and disposal facilities has generally lagged behind the development of sewerage systems.

Public awareness of the possible detrimental impact of inadequate sewage disposal led to a major shift in national policy in the 1970's. Recent years have, therefore, seen the allocation of major investments to programs for the improvement of wastewater treatment and disposal, in the form of a National Sewage Plan.

Except for the Dan Region Metropolitan Area, for which a separate reclamation scheme, including tertiary treatment and groundwater recharge, is implemented (Idelovitch 1983), the entire plan is based on the recycling of secondary effluents.

The plan is based on the construction of regional sewage treatment plants serving several adjacent towns and cities. These plants employ conventional biological treatments in anaerobic ponds and a series of aerobic ponds with or without mechanical aerating means. Secondary effluents are utilized through specially designed irrigation schemes for the irrigation of nonedible crops, such as cotton.

At present, about 58% of the potential flow of wastewater from cities and towns throughout the country (180 MCM, not including the Dan Region) undergoes a certain degree of treatment and about 37% is used for irrigation, either directly or after impoundment over the winter period. The corresponding figures for the five main regions are given in Table 5.

The main component of these projects is the seasonal storage element, consisting of one or more earth reservoirs. A typical reservoir would have a surface water area of about 15 ha, a water column 7–9 m deep, and a capacity of about 1 MCM. The bottom of the reservoirs is treated against infiltration by compacted clay and where necessary with plastic sheets covered with a layer of sand.

The largest wastewater treatment and reuse scheme of this type is the Kishon Reclamation Scheme completed in 1982. The scheme was designed to renovate

Table 5. Wastewater treatment and reuse of secondary effluents (in MCM yr^{-1})

Region	Total	Treated	Impounded	Utilized	Cotton irrigation requirement
Kinneret	13	6	1	2	88
Western Galilee and Carmel	52	25	15	29	59
Coastal Plain	57	44	4	14	53
Mountains	38	12	8	14	25
Negev	20	18	1	7	19
Total[a]	180	105	29	66	244

[a] Excluding 80 MCM yr^{-1} from the Dan Region for which a separate reclamation scheme, including groundwater recharge, is currently being implemented

wastewater generated in the Haifa Metropolitan Area (25 MCM yr^{-1}). Secondary effluents are impounded in a 12 MCM reservoir in the Jezreel Valley, 30 km to the east of Haifa, from which they are to be pumped for irrigation of nonedible crops in the summer months.

Recent research indicates that impoundment after primary treatment causes marked improvement in effluent quality as measured by such indicators as reduced biological oxygen demand (BOD), suspended solids, and ammonium nitrogen (Abeliovitch 1980, Dor 1980). Similarly, storage of secondary effluents for 2 months causes a 10 to a 100-fold decrease in the number of coliforms, fecal coliforms, and fecal streptococci, such that with effective disinfection at the end of the storage period, effluents were rendered effectively free of enteroviruses (Kott 1974, Kott 1980).

17.3.3 Solid Waste Disposal

The disposal of municipal solid waste in Israel is achieved almost exclusively by dumping in landfill sites. At present, there are about 160 sites, ranging in size from less than 1 ha to 20 ha or more, located on the outskirts of urban areas.

These sites cover a total area of about 395 ha and bury about 1 million tons of solid waste a year, which is equivalent to an average of 2,400 ton ha^{-1}. The regional distribution in terms of the number of sites and total area is given in Table 6.

Table 6. Landfill sites by region

Region	No. of sites	Total area (ha)
Kinneret	19	22
Western Galilee	66	77
Coastal Plain	55	159
Mountains	6	60
Negev	14	77
Total	160	395

244   M. Waldman and Y. Shevah

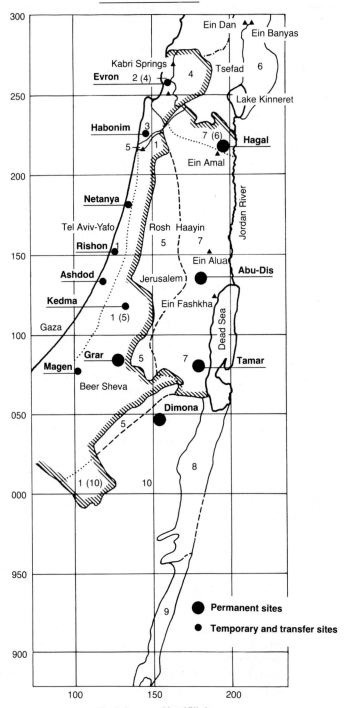

Fig. 3. Proposed landfill sites

Environmental impact studies conducted as part of the preparation of a National Master Plan for Solid Waste Disposal in 1976 indicated that most of the existing dumping sites pose a real pollution hazard to the soil profile and groundwater aquifers. The Master Plan, therefore, calls unequivocally for the existing sites to be abandoned, to be replaced by five new sites at strategic locations supplemented by six temporary and transfer sites located as shown in Fig. 3. Special measures are to be incorporated to monitor the accumulation of the various pollutants and to prevent the deterioration of environmental quality. However, because the vast capital investment required for the implementation of the proposed scheme and the major changes it would require in the whole solid waste disposal procedure, the plan is still awaiting government approval.

17.4 Discussion

The data presented demonstrate that the pollution hazard in the most densely populated Coastal Plain and in other regions underlain by active aquifers supplying fresh water to the nation is considerable. However, proper planning and management can ensure that pollution does not exceed acceptable levels.

Reduction of the quantity of pollutants leached through the soil profile in cultivated land can be achieved by economic use of water and by the use of irrigation techniques that prevent excess drainage and rapid percolation.

Similarly, proper planning for the treatment and reuse of wastewater so as to avoid on-site pollution and provide for the spread of the pollution load over a large area in a controlled environment can retard most of the undesirable processes of pollution as are currently known.

However, this approach has certain drawbacks as it is limited to the treatment of visible sources that can be identified, quantified, and traced to the source of origin. At the same time, other sources of pollutants that are less easily identified, but may have a far greater negative impact on our environment, are left untouched. Moreover, the engineering solutions of this approach rely entirely on the soil profile for the treatment, re-use, and ultimate disposal of the various wastes, while the long-term effects are not fully understood.

Accordingly, it is advisable that widespread use of such programs on a national scale should be supported by comprehensive research to investigate the fate of pollutants before and after they reach the unsaturated zone. Research topics should include: biological and chemical breakdown of resistant pollutants, including detergents and pesticides; monitoring the movement of pollutants through the unsaturated zone to quantify the effect of hydraulic properties and the purification achieved by the movement through the soil profile; simulation modelling to forecast cumulative effects of long-term disposal; and the development of preventive techniques to eliminate or retard the passage of pollutants through the unsaturated zone. Research into methods of recycling wastes so as to minimize the quantities of residues for land disposal should also be intensified.

References

Abeliovitch A (1980) Biology of Impounded Effluent. Desert Institute Report, Ben Gurion University, Beer Sheva, Israel

Bole JB, Carefoot JM, Chang C, Oosterveld M (1981) Effect of wastewater irrigation and leaching percentage on soil and groundwater chemistry. J Environ Qual 10:177–183

Bolton EF, Aylesworth JW, Hore FR (1970) Nutrient losses from tile drains under three cropping systems and two fertility levels on a Brookston clay soil. Can J Soil Sci 50:275–279

Bouwer H, Chaney RL (1974) Land treatment of wastewater. Adv Agron 26:133–176

Day AD, McFadyen JA, Tucker TC, Cloff CB (1981) Effects of municipal waste water on the yield and quality of cotton. J Environ Qual 10:47–49

Dor I (1980) Naan Reservoir Studies. Report of the Hebrew University, Jerusalem, Israel

Feigin A (1975) The effect of irrigation with effluents on soil and crops. In: Proc Recycling of Wastewater for Agricultural and Industrial Uses, Herzlia, Nov 1975

Hergert GW, Glausner SD, Bouldin DR, Zwerman PJ (1981) Effects of dairy manure on phosphorous concentrations and losses in tile effluents. J Environ Qual 10:345–349

Hillel D, Rawitz E, Pratt PF (1980) Nitrates leakage toward groundwater in agricultural field. P 14 in Report to United States – Israel Binational Science Foundation, (abstracts) 1974–1978 Jerusalem

Idelovitch E (1983) Use of soil-aquifer system for sewage reclamation and pollution control. In: Int Workshop on Behaviour of Pollutants in the Unsaturated Zone. Bet-Dagan, Israel

Khaleel R, Reddy KR, Overcash MR (1981) Changes in soil physical properties due to organic waste applications: a review. J Environ Qual 10:133–141

Kipnis TA, Feigin AA, Dovrat A, Levanon D (1979) Ecological and agricultural aspects of nitrogen balance in perennial pasture irrigated with municipal effluents. Prog Water Technol 11(415):127–138

Kott Y (1974) Unrestricted use of wastewater in irrigation. Report of Mekorot Water Co, Tel Aviv, Israel

Kott Y (1980) The inactivation of micro-organisms by ozonation. In: Current Research in Water and Wastewater Technology. NRCD Public, Jerusalem, Israel

Marchaim U, Criden J (1981) Research and development in the utilization of agricultural wastes in Israel for energy, feedstock fodder and industrial products, in Fuel Gas Production from Biomass, edited by DL Wise Vol I Bocakaton, Fa, CRC Press

Mercado A (1983) Modelling of groundwater pollution from diffused sources: the coastal aquifer of Israel as a case study. In: Int Workshop on Behaviour of Pollutants in the Unsaturated Zone. Bet-Dagan, Israel

Ronen D, Kanfi Y, Margaritz M (1983) Nitrogen presence in groundwater as affected by the unsaturated zone. In: Int Workshop on Behaviour of Pollutants in the Unsaturated Zone. Bet-Dagan, Israel

Stewart BA, Viets FG, Hutchinson GL (1968) Agriculture's effect on nitrate pollution of groundwater. J Soil Water Conserv 23:13–15

Viets FG (1971) The mounting problem of cattle feedlot pollution. Agric Sci Rev 9:1–8

Waisman I, Shalhevet J, Kipnis T, Feigin A (1981) Reducing groundwater pollution from municipal waste water irrigation of Rhodes grass grown on sand dunes. J Environ Qual 10:434–438

Waisman I, Shalhevet J, Kipnis T, Feigin A (1982) Water regime and nitrogen fertilization for Rhodes grass irrigated with municipal waste water on sand dune soils. J Environ Qual 11:230–232

Winant WM, Menser HA, Bennett OL (1981) Effects of sanitary landfill leachate on some soil chemical properties. J Environ Qual 10:318–322

F. Unsaturated Zone and Groundwater: Pollution Control

Introductory Comments

G. DAGAN

The unsaturated zone separates the soil surface, where pollutants generally originate, from the groundwater. Thus, the unsaturated zone plays an important role in the transfer of pollutants and it can be used conveniently as a buffer to modify their impact upon groundwater.

The chapters by Bouwer (Chap. 18) and Idelovitch (Chap. 19) are similar in content and deal precisely with a few operating projects of infiltration of partially treated sewage effluents by spreading basins. The three schemes, two in Arizona and one in Israel, have already been in operation for a few years and both chapters present field results collected so far. These results comprise comparisons between the quality of the sewage water applied to the infiltration ponds and that of the water pumped by adjacent wells from the underlying phreatic aquifer for reuse. Whereas the preceding parts of this book have analyzed the various physical and chemical processes which take part in the unsaturated zone, the chapters here present the overall effect of these processes upon the quality of water. Although the degree of purification attained by the spreading basins – unsaturated zone – aquifer system varies with the type of pollutant (inorganic, organic, or biological) and with local conditions, the common conclusion for the three operations is that such projects are effective and successful in purifying secondary sewage and upgrading it to an effluent suitable for unrestricted irrigation, a variety of industrial and nonpotable municipal uses as well as a raw water source for drinking water supply. The reader may find in the articles of Bouwer and Idelovich valuable guidelines for design of such wastewater renovation schemes, especially in areas of dry climate similar to that of Arizona and Israel.

The chapter by Mercado (Chap. 20) is somehow different in content and approach. It summarizes the author's work in developing a quantitative model to predict pollutant concentration in the soil surface-unsaturated zone-groundwater system and to use it for management purposes. The scale is regional, i.e., much larger than in the other chapters and the aim is the long-range protection of the quality of groundwater. Mercado suggests the use of a simplified cell model to predict the change of pollutant concentration with time and the effect of the unsaturated zone manifests then in the travel time from surface to groundwater and its retardation caused by adsorbtion. The use of the model as a decision-making tool is illustrated for the long-range program of sewage management in the coastal plain of Israel.

Summarizing Part F, it can be said that in a sense, it is the culmination of the entire book as it illustrates the role played by the unsaturated zone in transfer of pollutants in engineering projects and design. The integration of the analysis and models of the fundamental processes which are described in Parts A–D into such applications is a matter of further study, which will hopefully be facilitated by the present book.

18. Wastewater Renovation in Rapid Infiltration Systems

H. Bouwer

18.1 Introduction

The ability of vadose zones and aquifers to remove pollutants from low quality water flowing through such media is utilized in rapid-infiltration or soil-aquifer treatment systems. With these systems, partially treated sewage effluent or similar low quality water (processing plant effluent, urban runoff, etc.) is infiltrated into the soil (normally with basins or furrow systems), from where it percolates down to the groundwater. It then moves laterally through the aquifer for some distance before it leaves the aquifer or is collected from the aquifer as "renovated" water. The renovated water can be collected with wells or subsurface drains for reuse or disposal or it can seep naturally into streams or other surface water (Fig. 1). The latter systems are used to minimize pollution of streams or other receiving waters by giving the wastewater soil-aquifer treatment rather than discharging it directly into the surface water.

18.2 Basin Management

The infiltration basins or furrows must be regularly dried to allow atmospheric oxygen to enter the soil for decomposition of organic matter, nitrification of ammonium, and other aerobic processes. Drying also helps restore infiltration rates, which normally decrease during infiltration due to accumulation of solids, algae, etc. on the soil surface. Infiltration schedules may range from 8 h wet, 16 h dry each day to flooding and drying periods of several weeks each. The optimum schedule depends on the local conditions of soil, climate, and wastewater characteristics (suspended solids, biochemical oxygen demand) and whether the system is to be operated for maximum hydraulic capacity or for maximum removal of nitrogen from the water. Hydraulic loading rates of rapid-infiltration systems typically are in the range of $50-500 \, \text{m yr}^{-1}$ ($0.5-5$ million $\text{m}^3 \, \text{yr}^{-1} \text{ha}^{-1}$ of basin area). Mechanical removal of accumulated solids from the soil surface may be periodically required to maintain high infiltration rates.

Because the bottom of infiltration basins is usually covered by a clogging layer of accumulated solids, infiltration rates are reduced, soil-water pressure heads beneath the clogged layers are negative, and the soil below the clogged layer is unsaturated. Infiltration rates for soils flooded with secondary sewage effluent

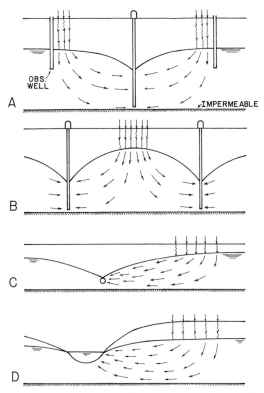

Fig. 1 A–D. Soil-aquifer treatment systems with infiltration areas in two parallel rows and line of wells midway in between (**A**), infiltration areas in center surrounded by a circle of wells (**B**), collection of renovated water by subsurface drain (**C**), and natural drainage into surface water (**D**)

typically are about one-half of the rates for clear water, which in turn are about equal to the hydraulic conductivity of the upper soil layers. In warm, arid climates, drying periods are about as long as flooding periods, so that the long-term infiltration rate or hydraulic loading rate, will be about one-fourth the hydraulic conductivity of the top soil expressed for the same time period. For primary effluent and/or more humid, colder climates, the ratio between hydraulic loading rate and saturated hydraulic conductivity of the soil can be much less, for example 0.1.

To allow flexibility in scheduling flooding and drying in a rapid-infiltration system, a relatively large number of basins is better than only a few basins. Since overloading of the infiltration system must be avoided, a few reserve basins should always be included to take the wastewater during times of low infiltration rates (rainy or cold periods, for example) or when part of the system is undergoing repair. Draining the basins by gravity at the end of a flooding period hastens the drying and gives greater hydraulic loading rates than when the basins are dried by stopping the inflow and letting all the water go into the ground by infiltration.

Suspended algae (for example, *Carteria klebsii*) in the wastewater should be minimized. Such organisms can clog the soil and cause calcium carbonate to pre-

cipitate due to the pH increase of the wastewater as the algae absorb CO_2 from
the water for photosynthesis. Growth of filamentous algae (for example, *Oscilla-
toria* sp.) on the bottom of the infiltration basins is not as serious and can actually
increase infiltration rates as oxygen bubbles form in the algal mat. This causes
portions of the algal mat to float to the surface, carrying suspended materials up
with them and exposing fresh, "rejuvenated" basin soil.

18.3 Quality Improvement

As the wastewater (partially treated sewage effluent, for example) moves
through the soil and down to the groundwater, suspended solids, bacteria,
viruses, and biodegradable carbon will be essentially completely removed. Some
residual organic carbon, including synthetic materials, can be expected in the
renovated water. Heavy metals and phosphate concentrations will be greatly reduced during passage through the unsaturated zone and aquifer, especially in calcareous materials. For secondary sewage effluent where nitrogen is mostly in the
ammonium form, the nitrogen will be essentially completely converted to nitrate
if short, frequent infiltration periods are held. It will stay as ammonium if infiltration periods are long and the cation exchange complex in the soil has become
saturated with ammonium and it will be partially converted to nitrogen gas and
nitrogen oxides if infiltration periods are intermediate, so that ammonium is adsorbed in the soil during flooding and nitrified and denitrified during drying
(Bouwer and Chaney 1974, Bouwer et al. 1974b, 1980, Bouwer 1982). Most of
these quality improvements take place in the top meter of the unsaturated zone.
However, for additional polishing treatment and peace of mind, the travel distance of the water through the vadose zone and aquifer should always be taken
as large as possible.

18.4 Flushing Meadows Project

Since hydraulic capacity of a rapid infiltration system and quality improvement of the wastewater by soil-aquifer treatment are site specific, local experimen-
tation is usually desirable to see what results can be expected and how the rapid
infiltration system should be best designed and managed. An example of such a
pilot project is the Flushing Meadows Project in the Salt River floodplain west
of Phoenix, Arizona. This project, installed in 1967, consisted of six parallel infiltration basins 6.1×213 m each that were intermittently flooded with secondary
sewage effluent (activated sludge process, nonchlorinated). The soil was fine loamy sand for the top 1 m underlain by sand and gravel in various layers to a depth
of about 75 m where a clay layer began. The water table was at a depth of 3 m
and renovated water was sampled from depths of 6–9 m beneath the basins and
at various distances from the basins to study the effect of additional lateral move-

ment through the aquifer on quality of renovated water. The monitoring wells consisted of 15 cm diameter cased wells open at the bottom. The Flushing Meadows Project was operated from 1967 until 1978. In the first 5 years, the basins were flooded and managed to maximize hydraulic loading rates. In the second 5 years, the basins were managed to maximize denitrification. Main results obtained from the Flushing Meadows Project were as follows (Bouwer et a. 1974 a, b, 1980):

1. The largest hydraulic capacity of the system was obtained with flooding periods of 2–3 weeks alternated with drying periods of 10–20 days. At this schedule and using a water depth of 0.3 m in the basins, the hydraulic loading rate was 122 m yr^{-1}. Vegetation or gravel layers in the basins offered no particular advantages. Thus, the best bottom condition was bare soil. Volunteer vegetation seemed to have no undesirable effects, so weed control was not necessary. Sludge-like accumulation on the basin bottoms was removed every 1 or 2 years. After 10 years of operating the project and a total infiltration of 754 m, there were no signs of reductions in hydraulic loading or in hydraulic conductivity of the underlying aquifer.

2. Nitrogen removal from the effluent water as it seeped through the ground to become renovated water was about 30% at maximum hydraulic loading, but 65% if the loading rate was reduced to about 70 m yr^{-1} by using 9 day flooding and 12 day drying cycles and by reducing the water depth in the basins to 0.15 m. The form and concentration of nitrogen in the renovated water sampled from the aquifer below the basins were slow to respond to a reduction in hydraulic loading (Bouwer et al. 1980). In the tenth year of operation (1977), the renovated water contained 2.8 mg l^{-1} ammonium nitrogen, 6.25 mg l^{-1} nitrate nitrogen, and 0.58 mg l^{-1} organic nitrogen, for a total nitrogen content of 9.6 mg l^{-1}. This was 65% less than the total nitrogen content of the secondary sewage effluent, which averaged 27.4 mg l^{-1} (mostly as ammonium nitrogen) in that year.

3. Phosphate removal increased with increasing distance of underground movement of the sewage water. After 9 m of downward movement to and in the underlying aquifer, removal was about 40% at high-hydraulic loading and 80% at reduced hydraulic loading. Additional lateral movement of 61 m through the aquifer increased the removal to 95% in 1977 (i. e., a concentration of 0.51 mg l^{-1} phosphate phosphorus in the renovated water versus 7.9 mg l^{-1} in the effluent). After 10 years and a total infiltration of 754 m, there were no signs of a decrease in phosphate removal.

4. Fluoride removal paralleled phosphate removal. Fluoride concentrations in 1977 were 2.08 mg l^{-1} for the effluent, 1.66 mg l^{-1} for the renovated water sampled from wells between the basins, and 0.95 mg l^{-1} for renovated water sampled from a well 30 m away from the basins.

5. Boron was not removed and was present at concentrations of 0.5 to 0.7 mg l^{-1} (0.59 mg l^{-1} in 1977) in both effluent and renovated water.

6. Concentrations of zinc, copper, cadmium, and mercury in effluent and renovated water were below the maximum limits for drinking and permanent irrigation. The concentration of lead in the renovated water was 0.066 mg l^{-1}, which exceeded the 0.05 mg l^{-1} limit for drinking water, but was well below the 5 mg l^{-1} limit for permanent irrigation (Nat. Acad. of Sci., Nat. Acad. of Eng., 1972).

7. The biochemical oxygen demand (BOD) was reduced from a range of 10–20 mg l^{-1} for the effluent to essentially zero for the renovated water. Corresponding ranges for chemical oxygen demand (dichromate technique) were 30–60 mg l^{-1} in the effluent and 10–20 mg l^{-1} in the renovated water. Not all organic carbon was removed from the effluent water as it seeped through the soil and aquifer. The total organic carbon (TOC) contents of the renovated water averaged 5.2 mg l^{-1} for the wells between the basins and 3.3 mg l^{-1} for the wells 30 m away from the basins. The average TOC content of the secondary sewage effluent was 19.2 mg l^{-1}.

8. The secondary sewage effluent contained 21 virus units (PFU's) per liter, but viruses could not be detected in renovated water sampled below the basins (Gilbert et al. 1976). Fecal coliforms were present at concentrations of up to several hundred per 100 ml in the renovated water sampled below the basins, but could not be detected in renovated water sampled 91 m away from the basins. Hence, a lateral movement of about 100 m through the aquifer apparently produced renovated water free from fecal coliforms. Laboratory studies showed that distilled water (rain!) could remobilize previously adsorbed viruses in the soil only during the first 24 h of a drying period, but that addition of calcium chloride to the water prevented most of this desorption (Lance et al. 1976). This finding is important for managing rapid infiltration systems in humid climates.

18.5 23rd Avenue Project

Following the successful results of the Flushing Meadows Project, the City of Phoenix became interested in wastewater renovation by soil-aquifer treatment on a commercial scale. This led to the construction in 1975 of the 23rd Avenue Project, where the lower, 16-ha pond of a 2-ponds-in-series stabilization lagoon system was converted into four parallel rapid infiltration basins of 4 ha each (Fig. 2). The renovated water, which is to be pumped from three wells on the center dike of the basin area, could be delivered to an irrigation district for direct sale of the renovated water or for exchange for high quality groundwater pumped from district wells. This groundwater would then be used to augment the city's water supply. For a complete report on this project, reference is made to Bouwer and Rice (1984).

The soils at the 23rd Avenue Project are sandier and more gravelly with more boulders than at the Flushing Meadows Project. The groundwater is also deeper, i.e., between 15 and 25 m before 1978, and between 5 and 15 m after 1978. The groundwater table rise after 1978 was due to a series of wet years, resulting in releases of water through the Salt River, which flows south of the infiltration basins and which recharged the groundwater. The vadose zone below the 23rd Avenue Project consists of an irregular succession of sand, gravel, and boulder layers with occasional silty strata (Fig. 3).

For the first 2 years of operation of the project, the secondary effluent (activated sludge, nonchlorinated) from the 23rd Avenue sewage treatment plant flowed through a 32-ha lagoon before it entered the infiltration basins (Fig. 2).

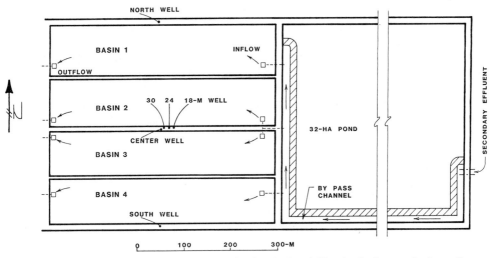

Fig. 2. 23rd Avenue Rapid-Infiltration Project showing the four infiltration basins, monitoring wells, production well (*center well*), and 32 ha lagoon with bypass channel

Fig. 3. Wall of gravel pit near 23rd Avenue Project as indicator of vadose zone below infiltration basins

This resulted in a heavy growth of suspended algae (*Carteria klebsii*) in the effluent water before it entered the infiltration basins. Consequently, hydraulic loading rates were only 21 m yr^{-1}. To reduce the growth of algae in the secondary effluent, a bypass channel was constructed around the 32-ha pond so that the effluent could flow directly to the infiltration basins. In addition, the water depth in the infiltration basins was reduced from about 1 m to 0.2 m to increase the rate of turnover of the water in the infiltration basins, thus reducing the time that a given body of water was exposed to sunshine and, hence, the opportunity for growth of suspended algae. After the bypass channel was put into operation and the surface soil in the infiltration basins was ripped to break up possible crusts of calcium carbonate formed by the previous high algal activity, hydraulic loading rates were about 100 m yr^{-1}. At this rate, the 16 ha infiltration system has a capacity of about 44,000 m^3 d^{-1}.

The aquifer below the infiltration basins is unconfined, very permeable (sands and gravels, similar to Fig. 3), and at least 70 m thick (from project well logs). The actual thickness probably is several hundred meters more as indicated by other well logs and the general geology of the area.

The project has several observation wells 18–30 m deep and one production well which is perforated from 30–54 m and has a capacity of about 13,000 m^3 d^{-1} (center well in Fig. 2). Thus, two more wells symmetrically located on the center dike would have to be drilled to pump renovated water out of the aquifer at the same rate as it arrives at the aquifer from the infiltration basins. The new wells would be drilled somewhat deeper than the center well to obtain the needed additional capacity. Pumping the renovated water out at the same rate as the infiltration rate should establish an equilibrium system in the aquifer and should prevent movement of renovated water into the aquifer outside the infiltration system.

The sewage water used for the 23rd Avenue infiltration project was secondary effluent, activated sludge process. Initially, this effluent was not chlorinated. However, in November 1980, a chlorination facility was put into operation at the treatment plant that dosed the effluent with chlorine at the rate of 1.5 mg l^{-1}. This offered an opportunity to study how the performance of the soil-aquifer treatment system was affected by chlorination of the effluent. Therefore, in discussing some of the water quality aspects, distinction is made between the pre- and post-chlorination periods. Quality parameters of the secondary effluent prior to infiltration and of the renovated water from the center well (unless mentioned otherwise) were as follows [see Bouwer and Rice (1984) for complete report]:

1. The total dissolved salts content of the secondary effluent entering the ground in the last few years of the project averaged about 750 mg l^{-1} and that of the renovated water about 790 mg l^{-1}. The increase of 40 mg l^{-1} can only be partly explained by evaporation from the basins and the soil. Mobilization of calcium carbonate in the soil and aquifer as the pH of the effluent dropped from slightly alkaline to neutral due to biological reactions could also have contributed to the increase in dissolved salts.

2. The suspended solids content of the secondary effluent going into the basins before the bypass channel was constructed was often in the 50–100 mg l^{-1} range in the summer and about 15 mg l^{-1} in the winter. The high summer values were due to algae growth in the 32 ha lagoon. After the bypass channel was put

Table 1. Fecal coliform concentrations of secondary effluent and renovated water in three different periods (colonies/100 ml)

		Prechlorination pre-bypass	Prechlorination post-bypass	Postchlorination post-bypass
Secondary effluent entering basin		10^4	1.8×10^6	3.5×10^3
Renovated water from center well	Average	2.3	22	0.27
	range	0–40	0–160	0–3

into operation, the suspended solids content of the secondary effluent entering the infiltration basins averaged about 11 mg l^{-1} with a range of 5–20 mg l^{-1}. The suspended solids content of the renovated water was about 1 mg l^{-1}, mostly fines from the aquifer.

3. Nitrogen was mostly in the form of ammonium in the secondary effluent (average about 16 mg l^{-1} NH$_4$-N). There were also about 2 mg l^{-1} organic nitrogen and traces of nitrite and nitrate. The 2 week flooding and drying cycles must have been conducive to denitrification in the soil, because the total N content of the renovated water averaged about 5.6 mg l^{-1} of which 5.3 was as nitrate, 0.1 as ammonium, 0.1 as organic, and 0.02 as nitrite. The nitrogen removal percentage thus was about 70%. This removal was the same before and after the secondary effluent was chlorinated, indicating that the low residual chlorine of the effluent by the time it infiltrated into the ground apparently had no effect on the nitrogen transformations in the soil.

4. Phosphate-phosphorus concentrations in the last few years averaged 5.5 mg l^{-1} for the secondary effluent and 0.37 mg l^{-1} for the renovated water pumped from the center well. The shallower wells showed a higher phosphate content, indicating that precipitation of calcium phosphate continued in the aquifer.

5. From a microbiological standpoint, there were three periods, (a) the prechlorination, prebypass channel period when nonchlorinated secondary effluent flowed through the 32 ha lagoon before entering the infiltration basins; (b) the prechlorination, postbypass channel period when nonchlorinated secondary effluent flowed around the lagoon and directly into the infiltration basins; and (c) the postchlorination, postbypass channel period when chlorinated secondary effluent flowed directly into the infiltration basins. The fecal coliform densities in the secondary sewage effluent entering the basins and in the renovated water from the center well are shown in Table 1. The high fecal coliform concentrations in the renovated water generally occurred when water that had infiltrated at the start of a new flooding period arrived at the intake of the well. Apparently, the absence of a clogged soil surface in the basins when flooding was resumed and the decreased bacterial activity in the soil during drying allowed fecal coliforms to move through the unsaturated zone to the groundwater. After a few days of continued flooding, however, fecal coliforms were again completely removed by the clogged surface and in the unsaturated zone and did not show up in the renovated water from the wells. The data in Table 1 show that the higher the fecal coliform concentration in the secondary effluent going into the ground, the more fecal coliforms show up in the renovated water. Virus concentrations in the renovated wa-

ter, determined on samples of 1,000 to 2,000 l, were 1.3 pfu/100 l before the secondary effluent was chlorinated and zero after it was chlorinated. Chlorination thus had a beneficial effect on fecal coliform and virus concentrations in the renovated water.

6. The total organic carbon (TOC) concentration of the secondary effluent averaged 12 mg l^{-1} where it entered the infiltration basins and 14 mg l^{-1} at the opposite ends. This increase was probably due to biological activity in the water as it moved through the basins. The renovated water had a TOC content of 2 mg l^{-1}. The TOC removal was the same before and after chlorination of the secondary effluent, indicating that chlorination had no effect on the microbiological processes in the soil. The concentration of organic carbon in the renovated water of 2 mg l^{-1} was higher than the 0.7 mg l^{-1} typically found in unpolluted groundwaters and which is mostly due to humic substances like fulvic acid (Thurman 1979). Thus, the renovated sewage water from the soil-aquifer treatment process probably contained a number of synthetic organics, some of which could be carcinogenic or otherwise toxic.

18.6 Identification of Trace Organics

The nature and concentration of trace organics in the secondary sewage effluent and renovated water were determined by Stanford University's Environmental Engineering and Science Section, using gas chromatography and mass spectrometry. The studies were carried out for 2 months with nonchlorinated effluent and then for 2 months with chlorinated effluent, taking weekly and biweekly samples. As could be expected, the results showed a wide variety of all sorts of organics (including priority pollutants), many in concentrations on the order of µg l^{-1}. For a full report on this work, reference is made to Bouwer et al. (1983). The main results will be summarized in the following paragraphs.

A comparison between halogenated aliphatic hydrocarbons and chlorinated aromatics in the secondary effluent as it entered the infiltration basins before and after chlorination showed that chlorination resulted in higher chloroform concentrations and produced three additional brominated trihalomethanes (Table 2). Otherwise, chlorination had no significant effect on the chlorinated organics in the secondary effluent. This was also true for the nonhalogenated aliphatic and aromatic compounds where differences between the pre- and postchlorination periods were small and most likely due to normal concentration fluctuations in the effluent. Nonhalogenated compounds identified in the secondary effluent included the aliphatic hydrocarbons 5-(2-methylpropyl) nonane, 2,2,5-trimethylhexane, 6-methyl-5-nonene-4-one, 2,2,3-trimethylnonane, and 2,3,7-trimethyloctane; and the aromatic hydrocarbons o-xylene, m-xylene, p-xylene, C_3 benzene isomer, C_3 benzene isomer, styrene, 1,2,4-trimethyl benzene, ethylbenzene, naphthalene, phenanthrene, and diethylphthalate. The concentrations of these materials were less than 1.5 µg l^{-1} and mostly below 1 µg l^{-1} except for the diethylphthalate which had a concentration of 20 µg l^{-1} in the prechlorination period and 15 µg l^{-1} in the postchlorination period.

Table 2. Halogenated organic compounds in secondary effluent during pre- and postchlorination period ($\mu g \, l^{-1}$)

Constituent[a]	Secondary effluent (basin inflow)	
	Prechlorination	Postchlorination
Halogenated aliphatic hydrocarbons		
Chloroform	2.88	4.79
1,1,1-Trichloroethane	2.45	1.79
Carbon tetrachloride	0.13	0.15
Bromodichloromethane	$-$[b]	0.51
Trichloroethylene	0.91	0.53
Dibromochloromethane	$-$	0.46
Tetrachloroethylene	2.21	1.82
Bromoform	$-$	0.13
Chlorinated aromatics		
o-Dichlorobenzene	4.11	3.18
m-Dichlorobenzene	1.15	0.53
p-Dichlorobenzene	2.70	2.82
1,2,4-Trichlorobenzene	0.33	0.44
Trichlorophenol	0.01	0.02
Pentachlorophenol	0.02	0.04
Pentachloroanisole[c]	0.63	0.26

[a] Identification confirmed by comparison with standards
[b] $-$ Not detected
[c] Only compound that is not a priority pollutant

Of the volatile organics, about 30%–70% volatilized from the effluent as it flowed through the infiltration basins. These compounds included the chlorinated aliphatic hydrocarbons chloroform, 1,1,1-trichloroethane, trichloroethylene, and tetrachloroethylene; the chlorinated aromatic hydrocarbons o-dichlorobenzene, m-dichlorobenzene, p-dichlorobenzene, 1,2,4-trichlorobenzene, and (chloromethyl)-benzene; the aliphatic hydrocarbons 2,2,5-trimethylhexane, 5-(2-methylpropyl) nonane, and 2,2,3-trimethylnonane; and the aromatic hydrocarbons o-xylene, m-xylene, 1,2,4-trimethyl benzene, C_3-benzene isomer, and naphthalene.

18.7 Removal of Trace Organics in the Unsaturated Zone

To study the removal of trace organics from the effluent water in the unsaturated zone, renovated water samples were obtained from an 18 m deep well in the center of the project (Fig. 2). This well, which was cased with steel pipe open at the bottom, yielded renovated water from the top of the aquifer, so that the results would give a good indication of the removal of trace organics in the unsaturated zone. Table 3 shows the concentrations of nonhalogenated organics in the secondary sewage effluent as sampled in the basins and the percentage reduc-

Table 3. Percentage decrease in concentration of nonhalogenated hydrocarbons during passage through unsaturated zone

Constituent	Prechlorination period		Postchlorination period	
	Geometric mean concentration of secondary effluent (27 samples)	Average decrease in renovated water from 18 m well (6 samples)	Geometric mean concentration of secondary effluent (27 samples)	Average decrease in renovated water from 18 m well (6 samples)
	(μg l^{-1})	(%)	(μg l^{-1})	(%)
Aliphatic hydrocarbons				
5-(2-Methylpropyl) nonane	0.35	>94	0.57	>96
2,2,5-Trimethylhexane	0.11	>82	0.18	>89
6-Methyl-5-nonene-4-one	0.41	93[a]	0.94	98[a]
2,2,3-Trimethylnonane	0.21	76[a]	0.25	>92
2,3,7-Trimethyloctane	0.12	50[a]	0.27	>93
Aromatic hydrocarbons				
o-Xylene	0.45	67[a]	0.50	88[a]
m-Xylene	0.76	78[a]	1.00	98[a]
p-Xylene	0.17	53[a]	0.12	92[a]
C$_3$-Benzene isomer	0.56	84[a]	0.34	>94
C$_3$-Benzene isomer	0.48	85[a]	0.53	96[a]
Styrene	0.26	>92	0.58	98[a]
1,2,4-Trimethyl benzene	0.80	78[a]	1.04	96[a]
Ethylbenzene	0.19	53[a]	0.15	67
Naphthalene	0.22	68[a]	0.63	91[a]
Phenanthrene	0.10	80	0.10	90
Diethylphthalate	19	20	10	90

[a] Level of significance for the difference between basin and well concentrations based on a t-test comparison is less than or equal to 0.1

tion in these concentrations after the water had passed through the unsaturated zone and was sampled as renovated water from the 18 m well. Table 4 shows the same data for the halogenated organics.

The results show that nonhalogenated hydrocarbons (Table 3) decreased 50% –99% during percolation through the soil with concentrations in the renovated water being near or below the detection limit. However, most of the compounds could still be detected in the renovated water. Reduction percentages were generally higher during the postchlorination period as a result of higher concentrations in the effluent water observed for many of the nonhalogenated compounds. These compounds are subject to microbial decomposition and, presumably, were removed during soil percolation by this process. Decreases in the concentrations of the nonhalogenated priority pollutants were comparable to those for the other nonhalogenated aliphatic and aromatic hydrocarbons. Concentration variations in the renovated water were less than those in the sewage effluent in the basins. Thus, percolation through the unsaturated zone had the effect of damping fluctuations in concentrations and eliminating extreme values.

Table 4. Percentage decrease in concentration of halogenated organic substances during passage through unsaturated zone

Constituent	Prechlorination period		Postchlorination period	
	Geometric mean concentration of secondary effluent (27 samples)	Average decrease in renovated water from 18 m well (6 samples)	Geometric mean concentration of secondary effluent (27 samples)	Average decrease in renovated water from 18 m well (6 samples)
	(μg l^{-1})	(%)	(μg l^{-1})	(%)
Chlorinated aliphatic hydrocarbons				
Chloroform	2.72	61[b]	3.46	88[b]
1,1,1-Trichloroethane	2.94	34	1.41	84[b]
Carbon tetrachloride	0.12	0	0.12	42[b]
Bromodichloromethane	–[a]	–	0.26	> 62
Trichloroethylene	0.91	– 180[b]	0.39	– 267[b]
Dibromochloromethane	–	–	0.23	> 57
Tetrachloroethylene	2.63	– 97[b]	1.69	31[b]
Bromoform	–	–	0.08	> 10
Chlorinated aromatics				
o-Dichlorobenzene	3.52	25	2.40	10
m-Dichlorobenzene	0.79	58[b]	0.38	5
p-Dichlorobenzene	2.25	33[b]	1.81	10
1,2,4-Trichlorobenzene	0.19	42[b]	0.38	71[b]
Trichlorophenol	0.01	0	0.02	0
Pentachlorophenole	0.02	0	0.04	0
Pentachloroanisole	0.43	– 150	0.18	– 120

[a] – Not detected
[b] Level of significance for the difference between basin and well concentrations based on a t-test comparison is less than or equal to 0.1

The halogenated organic compounds (Table 4) generally decreased to a lesser extent with passage through the unsaturated zone than the nonhalogenated compounds (Table 3). Of the halogenated aliphatic hydrocarbons, the renovated water concentrations of chloroform and 1,1,1-trichloroethane were lower than those in the basin water during both periods. The brominated trihalomethanes present in the secondary effluent with chlorination were not detected in the renovated water samples. This may have been the result of slow transport due to sorption or to chemical or biological transformation. The concentrations of trichloroethylene and pentachloroanisole were significantly higher in the renovated water than in the basin during both sampling periods. Tetrachloroethylene exhibited a similar concentration increase in the prechlorination period, but not in the postchlorination period.

The chlorinated aromatics appeared to be relatively refractory and mobile in the ground because they showed much less concentration decrease than the nonchlorinated aromatic hydrocarbons. Less decrease in the dichlorobenzenes

Table 5. Average TOX concentrations and TOX/TOC ratios for secondary effluent in basins and renovated water from 18 m well

	Prechlorination		Postchlorination	
	TOX $\mu g\ l^{-1}$	TOX/TOC mol Cl/mol C	TOX $\mu g\ l^{-1}$	TOX/TOC ml Cl/mol C
Secondary effluent (average for basin)	84	0.0031	142	0.0050
18 m Well	65	0.0069	55	0.0059

was observed after chlorination than before. Complete breakthrough appeared to occur for the chlorophenols, but concentrations were near detection limits so that positive conclusions could not be made. A combination of biodegradation and sorption processes might have been responsible for the decreases observed.

Total organic halogen (TOX) concentrations and ratios of TOX to total organic carbon (TOX/TOC) for the secondary sewage effluent and renovated water are shown in Table 5. TOX concentrations of the secondary effluent were significantly higher with chlorination than without. However, the renovated water TOX concentrations were similar for both periods. The ratio of TOX to TOC was higher in the renovated water than in the secondary effluent samples, implying that the halogenated organic compounds comprise the more refractory and mobile portion of the TOC.

18.8 Other Organic Micropollutants

In addition to the aliphatics and aromatics mentioned, other compounds tentatively identified in organic extracts of the samples of secondary sewage effluent and renovated water using gas chromatography/mass spectrometry were: fatty acids, resin acids, clofibric acid, alkylphenol carboxylic acids (APEC's), trimethylbenzene sulfonic acid, steroids, n-alkanes, caffeine, Diazinon, alkylphenol polyethoxylates (APE's), and trialkylphosphates. Several of the compounds were detected only in the secondary effluent and not in the renovated water. A few others, Diazinon, clofibric acid, and tributylphosphate, decreased in concentration with soil passage, but were detected in the renovated water. The APE's appeared to undergo rather complex transformations during ground filtration. They appeared to be completely removed with soil percolation during the prechlorination period, but after chlorination two isomers were found following soil passage, while others were removed.

18.9 Conclusions

The results of the Phoenix studies and of similar work on other projects (McKim 1978, Idelovitch and Michail 1982) show that the unsaturated zone can serve as an effective treatment system for partially treated wastewater. The renovated water from the Phoenix projects, for example, meets the health, agronomic, and aesthetic requirements for unrestricted irrigation, including vegetable crops that are consumed raw (Bouwer 1982). The water also meets the standards for lakes with primary contact recreation (Matters 1981). Potable use of the renovated water will require additional treatment, for example, activated carbon adsorption, reverse osmosis, and disinfection. Such treatment, however, will be more effective and economical for renovated water from a soil-aquifer treatment system than for effluent from a conventional sewage treatment plant.

In the Phoenix studies, secondary effluent was used because that was what the treatment plants provided. The secondary (biological) treatment step, however, is not necessary because the soil-aquifer treatment system can handle relatively large amounts of organic carbon. Thus, where sewage effluent is to be used for a rapid-infiltration system, primary treatment may suffice (Rice and Gilbert 1978, Lance et al. 1980, Leach et al. 1980, Carlson et al. 1982, Rice and Bouwer 1983). Some additional clarification or lime precipitation of the primary effluent may be desirable, however, to reduce suspended solids and improve the quality of the primary effluent.

Where posttreatment of the renovated water is needed, the soil-aquifer treatment system should be designed as in Fig. 1A because the wells discharge essentially 100% renovated water. In contrast, the wells in the system of Fig. 1B discharge a mixture of renovated wastewater and native groundwater. This provides the benefit of blending, but it also increases the capacity and, hence, the cost of the posttreatment plant. Encroachment of renovated sewage water into the aquifer outside the soil-aquifer treatment system of Fig. 1A can be avoided by monitoring groundwater levels below the outer edges of the infiltration areas and by managing infiltration and pumping rates so that the water levels in the monitoring wells do not rise above the general groundwater level outside the system. By maintaining water levels in the monitoring wells at the same elevation as the water table outside the system, very little if any native groundwater will be drawn into the soil-aquifer treatment system. This is important where there are legal or other restrictions on pumping indigenous groundwater. Finally, the cost of putting partially treated wastewater underground and pumping it from wells as renovated water is low and soil-aquifer treatment systems do not require highly trained operators.

References

Bouwer H (1982) Wastewater reuse in arid areas. In: Middlebrooks EJ (ed) Water Reuse. Ann Arbor Sci Publ, Ann Arbor 6:137–180

Bouwer H, Chaney RL (1974) Land treatment of wastewater. In: Brady NC (ed) Advances in agronomy Vol 26 Academic Press, London New York pp 133–176

Bouwer H, Rice RC (1984) Renovation of wastewater at the 23rd Avenue rapid-infiltration project. J Water Pollut Control Fed 56:76–83

Bouwer H, Rice RC, Escarcega ED (1974a) High-rate land treatment I. Infiltration and hydraulic aspects of the Flushing Meadows Project. J Water Pollut Control Fed 46(5):835–843

Bouwer H, Lance JC, Riggs MS (1974b) High-rate land treatment. II. Water quality and economic aspects of the Flushing Meadows Project. J Water Pollut Control Fed 46(5):844–859

Bouwer H, Rice RC, Lance JC, Gilbert RG (1980) Rapid-infiltration research – The Flushing Meadows Project, Arizona. J Water Pollut Control Fed 52(10):2457–2470

Bouwer EJ, McCarty PL, Bouwer H, Rice RC (1983) Organic contaminant behavior during rapid infiltration of secondary wastewater at the Phoenix 23rd Avenue project. Water Res (in Press)

Carlson RR, Lindstedt KD, Bennett ER, Hartman RB (1982) Rapid infiltration treatment of primary and secondary effluents. J Water Pollut Control Fed 54:270–280

Gilbert RG, Gerba CP, Rice RC, Bouwer H, Wallis C, Melnick JL (1976) Virus and bacteria removal from wastewater by land treatment. Appl Environ Microbiol 32:333–338

Idelovitch E, Michail M (1982) Dan Region project. Recharge with municipal effluent. Period 1980–1981, Tahal-Water Planning for Israel, Tel-Aviv Israel

Lance JC, Gerba CP, Melnick JL (1976) Virus movement in soil columns flooded with secondary sewage effluent. Appl Environ Microbiol 32(4):520–526

Lance JC, Rice RC, Gilbert RG (1980) Renovation of wastewater by soil columns flooded with primary effluent. J Water Pollut Control Fed 52(2):381–388

Leach LE, Enfield CG, Harlin CC Jr (1980) Summary of long-term rapid infiltration system studies. US Environ Protec Agency, Ada OK Rep EPA-600/2-80-165, 48 pp

Matters MF (1981) Arizona rules for irrigating with sewage effluent. Proc Sewage Irrigat Symp Phoenix, Arizona, US Water Conservat Lab pp 6–12

McKim HL (ed) (1978) State of knowledge in land treatment of wastewater. Proc Int Symp US Army Corps Eng Vol 1 and 2. Hanover New Hampshire

National Academy Sciences and National Academy of Engineering (1972) Water quality criteria, 1972. Comm Water Qual Criteria Request US Environ Protect Agency, Washington DC

Rice RC, Bouwer H (1983) Soil-aquifer treatment using primary effluent. J Water Pollut Control Fed 56:84–88

Rice RC, Gilbert RG (1978) Land treatment of primary sewage effluent: Water and energy conservation. Hydrology and water resources in Arizona and the Southwest. Univ Arizona Press, Tucson 8:33–36

Thurman EM (1979) Isolation, characterization, and geochemical significance of humic substances from groundwater. PhD Thesis. Univ of Colorado

19. Use of Soil-Aquifer System for Effluent Purification and Reuse

E. IDELOVITCH

19.1 Introduction

In areas where groundwater basins have been depleted by overpumping and where there is danger of groundwater salinization by seawater intrusion, sewage effluents are used for groundwater replenishment. In such cases, the recharge of the effluent to a potable aquifer is the ultimate use of the reclaimed water (Fig. 1). A very high degree of treatment (virtually to drinking water quality) is provided prior to recharge. The recharge can be carried out either by spreading basins or by injection wells. The recharged effluent loses its identity and its movement in the aquifer can hardly be traced. Production wells located in the vicinity of the recharge zone pump an admixture of recharged effluent and native groundwater, which is supplied to unrestricted uses, including drinking. The main concerns connected with such a scheme are: the high cost, because of the high level of treatment required prior to recharge, and the possible long-term health hazards associated with the ingestion of effluent, even though of very high quality.

A different approach to groundwater recharge with effluent was adopted in the Dan Region Project – the largest and most advanced wastewater reuse scheme in Israel (Idelovitch et al. 1980). This approach, which is referred to as soil-aquifer treatment (SAT), consists of controlled passage of the effluent through the un-

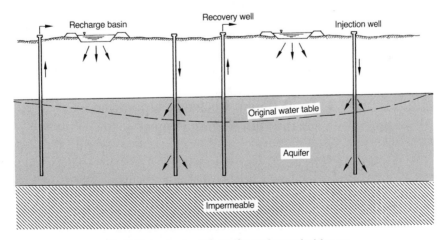

Fig. 1. Groundwater recharge for aquiver replenishment

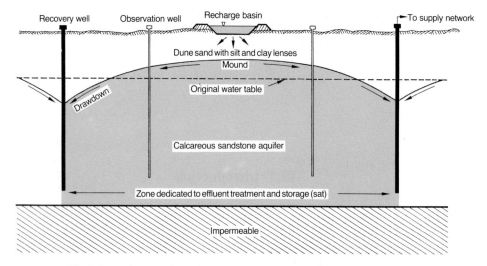

Fig. 2. Groundwater recharge for treatment and storage – Dan Region Scheme

saturated zone and a limited portion of the aquifer, which is separated from the rest of the aquifer by a ring of adequately spaced recovery wells surrounding the recharge area (Fig. 2). Only partial treatment is carried out prior to recharge, since the soil-aquifer system provides for additional treatment by a combination of physical, chemical, and biological purification processes. The recharge is carried out by spreading basins; injection wells are not suitable for this type of operation, because they would require high quality effluent in order to avoid well clogging. The recharged effluent, which can be traced and monitored in the subsoil and in the aquifer, is pumped by the recovery wells and supplied to uses compatible to its quality.

Generally these are nonpotable uses, such as industrial and unrestricted agricultural uses, including irrigation of vegetables to be eaten raw and livestock watering. Accidental drinking of the reclaimed water would not imply any health hazard. The reclaimed water could also serve as a raw source for drinking water supply (Idelovitch 1978).

19.2 The Effluent Recharge Operation in the Dan Region Project

The Dan Region is Israel's largest metropolitan area, which includes the city of Tel Aviv-Jaffa and several other neighboring municipalities located south, east, and north of Tel Aviv.

The Dan Region Sewage Reclamation Project – Stage One, which has been in full operation since January 1977, consists of facilities for treatment and groundwater recharge of the municipal wastewater discharged from the southern parts of the Dan Region (Holon, Bat Yam, Jaffa, and South Tel Aviv), and from the neighboring municipality of Rishon-Le-Zion.

The wastewater pumped to the treatment plant undergoes biological treatment in two parallel series of facultative ponds with recirculation and physicochemical treatment in two stages: high lime-magnesium treatment carried out in two reactor clarifiers, followed by detention of the high pH effluent in a series of polishing ponds, mainly for free ammonia stripping and natural recarbonation (Figs. 3 and 4).

A small amount of well water containing effluent which in the past seeped from the treatment ponds is also pumped to the last polishing pond by means of Dan wells Nos. 6, 7, and 8. This operation fulfills two objectives: to complete the first stage recarbonation prior to recharge in order to avoid scale deposits and to recycle seeped effluent in order to lower groundwater levels and to avoid water losses to the sea.

The treated effluent is recharged to the regional groundwater aquifer by means of spreading basins located in the vicinity of the wastewater treatment plant (Fig. 4).

The recharged water, after additional treatment and prolonged detention in the soil-aquifer system, will be pumped by means of recovery wells surrounding the recharge basins for reuse in the south of the country. In the final stage of the project, when the recovery wells will pump mostly recharged effluent, the reclaimed water will be supplied only for nonpotable uses (mainly unrestricted irrigation) by means of a separate supply network.

The recharge site is located in an area of rolling sand dunes near the Mediterranean coast, lying above the central part of the coastal aquifer (Pleistocene). The aquifer consists mainly of calcareous sandstone and is divided into subaquifers by silt and clay layers.

The surface soil of the recharge basins is a relatively uniform, fine sand (less than 0.3 mm).

The climate of the area is typically Mediterranean. Summers are warm and dry, whereas winters are mild, with rainy spells. The average annual precipitation is 500–600 mm. The average temperatures usually range between 20°–30 °C in summer and 10°–20 °C in winter.

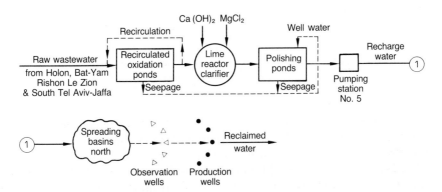

Fig. 3. Flow diagram of Dan Region Project – Stage One

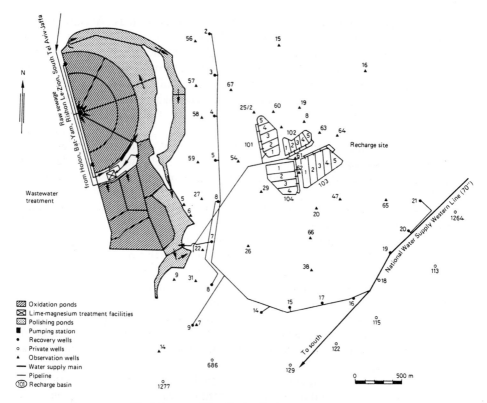

Fig. 4. Layout of Dan Region Project – Stage One

The recharge area consists of four basins (101, 102, 103, and 104), covering at present a total net area of about 24 ha. Originally, only basin 102 was divided into subbasins, for various research purposes. Later, all basins were divided into subbasins in order to enable higher hydraulic loading per unit area and greater operational flexibility.

A ring of recovery wells, spaced 300–400 m from one another surround the recharge area on the western, southern, and eastern side; they are located at distances varying between 800–1,800 m from the center of the recharge zone (Fig. 4). Additional wells will be constructed in future north of the recharge basins, in order to minimize the influence of the recharge operation on private wells. All the wells are pumping at present to the national supply network, with the exception of wells Nos. 6, 7, and 8, which have been pumping to the last polishing pond since 1978.

Installations for sampling the percolating effluent from the upper layer (2 m) of the unsaturated zone, which were specially designed for this purpose, were constructed in two subbasins of basin 102 and have been used for research purposes.

A monitoring network of 20 observation wells was established between the recharge basins and the recovery wells. Samples from the observation wells are taken after at least half an hour of pumping with compressed air, which is sufficient in order to exchange the water in the well several times. The observation wells are of 2 or 3 in diameter and are perforated at various depths. At some locations, two separate wells were drilled to different subaquifers or to different layers of the same subaquifer. The observation wells are located at distances varying between 30–800 m from the recharge basins.

About 80 million m^3 of effluent were recharged to the groundwater aquifer during 6 years of operation 1977–1982 (Idelovitch et al. 1983). The spreading basins were flooded intermittently in order to maintain high infiltration rates and to enhance effluent purification during percolation. A short recharge cycle was employed, consisting usually of 1 day flooding and 2–3 days dry, in order to ensure that aerobic conditions predominate in the unsaturated zone and in the aquifer.

Average infiltration rates varied between 1–1.5 m d^{-1} and average hydraulic loads varied between 100–150 m yr^{-1}, depending on effluent quality, climatic conditions, and the frequency of basin cleaning.

The vertical flow velocity in the unsaturated zone is estimated at 1.5–2 m d^{-1} and the depth of the unsaturated zone below the recharge basins varies between 15–30 m; thus, the travel time of the recharged water through the unsaturated zone varies between 10–20 days.

The horizontal flow velocity in the aquifer is estimated at 0.6–1 m d^{-1}; thus, the advance of the recharged water in the aquifer varies between 200–350 m yr^{-1}.

After 6 years of recharge, groundwater elevations in the center of the recharge zone rose by about 8 m. The area affected by the mound created below the recharge basins extended to about 1 km south and west and 1.5–2 km north and east of the recharge zone. The front of the recharged water advanced about 1 km from the center of the recharge area in all directions. Most of the recharged water is stored in the aquifer, where it displaces native groundwater toward the recovery wells.

19.3 Effluent Purification in the Soil-Aquifer System

The purification occurring by soil-aquifer treatment (SAT) was evaluated from the results of water quality analyses carried out in the effluent before recharge and in observation wells pumping 100% recharged water.

The recharge effluent has generally a low alkalinity and a low to moderate concentration of suspended solids, organic matter, nitrogen, and phosphorous. The effluent quality is usually higher in summer than in winter, because of the greater efficiency of both the lime clarification process and the ammonia stripping process.

The recharge water has a relatively high pH and Langelier saturation index, i. e., it has a slight tendency to form $CaCO_3$ scale.

The quality of the effluent is different from that of the native groundwater in the recharge zone (Table 1). The salinity of the recharge water, expressed as total dissolved solids or electrical conductivity is higher than that of the background water in the aquifer. With respect to major ions, the chloride and sodium concen-

Table 1. Quality of recharge effluent and of background aquifer water

Parameter	Unit	Recharge[a] effluent		Background[b] aquifer water	
Total dissolved solids	mg l^{-1}	500 –	700	200	–250
Electrical conductivity	μmhos cm^{-1}	750 –	1,050	300	–400
Chloride	mg l^{-1}	160 –	250	20	– 30
Sodium	mg l^{-1}	100 –	170	15	– 20
SAR	mg l^{-1}	3 –	8	0.6 –	1
BOD	mg l^{-1}	5 –	30	< 0.5	
COD	mg l^{-1}	60 –	120	< 0.5	
UV$_{254}$ absorbance	cm^{-1} × 10^3	200 –	275	< 3	
KMnO$_4$ consumption, as O$_2$	mg l^{-1}	10 –	18	0.2 –	0.5
Detergents	mg l^{-1}	0.5–	1.9	< 0.01	
Ammonia as N	mg l^{-1}	1 –	10	< 0.02	
Nitrogen Kjeldahl	mg l^{-1}	7 –	20	< 0.2	
Nitrate N	mg l^{-1}	0 –	1	2 –	5
Total nitrogen	mg l^{-1}	7 –	21	2 –	5
Phosphorus	mg l^{-1}	0.9–	3.8	0.01–	0.06

[a] Based on seasonal average data in the effluent pumped to the recharge basins
[b] Based on analyses of observation wells in the recharge zone before they were affected by effluent

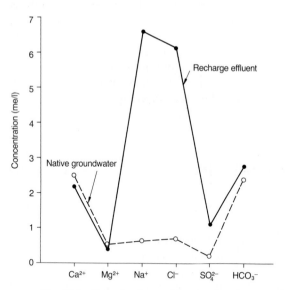

Fig. 5. Ionic composition (Schoeller) of recharge effluent and native groundwater

trations are much higher in the recharge water than in the groundwater, whereas calcium, magnesium, and bicarbonate concentrations are approximately the same, because of the softening effect of the high lime process (Fig. 5). The chloride ion, which is virtually unaffected by soil processes, serves as a reliable tracer of the movement of recharged effluent in the aquifer.

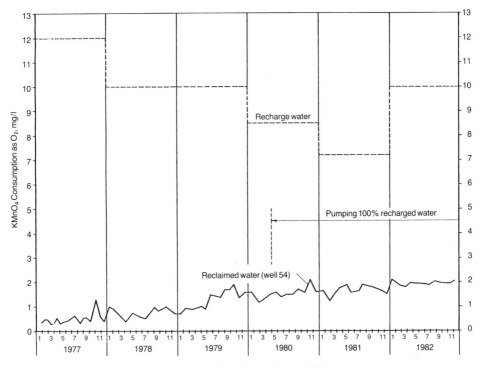

Fig. 6. Organics before and after SAT

In order to facilitate the understanding of the purification processes of the effluent during flow through the soil-aquifer system, the observation wells were grouped as follows (Idelovitch et al. 1983, Idelovitch and Michail 1983):

Group 1: Wells with a relatively short travel time from the nearest recharge basin (1–6 months): Nos. 29, 61, 62/2, 63, 64/1, and 64/2. Most of these wells have been pumping 100% recharged effluent for a long time. In two wells of this group (62/2 and 63), anoxic conditions became predominant lately (in 1982). In the other wells of this group, aerobic conditions were predominant.

Group 2: Wells with a longer travel time (9–12 months) or wells pumping from a deeper subaquifer: Nos. 19, 20, 47, 54, 60, and 62/1. These wells started pumping 100% recharged water more recently. In all these wells, aerobic conditions were predominant.

The removal of residual suspended solids (mostly organics) and of residual BOD was virtually complete in the soil-aquifer system. Good to very good removal of dissolved organic substances was obtained (Fig. 6), mainly by bacterial degradation which is favored by the long detention times in the aquifer (Idelovitch et al. 1982). Lower values of gross organic parameters, as well as of detergents, were obtained in Group 2 wells and in Group 1 wells, where aerobic conditions predominated, than in Group 1 wells where anoxic conditions pre-

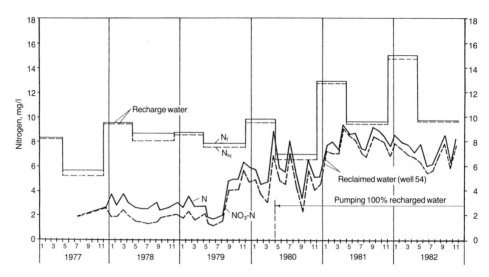

Fig. 7. Nitrogen before and after SAT

dominated (wells 62/2 and 63). The removal of dissolved organics by SAT under anoxic conditions was by about 15% lower than under aerobic conditions.

Phosphorous was removed very efficiently by SAT, from about 1–3 mg l^{-1} in the recharge water to less than 0.05 mg l^{-1} in the reclaimed water.

Nitrification was very efficient, especially in wells where aerobic conditions predominated. Most of the residual nitrogen in the effluent, which is found in unoxidized forms (Kjeldahl, i. e., ammonia and organic nitrogen), is converted to nitrate in the groundwater (Fig. 7). Denitrification was generally limited and fluctuating. In the wells where anoxic conditions predominated, denitrification was generally more efficient, but relatively high concentrations of ammonia have appeared lately (in 1982).

Denitrification can be enhanced if longer cycles of flooding-drying are employed (Bouwer et al. 1980), but ammonia breakthrough can occur under conditions of continuous flooding (Idelovitch and Michail 1980).

A reduction of pH, together with the conversion of carbonates to bicarbonates, occurred in the upper layer of the unsaturated zone by means of CO_2 and volatile acids produced by bacterial activity. Consequently, the Langelier Index of the reclaimed water was considerably improved and the reclaimed water pumped from the aquifer has no corrosive or scale-forming tendency.

The cation exchange process between sodium and potassium, on the one hand, and calcium and magnesium on the other hand, as well as the boron adsorption process, were exhausted in the area surrounding the recharge basins (Group 1 wells), but still acting to a certain extent in areas further away from the recharge basins (some of Group 2 wells).

The concentrations of some metals (Cd, Cr, Se, Cu, Mo, and Ni) and of phenolic compounds, which are higher in the recharge water than in the native groundwater, were reduced by SAT virtually to background levels (Table 2).

Table 2. Water quality before and after SAT

Parameter	Unit	Before SAT[a]	After SAT[b]
Suspended solids	mg l^{-1}	45	0
BOD$_f$	mg l^{-1}	6	< 0.5
COD$_f$	mg l^{-1}	50	10
KMnO$_4$ consumption$_f$, as O$_2$[c]	mg l^{-1}	8	2
UV$_{254}$ absorbance	cm^{-1} × 10^3	240	45
Detergents	mg l^{-1}	1	0.2
Nitrogen	mg l^{-1}	12	8.5
NH$_3$-N	mg l^{-1}	7	< 0.02
Organic-N	mg l^{-1}	4.9	0.5
NO$_3$-N	mg l^{-1}	0.1	8.0
Phosphorus	mg l^{-1}	2	< 0.02
Boron	mg l^{-1}	0.3	0.2
Sodium	mg l^{-1}	125	110
SAR	mg l^{-1}	4.4	2.7
pH	–	9.2	8.0
Langelier index	–	1.3	0.4
Cadmium	μg l^{-1}	1.2	0.5
Chromium	μg l^{-1}	6.5	3
Molybdenum	μg l^{-1}	8	3
Nickel	μg l^{-1}	21	12
Selenium	μg l^{-1}	6.5	1

[a] Average 1981–1982 in recharge water
[b] Average end of 1982 in Group 2 observation wells (underground travel time 9–12 months), pumping 100% recharged water
[c] The symbol f refers to filtered samples

The microbiological quality of the water pumped from the observation wells was very high in most cases: no coliforms, *E. coli, S. faecalis,* or enteroviruses were detected (Idelovitch 1982).

19.4 Conclusion

The indirect reuse of wastewater by means of groundwater recharge should be preferred, where feasible, because of its considerable merits related to effluent purification and storage. Knowledge of the purification processes occurring naturally underground (in the soil-aquifer system) can allow considerable savings in the treatment carried out artificially "above ground" (in the wastewater treatment plant), depending on the final objective of effluent reuse. In addition, the indirect reuse by SAT has an important psychological advantage from the point of view of the consumer, who is supplied from wells pumping groundwater and not from the outlet of a wastewater treatment plant.

Acknowledgements. The Dan Region Sewage Reclamation Plant is operated by Mekorot Water Company Ltd., the Jordan District. The author acknowledges the help and advice provided by Mrs. Medy Michail of Tahal in the preparation of this paper.

References

Bouwer H, Rice RC, Lance JC, Gilbert RG (1980) Rapid-infiltration research at Flushing Meadows Project. Arizona J WPCF

Idelovitch E (1978) Wastewater reuse by biological-chemical treatment and groundwater recharge. J WPCF

Idelovitch E (1982) Microbiological aspects of wastewater reuse by groundwater recharge. Presented at Int Symp on Enteric Viruses in Water, Herzlyia, Israel

Idelovitch E, Michail M (1980) Treatment effects and pollution dangers of secondary effluent percolation to groundwater. In: Prog Wat Tech, Vol 12. Pergamon, New York London, pp 949–966

Idelovitch E, Michail M (1983) Groundwater recharge for wastewater reuse in the Dan Region Project, summary of five – year experience 1977–1981. In: Asano T (ed) Artificial recharge of groundwater. Ann Arbor Sci (in press)

Idelovitch E, Terkeltoub R, Michail M (1980) The role of groundwater recharge in wastewater reuse: Israel's Dan Region Project. JAWWA

Idelovitch E, Michail M, Roth T (1982) Wastewater reclamation by groundwater recharge via spreading basins. Presented at Joint German – Israel Workshop in Water Technology, Dan Caesarea, Israel

Idelovitch E, Michail M, Goldman B (1983) Groundwater recharge with municipal effluent, Dan Region Project – sixth recharge year – 1982. Tahal – Water Planning for Israel Lid, Wastewater Reuse and Water Quality Division, Tel Aviv 01/83/31

20. Utilization of the Buffering Properties of the Unsaturated Zone in Groundwater Quality Management on a Regional Scale: The Coastal Aquifer of Israel as a Case Study

A. Mercado

20.1 Introduction

The coastal aquifer is one of the three major sources of the Israeli water system, supplying about 30% of the water consumption and planned to serve as the major long-term storage reservoir. The aquifer is phreatic, with depths to water levels ranging between few meters near the coast, up to 80–90 m at its southeastern parts. A thick bluish clay, named Saqya, forms the impervious bottom of the aquifer system; it rises gradually from 100–200 m below sea level at the seashore to nearly ground level at the Judean foothills in the east. The aquifer itself consists of sand and calcareous sandstone layers of Plio-Pleistocene age intersected with clay and loam lenses. The aquifer is replenished directly by percolating rainfall and return flow of irrigation and wastewater. The Mediterranean Sea is the natural outlet of the aquifer system.

The coastal plain overlying the aquifer has a total area of about 1,900 km^2. It includes some of the major urban communities in Israel, accounting for a total population of 1.7 million or about one-half of the total population. The projected population of the coastal plain for the year 2000 is about 3 million.

According to soil types, the coastal plain can be divided into three strips running parallel from north to south: the sand dunes are found along the shore, next to it eastward are light to medium soils, and further east is a strip of heavy soils and loess. The major agricultural crop is citrus orchards covering most central parts of the aquifer. Agriculture within the sand dunes is rather limited. This division of soils has a direct influence on the distribution of natural recharge and, in a way, also on pollution mechanism.

The combination of shallow phreatic water table, sandy soils, dense population, and intensive agriculture is highly unfavorable whenever groundwater is considered. Continuous water quality records show indeed that considerable parts of the aquifer area are polluted by nitrates, accompanied by chloride salinization, increase in total hardness, and alarming appearances of traces of heavy metals in some urban wells. It is estimated (Mercado 1980) that if nitrate levels follow current trends, increasing at the rate of about 1.3 ppm/year, a major portion of the water supply wells of the coastal aquifer will by the year 2000 pass the maximum allowable drinking water standard of 90 ppm NO_3^-. Chloride concentrations are increasing at an even faster rate of 1.5 ppm per year, which will eventually have serious implications for salt-sensitive soil-crop systems.

In view of these salinization and pollution trends, we believe that urgent protection measures should be taken in order to conserve groundwater quality without interfering with the population growth and economic development of the area. Appropriate engineering and economic decisions should be made now, although we still might lack a conclusive description of some of the physicochemical phenomena in the natural aquifer systems. The gap between the need for decisions and the lack of adequate physicochemical knowledge can be filled partly with the introduction of systems approach and simplified models oriented especially toward groundwater quality management on a regional scale.

20.2 Systems Approach to Groundwater Quality Management

Development and evaluation of measures for environmental protection of groundwater sources, to ensure the quantity and quality of present and future water supply demands, requires a knowledge of the quantitative relationships among human activities, hydrogeological parameters of the aquifer system, and aquatic chemistry. This knowledge is essential to elucidate some of the cause and effect pathways, to estimate pollution trends in the future, and to help in identifying the most suitable means of attacking major groundwater pollution problems. A good starting point is to define the major components of the groundwater quality system and their interrelationships.

Major paths of water and solutes in a given regional aquifer system can be described schematically in a flow chart consisting essentially, of three major sections (Fig. 1).

1. The land surface where most human activities, associated with the release of contaminants, are occurring.
2. The unsaturated zone (including the root zone of agricultural crops), which is considered to be the major "chemical reactor" for the whole system, is responsible also for the considerable time lag between the release of contaminants on the land surface and their arrival to the groundwater table.
3. The aquifer which is responsible for the dilution of contaminants and their transport to pumping wells and springs.

This conceptual description of the groundwater quality system is especially suitable for the case of diffused pollution sources overlying free-surface aquifers. The selection of other groundwater quality models for different field problems is discussed by Schwarz and Mercado (1974).

State variables of this system are usually the water quality parameters included in water quality criteria and standards. Maximum acceptable levels of these state variables are dictated by their use (domestic industrial, or agricultural), subject to existing standards and criteria, by flow patterns in the aquifer, and by forecasted exploitation plans. Decision variables frequently encountered in groundwater quality studies are location and timing of pumpage and artificial recharge, maximum tolerable load of various surface contaminants, concentration of toxic substances in sewage effluent, their control in pollution sources and the desired treatment level in plants, fertilizer dosage to crops, etc.

Groundwater quality models are always conjuncted with flow models; the latter can be regarded as a subsystem of the hydrological cycle and usually it can

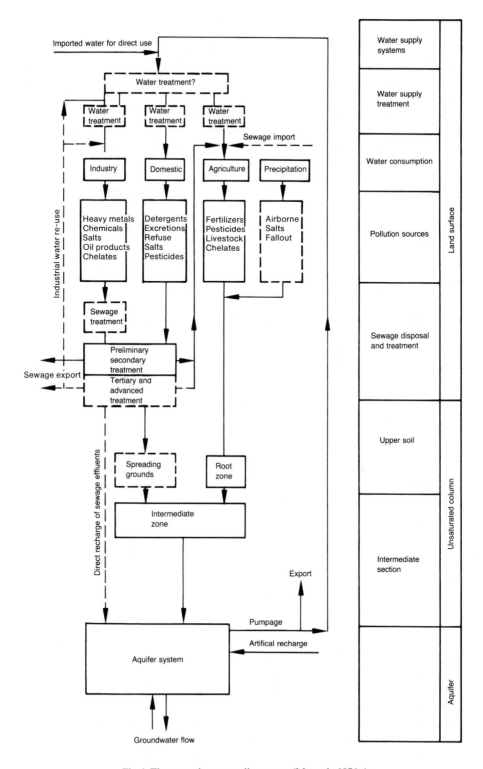

Fig. 1. The groundwater quality system (Mercado 1976 a)

be used separately. In groundwater quality models, the effects of surface activities and the processes occurring at the unsaturated zone should always be considered. This can be done by expanding existing groundwater models to include the relevant parameters of the land surface and the unsaturated zone or by defining the boundary conditions (varying with time) between the groundwater subsystem and the overlying surface. In such cases, the two models should be analyzed separately, reaching a solution by a repetitive iteration on the boundaries between them.

The laws of conservation of mass, for both water and aqueous species in the aquifer, provide a theoretical basis for developing a groundwater quality model of a given aquifer system. If the mass of aqueous species is to be conserved, the following relationship must be satisfied in any element of the system:

$$\left\{\begin{array}{l}\text{Rate of mass}\\\text{accumulation}\end{array}\right\} = \left\{\begin{array}{l}\text{Net rate of}\\\text{mass transfer}\end{array}\right\} + \left\{\begin{array}{l}\text{Net rate of appearance}\\\text{of mass by reaction}\end{array}\right\}$$
$$+ \left\{\begin{array}{l}\text{Contribution rate of}\\\text{mass by external sources}\end{array}\right\} \tag{1}$$

The mathematical statement of the above relationship, for both saturated and unsaturated zones of the aquifer system, are well-known and will not be repeated here. The first term on the right-hand side of Eq. (1) is described usually as a combination of hydrodynamic dispersion and convection. The latter is linked to the water balance. The form of the second term depends on the intrinsic characteristics and kinetics of the specific chemical reactions involved (Schwarz and Mercado 1974, Mercado and Billing 1975). The last term represents external inputs of contaminants to the specific section of the groundwater quality model. It should be remembered that the input function of the groundwater section is at the same time the output of the unsaturated zone model.

The combination of the separate differential equations which describe these relationships, and their solution according to a specified set of boundary and initial conditions, provides a model for studying groundwater pollution patterns in a given aquifer system. Examples for separate models of the unsaturated and saturated zones are numerous; however, only a few models describe simultaneously the overall system.

The use of such models for prediction purposes requires their calibration by fitting computed concentration variations to those measured in the field. In most instances, however, the scarcity of reliable field data, as compared to the complicated features of the terrestrial cycle of some contaminants, prevents the use of detailed models and alternative approximate tools must be developed instead.

Lumped models, of the single-cell type, are considered to serve this purpose; they are built upon extensive simplification of the conditions prevailing in reality and grouping of the pertinent characteristics and processes. Predefined parts of the aquifer are considered as fully mixed cells. Transport phenomena along the unsaturated column are expressed by the transit time of chemically inert solutes, corrected by the specific retardation factors due to adsorption and cation-exchange processes.

The use of such models is exemplified in the present communication for the preliminary assessment of the possible variations of some groundwater quality parameters in the coastal plain of Israel. They also provide the means for making economic and engineering decisions today, to be implemented immediately by planners.

The features of the single-cell model are outlined below, emphasizing the dominant role of the unsaturated column in buffering deterioration effects in over-exploited aquifers.

20.3 The Integrated Single-Cell Model

The single-cell model (Fig. 2) integrates pollution sources on the land surface, hydrological parameters of the aquifer and the unsaturated zone, and variations of average solute concentration. For the sake of brevity, two major sections of the model (the aquifer and the unsaturated zone) are discussed separately.

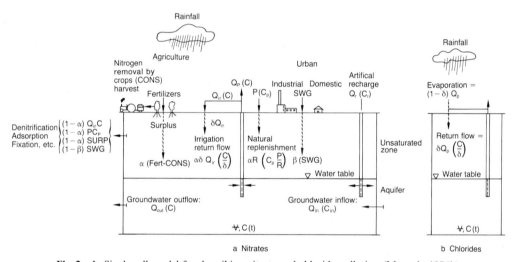

Fig. 2a, b. Single-cell model for describing nitrate and chloride pollution (Mercado 1976b)

The aquifer part of the single-cell model represents a portion of the aquifer volume within the investigated area. The whole aquifer is described by a single fully mixed element. The conservation of mass within one element is described by a simplified version of Eq. (1), where geochemical reactions within the aquifer and the dispersion mode of mass transfer were neglected.

> Rate of mass accumulation
> = net rate of convective mass transfer
> + vertical contribution rate of mass by pollution sources   (2)

or mathematically:

$$V\frac{dC}{dt} = Q_{in}C_{in} + Q_{un}C_{un} + Q_r C_r - Q_p C - Q_{out}C, \qquad (3)$$

where C is the solute concentration at time t, Q the discharge rate of various groundwater terms. The indices in, un, r, p, and out in Eq. (3), represent inflow, unsaturated, artificial recharge, pumpage, and groundwater outflow respectively, V is the effective groundwater volume used for dilution within the cell given by

$$V = Abn, \tag{4}$$

where A is the surface area of the model, b is the mean saturated thickness, and n is the aquifer porosity.

The flow terms in Eq. (3) obey the water balance equation

$$Q_{in} + Q_r + Q_{un} - Q_p - Q_{out} = AS \frac{d\phi}{dt}, \tag{5}$$

where S is the aquifer phreatic storativity ($S \leq n$), and ϕ the piezometric head. In numerous occasions, steady-state flow conditions nearly prevail, such that $d\phi/dt \approx 0$. Equation (5) is simplified, therefore:

$$Q_{in} + Q_r + Q_{un} - Q_p - Q_{out} \approx 0. \tag{6}$$

The average unsaturated flow term is approximated by the sum of natural replenishment and return flow from irrigation:

$$Q_{un} \cong AR + \delta Q_p, \tag{7}$$

where R is the natural replenishment rate and δ the return flow coefficient of pumped water, yielding finally an estimate for Q_{out}:

$$Q_{out} \cong Q_{in} + Q_r + AR - (1 - \delta) Q_p. \tag{8}$$

The concentration C_{un} of the vertical influx at the water table is approximated by

$$C_{un}(t) = \sum_j w_j W_j(t - \tau)/Q_{un}, \tag{9}$$

where W_j is the contribution rate of pollution source j at time $(t - \tau)$, τ is the mean transit time along the unsaturated column estimated from:

$$\tau = (RF) \frac{A\theta d}{AR + \delta Q_p}, \tag{10}$$

where θ the average moisture content of the unsaturated column, d its length, and RF the retardation factor (≥ 1) of adsorbable species in comparison to tagged water molecules. RF $= 1$ for inert species.

w_j in Eq. (9) is a linear proportion factor expressing the possible decrease of pollution loads due to various chemical transformations. Fixation and denitrification of nitrogen species is one of the possible examples, discussed further in this chapter. Obviously, $w_j = 1$ for chemically inert species, such as chlorides.

The evaluation of $\sum w_j W_j(t - \tau)$ in Eq. (9), depends on the specific conditions of the investigated area and the characteristics of the considered pollutant. To demonstrate the order of magnitude of the buffering properties of the unsaturated column, a simplified version of chloride recycling in phreatic aquifers is considered.

In our demonstration, chlorides are contributed by two sources only: (1) rainfall and (2) irrigation. In addition, the water balance terms Q_r and Q_{in} are set to zero.

Chloride flux at the water table is determined, therefore, from

$$Q_{un}C_{un}(t) = APC_p + Q_pC(t-\tau). \tag{11}$$

A further assumption is used that $C(t-\tau)$ can be approximated by:

$$C(t-\tau) = C(t) - \tau\frac{dC}{dt}. \tag{12}$$

Substituting above assumptions and approximations in Eq. (3), we get:

$$V\frac{dC}{dt} = APC_p + Q_pC - Q_p\tau\frac{dC}{dt} - Q_pC - [AR - (1-\delta)Q_p]C$$

or:

$$(V + Q_p\tau)\frac{dC}{dt} = APC_p - [AR - (1-\delta)Q_p]C \tag{13}$$

yielding as a solution:

$$C(t) = (1/\bar{B})[\bar{A} - (\bar{A} - \bar{B}C_0)\exp(-\bar{B}t)], \tag{14}$$

where the coefficients $\bar{A}$ and $\bar{B}$ are given by:

$$\bar{A} = APC_p/(V + Q_p\tau) \tag{15}$$

and

$$\bar{B} = -[AR - (1-\delta)Q_p]/(V + P_p\tau) \tag{16}$$

and C_0 is the initial concentration.

Above equations are valid only for $t > \tau$. For shorter time intervals, pumped water concentration will be about equal to the historical undisturbed concentrations determined simply by:

$$C_o = C_p\, P/R. \tag{17}$$

We should notice at this stage that the nonsaturated element has at least two properties which buffer the pollution process considerably: (1) The time retardation of pollutents by τ, which means that background concentrations are maintained for $t \leq \tau$. (2) The equivalent increase of the effective groundwater volume V by the product of $Q_p\tau$.

Figure 3 shows the estimated distribution of the transit time of chemically inert solutes along the unsaturated column in the Israeli coastal aquifer. Transit time range, according to this figure from a few years along the coast up to hundreds of years at its southeastern parts. This distribution is dictated in the major part by the combination of shallow groundwater table and sandy lithology along the coast, as compared to silt and clay profiles, as well as a deep water table at the southeast.

The transit-time of "real" pollutants, such as heavy metals for example, is much longer than predicted for chlorides because of anticipated cation exchange

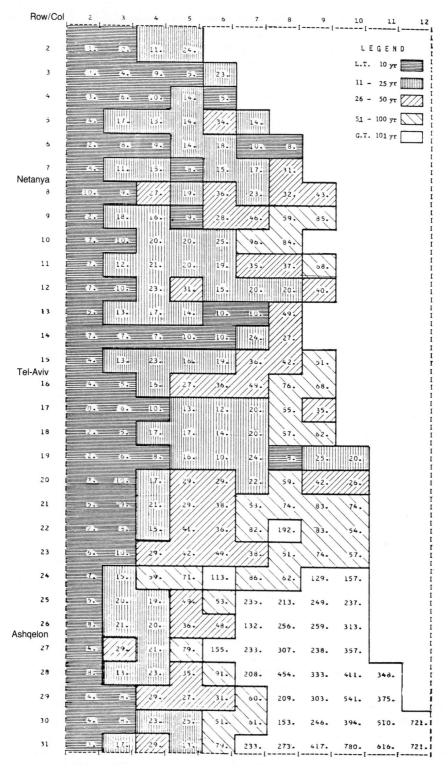

Fig. 3. Estimated transit time (years) of chemically inert solutes across the unsaturated column of the coastal aquifer (Mercado 1976a)

and adsorption processes. Above discussed buffering properties will increase, therefore, considerably.

The following section demonstrates the application of above concepts and engineering approximations for assessing the likelihood of groundwater contamination by sewage reclamation.

20.4 Large-Scale Sewage Reclamation for Irrigation – Risk Assessment of Possible Groundwater Contamination by Toxic Species

Forecasts for the industrial and domestic demands in Israel indicate that they might be tripled by the year 2000 to 2050. Sewage reclamation, to replace a portion of agricultural consumption, will be introduced then on a large scale to meet these requirements. Development programs made according to these forecasts are based partly on the assumption that replacing up to 50% of existing freshwater agricultural consumption by sewage effluents is feasible. Questions arise, however, of what will be the impact of this program on groundwater quality of the coastal plain and what precautions should be taken today in order to meet water quality criteria in the foreseeable future?

A study which was carried out by Tahal and sponsored by the Israel Water Commission, yielded a preliminary answer to these problems. The objective of this study was to develop quantitative tools and decision criteria for accepting (or rejecting) proposed plans for irrigation by sewage effluents in a given region. It will also contribute to the design criteria of treatment plants, sewage reclamation projects, pollution source control programs, and the implementation of dynamic regulations, to be issued by the Water Commissioner, for preventing groundwater pollution.

The assessment of potential pollution trends requires the development of semiquantitative tools to simulate reality. These tools, essentially of the above described single-cell type, are based on the routes of water and contaminants, discussed briefly below.

20.4.1 Recycling Schemes and Their Simulation

The water supply in a typical cell (Fig. 4) of the coastal plain aquifer consists of local pumpage (QP) and imported water (QIMP) through the National Water Carrier. A limited amount (QEXP) is exported to other regions. The total consumption $(QSIP=QP-QEXP+QIMP)$ is divided between urban (QUR) and agricultural (QIR) sectors. The division or replacement coefficient $(\alpha=QUR/QSUP)$ is a decision variable, depending partially on water quality parameters. The desired replacement, on a nationwide basis, is in the order of magnitude of 50%. A part $\delta (=65\%)$ of urban consumption is reclaimed and used as a part of the water inventory. Urban losses are compensated by an equivalent amount of imported water.

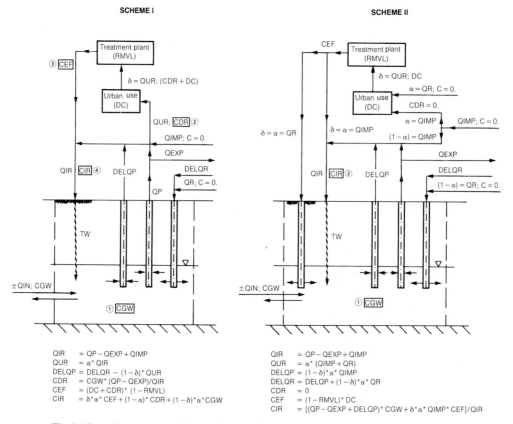

Fig. 4. Alternative conceptual schemes for recycling reclaimed effluents in the aquifer system (Mercado 1976 b, Mercado and Avron 1976)

Three alternative recycling schemes (Fig. 4) of reclaimed effluents conceptually defined in this study (Mercado and Avron 1976) for the purpose of the risk assessment. These schemes emphasize different "bottlenecks" of the groundwater quality system.

1. Direct use of sewage effluents for irrigation (Scheme I) provided their quality will be suitable for irrigation purposes. This alternative will ensure maximum protection of groundwater quality.
2. Indirect use of sewage effluents for irrigation (Scheme III). This alternative will be employed whenever sewage effluents will not meet the requirements for irrigation water or due to lack of sufficient surface storage to route winter effluents. The use of this alternative will obviously accelerate the deterioration of groundwater quality in specified recharge areas.
3. Partial replacement of imported water by sewage effluent (Scheme II), disconnecting urban consumption from the local aquifer system.

For the sake of brevity let us deal only with the first recycling scheme, shown in Fig. 4. Following this scheme, the concentration of contaminants (CDR) in

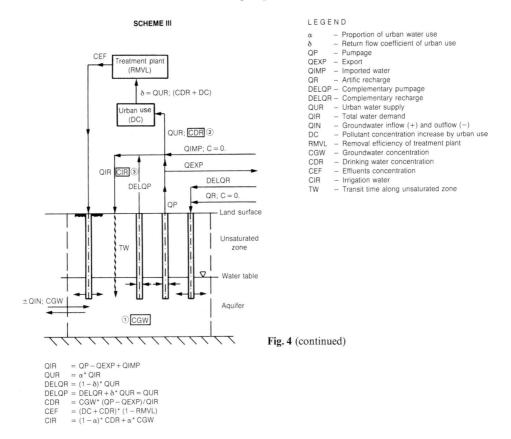

SCHEME III

LEGEND

α	– Proportion of urban water use
δ	– Return flow coefficient of urban use
QP	– Pumpage
QEXP	– Export
QIMP	– Imported water
QR	– Artific recharge
DELQP	– Complementary pumpage
DELQR	– Complementary recharge
QUR	– Urban water supply
QIR	– Total water demand
QIN	– Groundwater inflow (+) and outflow (−)
DC	– Pollutant concentration increase by urban use
RMVL	– Removal efficiency of treatment plant
CGW	– Groundwater concentration
CDR	– Drinking water concentration
CEF	– Effluents concentration
CIR	– Irrigation water
TW	– Transit time along unsaturated zone

Fig. 4 (continued)

$$QIR = QP - QEXP + QIMP$$
$$QUR = \alpha * QIR$$
$$DELQR = (1 - \delta) * QUR$$
$$DELQP = DELQR + \delta * QUR = QUR$$
$$CDR = CGW * (QP - QEXP)/QIR$$
$$CEF = (DC + CDR) * (1 - RMVL)$$
$$CIR = (1 - \alpha) * CDR + \alpha * CGW$$

municipal water supply networks varies with time according to the rise of average concentration (CGW) in the aquifer. Imported water (QIMP) will probably remain free of contaminants. Municipal waters in the course of their use are contaminated by a constant addition (DC) to their original concentration (CDR). Part δ of the urban consumption returns to the sewage system at a concentration of CDR + CD and is then conveyed to treatment plants. Treatment plants are characterized by their specific removal efficiency (RMVL) for a given contaminant. Sewage effluents at a concentration of CEF = $(1 - RMVL) * (CDR + DC)$ are reclaimed for irrigation at a concentration of CIR, determined as a weighted average of the various components.

Contaminants in irrigation water percolate through the unsaturated soil column and reach the water table after a time of $\tau = t_w * RF$; t_w is the transit time of chemically inert solutes and RF is the retardation factor of a specific pollutant due to cation exchange and adsorption processes. The contaminants are then mixed with the effective groundwater volume increased by $Q_p\tau$ causing a rise in the concentration of pumping wells tapping the aquifer.

The recycling scheme was simulated with the aid of a single-cell model (Mercado and Avron 1976). The only "sink" in the preliminary version of this model is the cell's treatment plant. Retarding processes in the unsaturated column, such as adsorption and cation exchange on clay minerals, were neglected, presuming chelation of organometal complexes to dominate.

Possible concentration of problematical contaminants in groundwater were determined with the aid of the above-mentioned model. Computations were repeated for 242 model cells, representing most of the coastal plain area.

20.4.2 Assessment of Possible Risks and Related Protection Measures

According to model computations (Mercado 1980), the northern part of the aquifer is most likely to be contaminated between the years 2000 and 2100 (Fig. 5). On the other hand, the southeastern part of the aquifer will remain intact probably for a period of more than 300 years. These computations are based on the assumptions that: (1) secondary effluents will predominate in the coastal plain; (2) there will be a recycling of one-half of the water consumption through urban consumers; and (3) adsorption and cation exchange will not play a significant role in determining the transit time of pollutants due to possible chelation of heavy metals with soluble organic substances in sewage. Similar computations were carried out for tertiary effluents, showing that most of the aquifer will remain intact up to the year 2050 and only moderate pollution is expected by 2100.

Besides determining possible concentrations in groundwater and other "check points" of the system, the computer analysis also yields some design guidelines with respect to alternative protection measures, such as maximum concentrations in raw sewage and treated effluents used for irrigation, recommended treatment levels (Fig. 6), necessary removal of contaminants at their source, and possible recycling of sewage water in particular cells. These recommendations are based on the dynamics of the groundwater system and the requirements to meet acceptable water quality criteria up to a given "target year". Choosing the target year is, so far, a subjective decision.

The recommended treatment levels shown in Fig. 6 are an example of translating simulation results to engineering alternatives that can be evaluated directly by the decision makers. According to this example, it seems that secondary treatment might not be sufficient to protect groundwater resources until the year 2050 and the introduction of tertiary treatment will be necessary to reclaim sewage effluents in the northern and central parts of the coastal plain. On the other hand, treatment level will not play a significant role in decisions concerning the future groundwater quality of the southeastern region due to the considerable retention capacity of its unsaturated zone.

20.4.3 Sensitivity Analyses

In view of the uncertainties in defining pollution mechanisms and estimating physicochemical parameters, it is important to analyze the sensitivity of pollution forecasts and proposed regulations to possible variations in such model parameters and management constraints. Sensitivity analysis is a common procedure in decision models; it enables verification of decisions based on limited information, screening of proposed research projects, and initiation of others according to the missing information required for the decision process. Sensitivity analysis is also

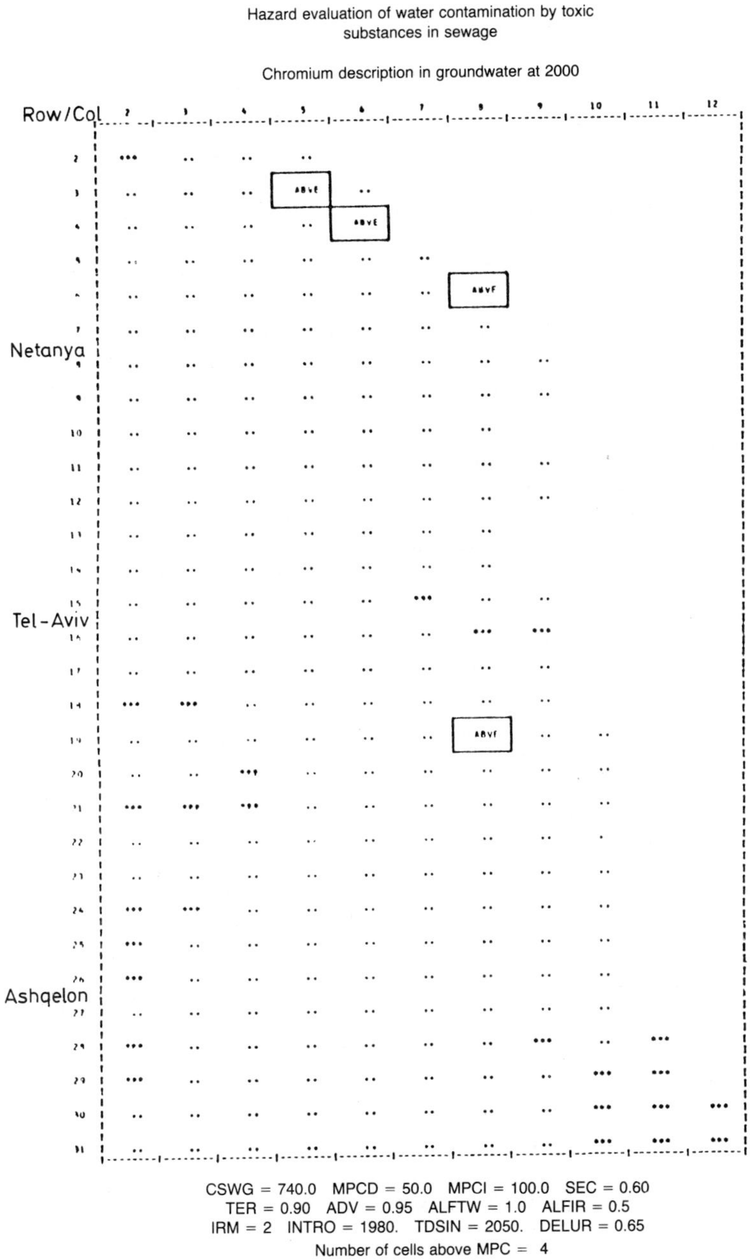

Hazard evaluation of water contamination by toxic substances in sewage
Chromium description in groundwater at 2000
Row/Col
Netanya
Tel-Aviv
Ashqelon
CSWG = 740.0 MPCD = 50.0 MPCI = 100.0 SEC = 0.60
TER = 0.90 ADV = 0.95 ALFTW = 1.0 ALFIR = 0.5
IRM = 2 INTRO = 1980. TDSIN = 2050. DELUR = 0.65
Number of cells above MPC = 4

Fig. 5. Descriptive maps of possible groundwater pollution by secondary effluents used for irrigation. *ABVE* indicate chromium concentrations exceeding drinking standards. Adsorption is neglected. Chromium has the highest relative toxicity of trace elements in secondary sewage (Mercado 1976a, Mercado and Avron 1976)

A. Mercado

Hazard evaluation of water contamination by toxic
substances in sewage

Chromium description in groundwater at 2050

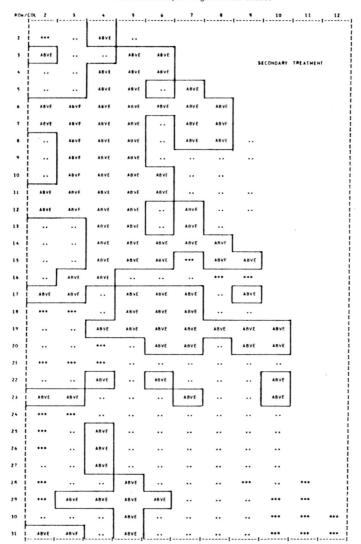

CSWG = 740.0   MPCD = 50.0   MPCI = 100.0   SEC = 0.60
TER = 0.90   ADV = 0.95   ALFTW = 1.0   ALFIR = 0.5
IRM = 2   INTRO = 1980.   TDSIN = 2050.   DELUR = 0.65
Number of cells above MPC = 99

Fig. 5 (continued)

Hazard evaluation of water contamination by toxic
substances in sewage

Chromium description in groundwater at 2100

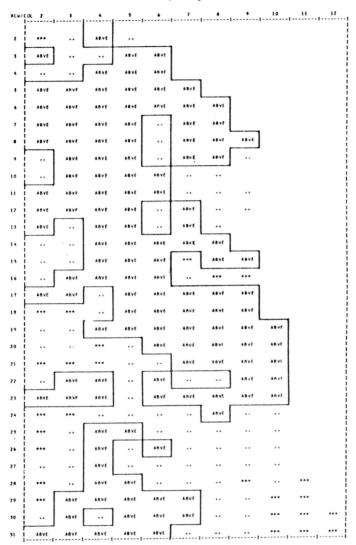

CSWG = 740.0 MPCD = 50.0 MPCI = 100.0 SEC = 0.60
TER = 0.90 ADV = 0.95 ALFTW = 1.0 ALFIR = 0.5
IRM = 2 INTRO = 1980. TDSIN = 2050. DELUR = 0.65
Number of cells above MPC = 133

Fig. 5 (continued)

Hazard evaluation of GW contamination by toxic
substances in sewage – average assumption

Recom. treatment to reach chromium MPC in groundwater

Row/Col	2	3	4	5	6	7	8	9	10	11	12
2	•••	RAW	TER	RAW							
3	TER	RAW	RAW	TER	ADV						
4	SEC	SEC	ADV	TER	TER						
5	SEC	SEC	TER	TER	SEC	TER					
6	TER	TER	TER	TER	TER	TER	ADV				
7	TER	TER	TER	TER	SEC	TER	TER				
Netanya 8	SEC	TER	TER	TER	SEC	TER	TER	SEC			
9	RAW	TER	TER	ADV	SEC	SEC	SEC	RAW			
10	SEC	TER	TER	TER	TER	RAW	RAW				
11	TER	TER	TER	TER	TER	SEC	RAW	RAW			
12	TER	TER	TER	TER	SEC	TER	RAW	RAW			
13	SEC	SEC	TER	TER	RAW	TER	RAW				
14	RAW	SEC	TER	TER	ADV	TER	TER				
15	RAW	RAW	TER	TER	TER	•••	TER	TER			
Tel-Aviv 16	SEC	TER	TER	SEC	SEC	RAW	•••	•••			
17	TER	TER	SEC	TER	TER	TER	SEC	TER			
18	•••	•••	RAW	TER	TER	TER	RAW	SEC			
19	SEC	RAW	TER	TER	TER	TER	ADV	TER	TER	TER	
20	RAW	RAW	•••	RAW	TER	TER	RAW	TER	TER	TER	
21	•••	•••	•••	SEC	SEC	RAW	RAW	RAW	RAW	RAW	
22	RAW	SEC	TER	RAW	TER	RAW	RAW	RAW	RAW	TER	
23	TER	TER	SEC	SEC	SEC	TER	SEC	RAW	RAW	TER	
24	•••	•••	RAW	RAW	RAW	RAW	RAW	RAW	RAW	RAW	
25	•••	RAW	TER	RAW	RAW	RAW	RAW	RAW	RAW	RAW	
Ashqelon 26	•••	RAW	TER	SEC	RAW	RAW	RAW	RAW	RAW	RAW	
27	RAW	SEC	TER	RAW	RAW	RAW	RAW	RAW	RAW	RAW	
28	•••	SEC	SEC	TER	RAW	RAW	RAW	•••	RAW	•••	
29	•••	TER	TER	TER	TER	RAW	RAW	RAW	•••	•••	
30	SEC	SEC	SEC	TER	SEC	RAW	RAW	RAW	•••	•••	•••
31	TER	TER	SEC	TER	RAW	RAW	RAW	RAW	•••	•••	•••

CSWG = 740.0   MPCD = 50.0   MPCI = 100.0   SEC = 0.60
TER = 0.90   ADV = 0.95   ALFTW = 1.0   ALFIR = 0.5
IRM = 2   INTRO = 1980.   TDSIN = 2050.   DELUR = 0.65

Fig. 6. Recommended treatment levels of sewage effluents necessary to conserve groundwater quality till 2050. Adsorption is neglected (Mercado 1976a, Mercado and Avron 1976)

important for identifying problematical pollutants, whose concentration and behavior in the aquifer system might endanger groundwater quality and they should, therefore, be studied more thoroughly.

References

Mercado A (1976a) The use of models in groundwater quality management. Proc Reading Water Res Center Conf Groundwater Qual Measurement, Predict and Protect, pp 653–688

Mercado A (1976b) Nitrate and chloride pollution of aquifers, regional study with the aid of a single cell model. Water Resour Res 5(12):731–747

Mercado A (1980) The coastal aquifer in Israel: some quality aspects of groundwater management. In: Shuval H (ed) Water Quality Management under Conditions of Scarcity: Israel as a Case Study. Ch V Academic Press, London New York pp 93–146

Mercado A, Avron M (1976) "Possible Contamination of the Coastal Plain Aquifer due to Irrigation by Sewage Effluents: Hazard Evaluation and Alternative Protection Measures". 01/76/04. Tahal, Tel Aviv (in Hebrew)

Mercado A, Billings GK (1975) The kinetics of mineral dissolution in carbonate aquifers as a tool for hydrological investigations: (I) Concentration-time relationships. J Hydrol 24:303–331

Schwarz Y, Mercado A (1974) System's approach to groundwater quality control and management. Belg Isr Symp Groundwater Qual Control Management, Brussels

Subject Index

Ecological Studies

Analysis and Synthesis

Editors: W.D.Billings, F.Golley, O.L.Lange,
J.S.Olson, H.Remmert

Springer-Verlag
Berlin
Heidelberg
New York
Tokyo

Ecological Studies

Analysis and Synthesis

Editors: W. D. Billings, F. Golley, O. L. Lange,
J. S. Olson, H. Remmert

Volume 38
The Ecology of a Salt Marsh
Editors: L. R. Pomeroy, R. G. Wiegert
1981. 57 figures. XIV, 271 pages
ISBN 3-540-90555-3

Contents: Ecosystem Structure and Function: Ecology of Salt Marshes: An Introduction. The Physical
and Chemical Environment. – Salt Marsh Populations: Primary Production. Aquatic Macroconsu-
mers. Grazers on *Spartina* and Their Predators.
Aerobic Microbes and Meiofauna. Anaerobic
Respiration and Fermentation. – The Salt Marsh
Ecosystem: The Cycles of Nitrogen and Phosphorus. A Model View of the Marsh. The Salt-Marsh
Ecosystem: A Synthesis. – References. – Index.

Volume 37
H. Jenny
The Soil Resource
Origin and Behavior
Corrected 2nd printing. 1983. 191 figures.
XX, 377 pages
ISBN 3-540-90543-X

Contents: Ecosystems and Soils. – Processes of Soil
Genesis: Water Regimes of Soils and Vegetation.
Behavior of Ions in Soils and Plant Responses. Ori-
gin, Transformation, and Stability of Clay Particles.
Biomass and Humus. Soil Colloidal Interactions
and Hierarchy of Structures. Pedogenesis of Hori-
zons and Profiles. – Soil and Ecosystem Sequences:
State Factor Analysis. The Time Factor of System
Genesis. State Factor Parent Material. State Factor
Topography. State Factor Climate. Biotic Factor of
System Genesis. Integration of Factors and Over-
view of Book. – Appendix 1: Names of Vascular
Species Cited. – Index.

Volume 36
C. B. Osmond, O. Björkman, D. J. Anderson
Physiological Processes in Plant Ecology
Toward a Synthesis with *Atriplex*
1980. 194 figures, 76 tables. XI, 468 pages
ISBN 3-540-10060-1

Contents: Physiological Processes in Plant Ecology:
the Structure for a Synthesis. – Systematic and
Geographical State of *Atriplex*. – Genecological Differentiation. – Genetic and Evolutionary Relationships in *Atriplex*. – *Atriplex* Communities: Regional
Environments and Their Ecological Analysis. –
Germination and Seedling Establishment. –
Absorption of Ions and Nutrients. – Water Movement and Plant Response to Water Stress. – Photosynthesis. – Productivity and Environment. – Epi-
log. – References. – Taxonomic Index. – Subject
Index.

Volume 35
J. A. C. Fortescue
Environmental Geochemistry
A Holistic Approach
1980. 131 figures. XVII, 347 pages
ISBN 3-540-90454-9

Contents: Introduction: General Overview. Outline
of Historical Development of Geochemistry. A Phi-
losophy for Environmental Geochemistry. Other
Approaches to Environmental Geochemistry. –
The Basics of the Discipline of Landscape
Geochemistry: Definitions of Concepts and Principles. Element Abundance. Element Migration in
Landscapes. Geochemical Flows in Landscapes.
Geochemistry Gradients. Geochemical Barriers.
Historical Geochemistry. Geochemical Landscape
Classification. Chemical Complexity and Landscape Geochemistry. Scientific Effort and Landscape Geochemistry. Space and Lanscape Geo-
chemistry. Time and Landscape Geochemistry.
Landscape Geochemistry as a Totality. – Applications of Landscape Geochemistry: Practical Applications of Landscape Geochemistry. A New Para-
digm for Environmental Geochemistry. – Summary
and Conclusions: Overview. Synopsis. References.
Author Index. Subject Index.

Springer-Verlag Berlin Heidelberg New York Tokyo